APPLICATIONS OF LEARNING & PLANNING METHODS

WORLD SCIENTIFIC SERIES IN COMPUTER SCIENCE

Published

7: Syntactic and Structural Pattern Recognition — Theory and Applications (Eds. *H Bunke & A Sanfeliu*)

8: Information, Randomness & Incompleteness — Papers on Algorithmic Information Theory, Second Edition (*G J Chaitin*)

9: P-Prolog — A Parallel Logic Programming Language (*Rong Yang*)

10: Intelligent Chinese Language Pattern and Speech Processing (Ed. *P S P Wang*)

11: Control Theory of Robotic Systems (*J M Skowronski*)

12: An Introduction to Chinese, Japanese, and Korean Computing (*J K T Huang & T D Huang*)

13: Mathematical Logic for Computer Science (*Z W Lu*)

14: Computer Vision and Shape Recognition (Eds. *A Krzyzak, T Kasvand & C Y Suen*)

15: Stochastic Complexity in Statistical Inquiry (*J Rissanen*)

16: A Perspective in Theoretical Computer Science — Commemorative Volume for Gift Siromoney (Ed. *R Narasimhan*)

17: Computer Transformation of Digital Images and Patterns (*Z C Li, T D Bui, Y Y Tang & C Y Suen*)

18: Array Grammars, Patterns and Recognizers (Ed. *P S P Wang*)

19: Structural Pattern Analysis (Eds. *R Mohr, Th Pavlidis & A Sanfeliu*)

20: A Computational Model of First Language Acquisition (*N Satake*)

21: The Design and Implementation of ConcurrentSmalltalk (*Y Yokote*)

23: Introduction to Theoretical Computer Science (*Ma Xiwen*)

24: A Digital Optical Cellular Image Processor — Theory, Architecture and Implementation (*K-S Huang*)

25: Computer Epistemology — A Treatise on the Feasibility of the Unfeasible or Old Ideas Brewed New (*T Vámos*)

26: Applications of Learning and Planning Methods (Ed. *N G Bourbakis*)

Forthcoming

22: From Humans to Computers — Cognition Through Visual Perception (*V V Alexandrov & N D Gorsky*)

27: Advances in Artificial Intelligence — Applications and Theory (Ed. *J Bezdek*)

28: Introduction to Database and Knowledge-Base Systems (*S Krishna*)

29: Pattern Recognition: Architectures, Algorithms and Applications (Eds. *R Plamondon & H D Cheng*)

For a complete list of published titles in the series, please write in to the publisher.

World Scientific Series in Computer Science – Vol. 26

APPLICATIONS OF LEARNING & PLANNING METHODS

Editor

Nikolaos G. Bourbakis

IBM, San Jose, California

World Scientific

Singapore • New Jersey • London • Hong Kong

Published by

World Scientific Publishing Co. Pte. Ltd.

P O Box 128, Farrer Road, Singapore 9128

USA office: 687 Hartwell Street, Teaneck, NJ 07666

UK office: 73 Lynton Mead, Totteridge, London N20 8DH

Library of Congress Cataloging-in-Publication data is available.

APPLICATIONS OF LEARNING AND PLANNING METHODS

ISBN 981-02-0546-5

Printed in Singapore by JBW Printers & Binders Pte. Ltd.

PREFACE

The goal of this book is to address some important issues and to present significant applications in the Machine Learning and Planning research areas. It discusses both symbolic and connectionist learning schemes using very recent results in various applications, such as intelligent recognition, design synthesis and production memory. It also presents planning and searching strategies with applications in the robotics area. This book can be used as a reference book in graduate level courses related to learning and planning. It is also appropriate for researchers and R&D engineers. The chapters have been written by experts in the learning and planning research fields. Each chapter provides a brief overview on a selected topic, addresses one or more problems in it and discusses in depth their solution. I would like to express many thanks to the authors for their fruitful and valuable contributions, the reviewers for their useful comments, and Professors P. Wang and H. Bunke, Dr. K. K. Phua, Ms. J. Gan, Ms M. -L. Foo, Ms C. S. Theng and Mr T. Moore for their support on this project.

Dr Nikolaos G. Bourbakis
IBM, San Jose, CA

CONTRIBUTORS

N. G. Bourbakis, *IBM, San Jose, CA*

R. J. Calistri-Yeh, *Odyssey Research Associates, Ithaca, NY*

K. Chung, *Soong-Sil University, Seoul, Korea*

D. D. Gajski, *University of California, Irvine, CA*

J.-L. Gaudiot, *University of Southern California, Los Angeles, CA*

J. R. Kipps, *University of California, Irvine, CA*

S. Lee, *University of Southern California, Los Angeles, CA*

H. Madala, *Clarkson University, Potsdam, NY*

T. M. Mitchell, *Carnegie Mellon University, Pittsburgh, PA*

A. Namatame, *National Defense Academy, Yokosuka, Japan*

P. W. Pachowicz, *George Mason University, Fairfax, VA*

N. S. V. Rao, *Old Dominion University, Norfolk, VA*

J. W. Shavlik, *University of Wisconsin-Madison, Madison, WI*

P. C.-Y. Sheu, *Rutgers University, Piscataway, NJ*

A. Sohn, *University of Southern California, Los Angeles, CA*

H. E. Stephanou, *Rensselaer Polytechnic Institute, Troy, NY*

T. Tanaka, *Toshiba Corporation, Kawasaki, Japan*

G. G. Towell, *University of Wisconsin-Madison, Madison, WI*

K. I. Trovato, *Philips Laboratories, Briarcliff Manor, NY*

Q. Xue, *Indiana-Purdue University, Indianapolis, IN*

F. Yegenoglu, *George Mason University, Fairfax, VA*

W. Zhang, *University of California, Los Angeles, CA*

CONTENTS

Chapter 6 **Automating Technology Adaptation in Design Synthesis** **129**

by J. R. Kipps and D. D. Gajski

Applications of Learning and Planning Methods
edited by N. G. Bourbakis

AN INTRODUCTION TO LEARNING AND PLANNING

NIKOLAOS G. BOURBAKIS
IBM, San Jose, CA

Artificial Intelligence is the science which studies the intelligent behavior of systems.

The Artificial Intelligence field has been growing exponentially the last decade, widening the spectrum of AI applications. Various methodologies and schemes have been developed in an attempt to contribute to the two main AI goals — theory of intelligence and design systems with intelligent behavior.

Learning and Planning are two scientific interrelated areas with significant contribution to the Artificial Intelligence domain. Learning deals with processes that represent, create, interrelate, and generalize rules for information and knowledge manipulation, making a system autonomous and/or adaptable to a particular environment or problem. Planning deals with the strategies required to achieve a goal or to solve a particular problem through efficient decomposition and/or composition of various tasks.

This book combines both learning and planning schemes and their applications in different domains. The first part, *Learning*, has seven chapters on the ongoing research work in symbolic and connectionist learning. The second part, *Planning*, contains also seven chapters which provide the current research efforts in planning and searching methodologies and their applications. Special emphasis is given to the path planning of robots in known, uncertain or unknown environments.

The chapters in Part I deal with important issues and their applications in intelligent recognition, design synthesis, etc. In particular, the first chapter discusses an embedding learning scheme in a general frame-based architecture. It describes CHUNKER, an explanation-based chunking mechanism built on top of THEO, a software framework to support development of self-modifying problem solving systems. The second chapter proposes a new architecture, Chebychev networks with Chebychev polynomials as input units, for supervised neural network learning. It shows that the inductive capability or generality of supervised learning depends on how the continuous mappings are represented and obtains the necessary conditions of the networks' architecture for learning the structures of the continuous mappings. The third chapter describes a variety of inductive learning algorithms, such as basic, multilayer, combinatorial and harmonical ones, and discusses their computational aspects. The fourth chapter presents a hybrid system that combines the symbolically-oriented explanation-based learning paradigm with the neural back-propagation algorithm. It provides empirical results which show that the hybrid system learns a concept more accurately than the explanation-based system by itself and also learns faster and generalizes better than the neural learning system by itself. The fifth chapter presents a methodology developed to acquire and recognize noisy texture concepts. In particular, it deals with the acquisition of a texture concept incorporated with symbolic inductive learning from provided examples. It manipulates the acquired concept descriptions in order to remove the noise and deals with the application of concept optimization performed within a closed system. The system makes decisions dynamically for the recognition of patterns and curves. The sixth chapter describes a knowledge acquisition approach to automating technology adaptation with respect to the DTAS system. The DTAS system synthesizes generic logic components from a library of technology-specific cells. Using knowledge of fundamental principles for logic design, the Logic Learning Assistant generates design rules that map generic components into library cells. The seventh chapter demonstrates a multilayer Artificial Neural Network approach applied to the AI production systems paradigm, by describing two types of representations (local and hierarchical). To demonstrate the use of a neural network approach, the mapping of a generic production system is explained.

Part II has chapters which deal with planning and searching methodologies with applications to robotics. In particular, the first chapter discusses a resource-level conflict detection and conflict resolution scheme combined with a state-level planning algorithm for non-linear planning. The proposed scheme keeps track of the usage of individual resources during planning, and constructs a resource-usage-flow structure. It is also easily implementable in parallel distributed environments. The second chapter deals with heuristic tree searching. It proposes an approach based on non-parametric statistical inference methods which do not require precise information about the distribution. The third chapter presents four algorithms, starting with a direct adaptation of A* for recognizing correct plans, and progressing to an optimized algorithm for recognizing violated, ill-formed plans. The fourth chapter describes a differential A* planning method built on the A* configuration space approach to adapt quickly to the changes in the space by determining and updating the localized regions affected by those changes rather than regenerating the entire space. The fifth chapter deals with a real-time algorithm for planning collision-free navigation or manipulation paths in dynamic, cluttered environments. A Newton-like iteration is used to determine a sequence of knot points that guide the motion of several robots in a velocity vector field toward their targets while avoiding collisions with a set of moving obstacles. The sixth chapter presents a knowledge-based acquisition scheme in a real-time path planning process in an unknown navigation environment. The knowledge acquisition scheme is based on the extraction of the shapes of the current free navigation spaces and their synthesis by using attributed graphs. The seventh chapter investigates the problem of finding the collision-free paths of two planar robot manipulators with three revoluting joints, which cooperate to carry a rectangular object in a 2-D environment.

PART I: LEARNING

Applications of Learning and Planning Methods
edited by N. G. Bourbakis

CHAPTER 1

Embedding Learning in a General Frame-Based Architecture

Toshikazu Tanaka
Systems and Software Engineering Laboratory, Toshiba Corporation
70 Yanagicho, Saiwaiku, Kawasaki, Kanagawa 210, Japan
Tom M. Mitchell
School of Computer Science, Carnegie Mellon University
Pittsburgh, PA 15213, USA

Abstract

This research seeks to incorporate machine learning capabilities within a general-purpose frame-based architecture. We describe CHUNKER, an explanation-based chunking mechanism built on top of THEO, a software framework to support development of self-modifying problem solving systems. CHUNKER forms rules that improve problem solving efficiency, by generalizing and compressing the chains of inference which THEO produces during problem solving. After presenting the learning algorithm used by CHUNKER, we illustrate its application to learning search control knowledge, present results of simple experiments, discuss its relationship to THEO's other three learning mechanisms, and consider the relationship between architectural features of THEO and the effectiveness of CHUNKER.

1 Introduction

We describe an effort to embed a general learning mechanism within a frame-based problem solving architecture. More specifically, we are developing an architecture called THEO[8] in which all problem solving corresponds to inferring values of slots of frames. Each inferred slot value in THEO is stored along with its explanation, or derivation, in terms of other slot values. This paper discusses a learning mechanism, called CHUNKER, which utilizes these explanations to form general rules which can speed up the subsequent inference of similar slot values. CHUNKER is intended to organize every aspect of problem solving experience into recognition rules, and is similar in this respect to SOAR[5]. In order to generalize from a single instance of problem solving, it employs a variant of explanation-based learning[9][2], with the regression process specially adapted to THEO's knowledge representation and inference conventions. Embedding CHUNKER within THEO in this way opens the possibility of learning efficient rules for any slot of any frame in the system. We discuss the current implementation, examples of rules formed by the system, its application to learning control knowledge in state space search, and current limitations.

The basic motivation for this research is to build a self-improving frame-based architecture to support development of knowledge-based systems. Within the THEO architecture, we are currently exploring several learning mechanisms: explanation-based generalization, chunking, caching, and statistical learning. Motivations for developing a chunking mechanism in THEO include:

- To facilitate organizing and remembering all aspects of THEO's problem solving experience.
- To investigate the relative performance and interactions with the other learning mechanisms in THEO.
- To understand which architectural features such as uniform representation, truth maintenance, etc. are needed to facilitate effective learning.
- To investigate when to learn and what to learn, so that the benefits of learning outweigh the costs[6].
- To directly support the application of THEO to control a learning mobile robot[1][7].

Section 2 overviews the THEO architecture in which CHUNKER is embedded. Section 3 describes how learning tasks are formulated, and CHUNKER's learning algorithm. Section 4 introduces a case study in which CHUNKER is applied to learning search control knowledge in state space search using the Blocks World domain. In section 5, results of simple experiments are presented. Finally, in section 6, we compare CHUNKER with three other learning mechanisms in THEO, and discuss architectural features which make possible CHUNKER's functionality.

2 Overview of THEO

THEO is an integrated architecture intended as a research platform for knowledge representation, problem solving, learning, and their combination.

2.1 Knowledge Representation

THEO has a frame-based representation with its system knowledge and domain dependent knowledge represented in a uniform manner, similar in some respects to the RLL system[4]. Conceptually, every belief THEO can hold is an assertion of the form (*entity slot*) = *value*. For example, (`cube1 weight`) = `5000` represents the belief that the `weight` of `cube1` is `5000`. All problems in THEO have the form (*entity slot*), which represents the task of determining a justifiable belief of the form (*entity slot*) = *value*. One important feature is that such problems are themselves entities. Thus, THEO can hold beliefs about problems and pose problems about problems just as it can for any other entity. For example, ((`cube1 volume`) `solvable?`) = `t` represents the belief that the problem (`cube1 volume`) is solvable. Since meta-beliefs are indexed by the ground beliefs, their retrieval cost does not grow significantly as unrelated beliefs are added to the knowledge base. For convenience, we will henceforth drop the parentheses in describing such problems and beliefs about problems; we will write, for example, (`cube1 volume solvable?`) in place of ((`cube1 volume`) `solvable?`).

Figure 1 illustrates an example THEO knowledge base. As illustrated in this Figure, a THEO knowledge base consists of frame descriptions, each of which is recursively defined as a datastructure of the form *description* ::= (*name value description**), where *name* can be any Lisp atom, *value* can be any Lisp expression, and * denotes 0 or more iterations. For top level descriptions (i.e., those not nested within other descriptions), *name* stands for the name of the Theo object. For nested descriptions, *name* refers to the name of some slot of the object within which it is nested. For example, in Figure 1, the fourth object shown has the name `cube1`, and contains slots (nested descriptions) with named `generalizations`, `density`, `size`, and `color`.

2.2 Inference

When a problem is posed to THEO (e.g., to determine the value of (`cube1 weight`)), THEO accesses its memory to retrieve an explicit belief, or a cached slot value as a solution to this problem. Inference is invoked only when no such explicit belief is found. In this sense THEO's inference is demand-driven. THEO provides three levels of abstraction at which the programmer may specify procedures for inferring slot values. These three levels provide a range of sophistication and efficiency depending on the user's needs. The following summarizes the 3 levels of inference.

- Level 1 : *Togets*

  When a problem (*entity slot*) is given, determine its *value* by applying the Lisp function retrieved from the address (*entity slot* `toget`), to it. If no such function is specified, the default value of the `toget` slot implements Level 2.

- Level 2 : *Methods*

  Apply the list of methods retrieved from (*entity slot* `methods`) to the address (*entity slot*) until a value is obtained. For example, `use.rules` is a

```
(physobj *novalue*  (generalizations (frame)))
(box     *novalue*  (generalizations (physobj)))
(cube    *novalue*  (generalizations (box))
   (height *novalue*
      (definitions ((th size *de*))))
   (length *novalue*
      (definitions ((th size *de*))))
   (width *novalue*
      (definitions ((th size *de*)))))
(cube1 *novalue*  (generalizations (cube))
   (density 5) (size 10) (color red))
(cube2 *novalue*  (generalizations (cube))
   (density 5) (size 10) (color green))
(physobj.slot *novalue*  (generalizations (slot))
   (domain physobj)
   (available.methods (inherits defines
                          drop.context default.value)))
(weight *novalue*  (generalizations (physobj.slot))
   (definitions ((* (th density *de*) (th volume *de*)))))
(density *novalue*  (generalizations (physobj.slot)))
(volume *novalue*   (generalizations (physobj.slot))
   (domain box)
   (definitions ((* (th height *de*)
                    (th length *de*)
                    (th width *de*)))))
(height *novalue*   (generalizations (physobj.slot)))
(length *novalue*   (generalizations (physobj.slot)))
(width *novalue*    (generalizations (physobj.slot)))
(size *novalue*     (generalizations (physobj.slot)))
(color *novalue*    (generalizations (physobj.slot)))
(above *novalue*    (generalizations (physobj.slot))
   (inverse below))
(below *novalue*    (generalizations (physobj.slot))
   (inverse above))
```

Figure 1: Example of THEO Knowledge Base - Physical Object World

method which interprets rules learned by CHUNKER. If unspecified, the default value of `methods` is `(defines inherits drop.context default.value)` and the `defines` method implements Level 3.

- Level 3 : *Definitions*
  Interpret definitions found in (*entity slot* `definitions`) and in (*entity slot* `specialized.definitions`). The value of the `definitions` subslot may itself be inferred from knowledge of slots `inverse`, `plural`, `singular`, etc. The value of `specialized.definitions` may be inferred via explanation-based learning by TMAC.

Every inference step in THEO involves level 1, or the use of a *toget* function to compute a value for a given problem. If the default *toget* function is used, implementing level 2, then this invokes three additional types of *toget* functions to implement the methods, as follows:

- A *toget* function which obtains all the methods available for the given problem.
- *Toget* functions which implement the individual methods.
- Functions which collect values returned by the various methods and record explanations of their use. Some of these collect only the first value returned by the methods, while others collect all values. Some record explanations only for methods which successfully returned values while others collect explanations for all attempted methods independent of success.

2.3 Learning

Currently, four learning mechanisms are available in THEO, including CHUNKER. First, and simplest, is caching of inferred values. Second, SE[3] orders methods associated with individual slots so that the most promising and most cost-effective methods are tried earlier. Third, TMAC is an implementation of explanation-based learning[9][2] in THEO. It works only for slots which utilize inference level 3, and forms macro definitions by compressing the chains of definitions which were successful in previous computations. In contrast to TMAC, CHUNKER may be applied to any slot using any of the three levels of inference. CHUNKER is thus more broadly applicable than TMAC, though in some cases it produces weaker generalizations than TMAC. This is due to the fact that CHUNKER utilizes weaker information about the nature of the dependencies among slot values (i.e., the information that is available for slots using inference level 1).

3 Mechanism of CHUNKER

In this section, we describe in detail the implementation of the CHUNKER system.

3.1 Overview

CHUNKER is a learning mechanism which compiles Theo's problem solving experience into general rules which can help it infer related slot values in the future. It is based on a mechanism similar to the chunking mechanism of SOAR[5]. We intend for CHUNKER to be usable for every slot of every frame within THEO, and to be invoked whenever a new slot value is inferred successfully or even whenever the inference fails. More precisely, the problem faced by CHUNKER is as follows:

Input:

- *Belief:* A (*entity slot*) = *value* triple, with its associated explanation.
- *Operationality predicate:* A predicate used to judge whether or not a belief is operational. The default operationality predicate dictates that ground beliefs (i.e., beliefs with no explanations) are operational.
- *Variablization functions:* A list of functions that specifies which entities may be mentioned in the preconditions of the created rule. Each function corresponds to a relative address of some new entity in terms of the entity mentioned in the input belief. For example, the list `(*de* (th under *de*))` indicates that the rule may refer to both the entity in the input belief (which is named `*de*`), and the entity which is the value of the `under` slot of `*de*`.

Output:

- *Rule:* An *if-then* rule which directly maps operational supporting beliefs to a value for a given problem.
- *Assumptions:* A set of beliefs which support the rule, to be used by the TMS in maintaining the rule.
- *Generalization:* A generalization of the input *entity*, which describes the general class of entities to which the learned rule may be applied.

When invoked, CHUNKER traces back through the explanation tree until operational beliefs are reached, and collects such beliefs. Those beliefs which can be variablized form the rule left-hand side while the others are stored as assumptions. Figure 3 shows the rules formed by CHUNKER for the input belief of Figure 2, using both the default and user-supplied operationality and variablization inputs.

Features of the prototype include:

- It can in principle be applied to any slot instance in THEO.
- Learning is interleaved with problem solving. A learned rule can be used in an immediately succeeding inference.

- A learned rule may become compiled as part of another rule in subsequent learning.
- Operationality predicates can be made dynamic, depending on the current set of beliefs.
- Use of *assumptions* reduces the size of the rule and the cost of checking conditions.
- Since *assumptions* are stored as the explanation for the learned rule, rules are also maintained by the TMS. For example, a change of belief about operationality may retract rules, making rules keep up with a drifting computational environment.

Figure 2 shows an explanation structure recorded by the TMS while solving the problem (`cube1 totalweight`). `totalweight` is defined here to be the total weight of a given object plus all objects stacked on it. In this explanation tree, beliefs at the same indentation level support conjunctively a belief at next lower level of indentation. Beliefs used to derive inference methods are not shown, e.g. `|weight.definitions.1|`, although such beliefs are included in the actual explanation tree. This way of viewing each inference step is useful not only in providing a readable explanation, but also for formulating a generalization as described later. Figure 3 shows two rules learned from this explanation by CHUNKER using different operationality criteria and different variablization functions. Note that operationality and variablization drastically affects the form and the range of applicability of resulting rules.

```
(cube1 totalweight) = 8000
 <--|totalweight.definitions.1|--
 (cube1 weight) = 5000
   <--|weight.definitions.1|--
   (cube1 volume) = 1000
     <--|volume.definitions.1|--
     (cube1 height) = 10
       <--|cube.height.definitions.1|--
       (cube1 size) = 10
     (cube1 length) = 10
       <--|cube.length.definitions.1|--
       (cube1 size) = 10
     (cube1 width) = 10
       <--|cube.width.definitions.1|--
       (cube1 size) = 10
   (cube1 density) = 5
 (cube1 under) = pyramid1
 (pyramid1 totalweight) = 3000
   <--|totalweight.definitions.2|--
   (pyramid1 weight) = 3000
   (pyramid1 under) = nil
     <--inherit1--
     (pyramid under) = nil

User-defined operationality ..................
Default operationality .........................
```

Figure 2: Explanation for (cube1 totalweight)

Rule learned:

```
(iftrue
  (and (= (th density *de*) 5)
       (= (th size *de*) 10)
       (eq (th under *de*) 'pyramid1))
  8000)
```

Stored in: `(cube totalweight rules)`
Meaning: If the size of a given entity is 10, its density is 5, and it is under *pyramid*1, then its total weight is 8000.

(a) Rule learned using default operationality and default variablization

Rule learned:

```
(iftrue
  (and (= (th height *de*) 10)
       (= (th length *de*) 10)
       (= (th width *de*) 10)
       (= (th density *de*) 5)
       (= (th weight (th under *de*)) 3000)
       (equal (th generalizations
                  (th under *de*))
              '(pyramid)))
  8000)
```

Stored in: `(box totalweight rules)`
Meaning: If the height, length and width of a given entity are all 10, its density is 5, the entity is under an object with weight 3000 and this object is a *pyramid*, then the total weight of the entity is 8000.

(b) Rule learned using user-defined operationality and user-defined variablization

Figure 3: Example Rules in Physical Objects World

3.2 CHUNKER1 – Chunking at inference level 1

Chunking for slots which utilize only inference level 1 is performed by CHUNKER1. Every inference step in THEO involves the use of a *toget* function, which can be viewed as an inference rule. When a *toget* function is applied to obtain the value of a given slot, it may access the values of other related slots. Each such slot access can be viewed as a subproblem, and may lead THEO to invoke another *toget* function which may create its own subproblems. This recursive process terminates where THEO holds an explicit belief corresponding to a solution for a given subproblem. An inferred value may be cached for later query of the same problem. Thus, THEO's inference can be viewed as a process of computing the closure of a given set of beliefs with respect to a given set of *toget* functions.

If THEO is viewed in this way, its explanation trees can be viewed as proof trees for a given problem. CHUNKER performs a variant of explanation-based learning (EBL) [9] on this proof tree. The correspondence between CHUNKER1 and EBL is as follows: The goal concept for a given problem (*entity slot*) is (?e *slot*), where ?e is a universally quantified variable. CHUNKER regresses the goal concept (?e *slot*) through the explanation tree of a given training example. The domain theory is the set of *toget* functions. The EBL process consists of two phases: the explanation and goal regression processes. The explanation process is trivial because a proof tree for a given problem is provided automatically by THEO for every slot inference.

The goal regression process requires, however, some adaptation from that for standard Prolog-like rules. First, unlike such inference rules as in Prolog, a *toget* function has a domain for which it can be applied. Since a final top-level belief is conjunctively supported by inference steps involved, the intersection of such domains will be the domain of the learned rule. Second, also unlike such inference rules as in Prolog, *toget* functions in THEO do not bear the notion of variable and constant, nor unification. Therefore, in order to fully formulate CHUNKER as an EBL system, it is necessary to establish the corresponding notions in THEO. The following paragraphs discuss how this is done.

In order to make clear which subproblems should be variablized, we introduce the notion of constant and variable as follows. A subproblem is defined to be *constant* if it remains the same for any entity to which the *toget* function can be applied. A subproblem is defined to be *variable* if it is not *constant*. Since it is difficult to determine whether a subproblem is *variable* or *constant* simply by examining the explanation for a single slot query, CHUNKER must make assumptions about the structure of *toget* functions in THEO.

Consider a problem whose address is of the form $(a_1 \ldots a_n)$. CHUNKER assumes a subproblem is *variable* if and only if its address can be constructed by transforming the original problem address using the following address transformations:

- *Context:* $(a_1 \ldots a_j),\ j < n$
  An address which represents the context in which the problem $(a_{j+1} \ldots a_n)$ is

posed. If $j = n - 1$, it represents the entity for which a slot is posed.

- *Drop-Context:* $(a_i \dots a_n)$, $1 < i$
  An address which represents a generic problem with context $(a_1 \dots a_{i-1})$ dropped.
- *Meta-Problem:* $(a_1 \dots a_n\ b_1 \dots b_m)$
  An address which represents a problem about ... a problem about a given address.

A subproblem is assumed to be *constant* if it is not *variable.* If we restrict the way of constructing subproblems based on the address of a given problem $addr = (a_1 \dots a_n)$, subproblems will be of the form

$$(a_i \dots a_j),\ 1 \leq i \leq j \leq n,\ i \neq 1\ or\ j \neq n$$

or

$$(a_i \dots a_j\ b_1 \dots b_m),\ 1 \leq i \leq j \leq n$$

Therefore, if a subproblem takes the above form, CHUNKER1 regards the subproblem as *variable*, and otherwise, as *constant.*

If these assumptions are true, CHUNKER1 is guaranteed to produce truth- preserving generalizations. In fact, most of the actual coding of *toget* functions is done using the above address transformations. If these assumptions do not hold, the validity of the learned rules is not guaranteed because CHUNKER1 might include a constant condition into the left-hand side of a rule where a variable condition is required. On the other hand, when a constant subproblem coincidentally matches a portion of the given problem, CHUNKER1 will incorrectly variablize it. But this situation is somewhat artificial and rare in practice.

CHUNKER1 is implemented by a major function **chunker1** which recursively traces an explanation tree, beginning from the root node back to operational nodes. The function **chunker1** takes as input an address, *addr* or (*entity slot*), of a node in an explanation tree and determines how much the variablization target can be generalized. More specifically,

- If *addr* is an operational node then return the domain of the address as a constraint. This constraint may become a left-hand side condition or an assumption supporting the resulting rule. In Figure 4, this constraint is denoted by $?1 \subset$ D(*slot*), where D(*slot*) denotes the domain of *slot.*
- Otherwise, return the intersection of the following constraints
  - The domain of the given address. Denoted by $?1 \subset$ D(*slot*) in Figure 4.
  - The generality of the *toget* function used to compute the value for *addr.* Denoted by $?1 \subset$ *generality* to the left of a *toget* function in Figure 4.

– Return values from **chunker1**($addr_i$) where $addr_i$ are elements of the explanation supporting the given *addr*. Denoted by *?1* ⊂ *domain* to the left of a return path from lower level of branch in Figure 4.

Figure 5 shows an example of a rule learned by the **chunker1** program. This rule is learned for the problem (`cube1 weight`) with user-defined operationality which dictates `height`, `length` and `width` is operational, and the default-variablization which variablizes only `cube1`, the domain element of the given problem. Although the learned rule is stored into the `rules` slot of `physobj`, the last condition in the left-hand side constrains the applicability of the rule to only `cubes`.

3.3 CHUNKER2 – Chunking at inference level 2

CHUNKER1 has the advantage that it can be applied to any problem, or slot, in THEO, including slots for which users have written fairly arbitrary Lisp code for the attached *toget* function. CHUNKER1 does not analyze the structure of the Lisp code, and makes only the assumptions stated earlier regarding which address transformations can be used in the Lisp code. Based on this assumption and the training example trace showing which subproblems were generated for the specific example, it produces a generalized rule.

For some domains, chunking at inference level 1 suffices. In fact, learning of control rules for state space search (see Section 4) is implemented by CHUNKER1. In general, however, stronger generalizations may be produced if the learning algorithm is allowed stronger assumptions or more information regarding the inference methods involved. CHUNKER2 is a modification to CHUNKER1 which utilizes additional information for slots which are implemented using level 2 inference in THEO. The difference in the generalization abilities of CHUNKER1 and CHUNKER2 can be seen by comparing the rule produced by CHUNKER1 (see Figure 5) and the corresponding rule produced from the same training example by CHUNKER2 (see Figure 3). Whereas the rule produced by CHUNKER1 applies only to objects that are `cubes`, the rule produced by CHUNKER2 applies more generally to any `box`.

Each inference step at level 2, or *method*, is implemented via the combination of several *toget* functions. By knowing the structure with which these functions are combined, it is possible to produce stronger generalizations than those produced by CHUNKER1. CHUNKER2 is an improvement of CHUNKER1 in this sense. In addition to the basic function of CHUNKER1, CHUNKER2 performs a set of special interpretation functions for each method-related *toget* function. These interpretation functions know the structure of explanations for specific methods, modify the explanation associated with the method, and regress goal patterns through the modified explanation in a specialized manner. Since there are only a few method-related *toget* functions, these special interpretation functions can be developed for general use and do not degrade the generality of CHUNKER.

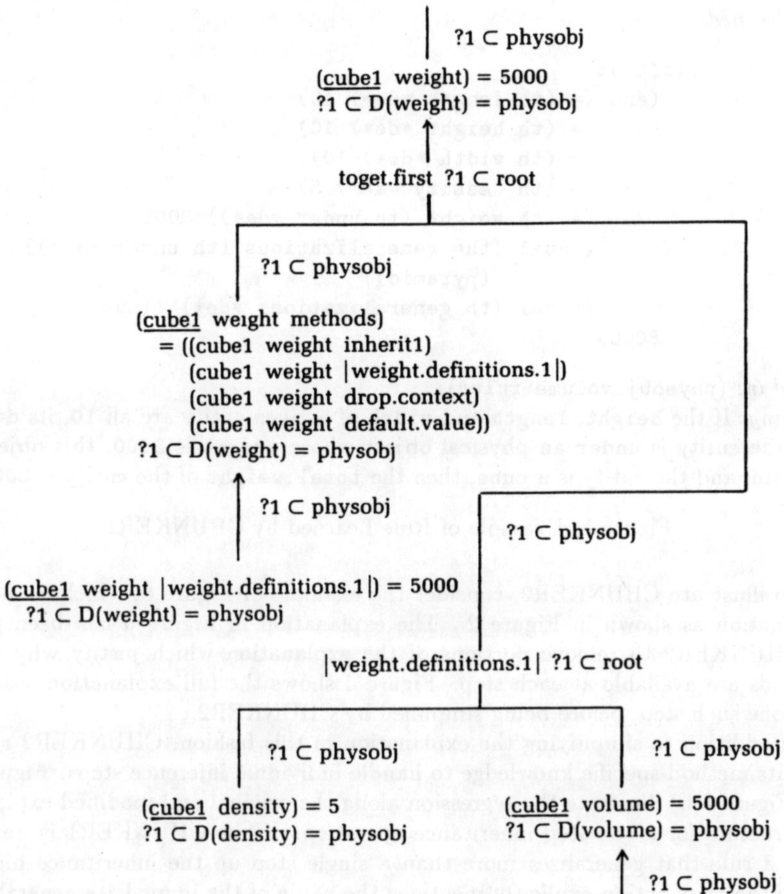

Underlined entity shows variablization target.

Figure 4: A trace of the **chunker1** function

Rule learned:

```
(iftrue
  (and (= (th length *de*) 10)
       (= (th height *de*) 10)
       (= (th width *de*) 10)
       (= (th density *de*) 5)
       (= (th weight (th under *de*)) 3000)
       (equal (the generalizations (th under *de*))
              '(pyramid))
       (equal (th generalizations *de*) '(cube)))
  8000)
```

Stored in: `(physobj volume rules)`
Meaning: If the `height`, `length` and `width` of a given entity are all `10`, its `density` is `5`, the entity is `under` an physical object whose `weight` is `3000`, this object is a `pyramid`, and the entity is a `cube`, then the `total.weight` of the entity is `8000`.

Figure 5: Example of Rule Learned by CHUNKER1.

To illustrate CHUNKER2, consider the example of Figure 6, which is the same explanation as shown in Figure 2. The explanation in Figure 6 has been pruned by CHUNKER2 to remove portions of the explanation which justify why various methods are available at each step. Figure 4 shows the full explanation associated with one such step, before being simplified by CHUNKER2.

In addition to simplifying the explanation in this fashion, CHUNKER2 also applies its method-specific knowledge to handle individual inference steps. Figure 7(a) and Figure 7(b) illustrate the regression along the original and modified explanation substructure for a two-step inheritance operation. While CHUNKER1 is unable to learn a rule that generalizes more than a single step up the inheritance hierarchy (since the explanation explicitly mentions the name of the immediate generalization of the problem instance), CHUNKER2 is able to generalize appropriately in this case.

In order for CHUNKER2 to generalize appropriately from inheritance, it uses its interpretation function corresponding to `toget.inherit`, the *toget* function that implements inheritance. First, information about the generalization hierarchy is eliminated from the explanation structure so that the resulting rule does not include it as a variablized condition. Second, a special substitution schema is employed so that a cascade of inheritance leads to a more general rule.

Figure 6 shows how the target concept (*?e* `totalweight`) = `8000` is regressed through the explanation structure. Note that in this example, there are two variablization targets: both the entity associated with the training example belief (i.e., `cube1`), and the value of the `under` slot of this entity (i.e., `pyramid`). Also note that the `inherit1` method is given a special interpretation and that beliefs about the

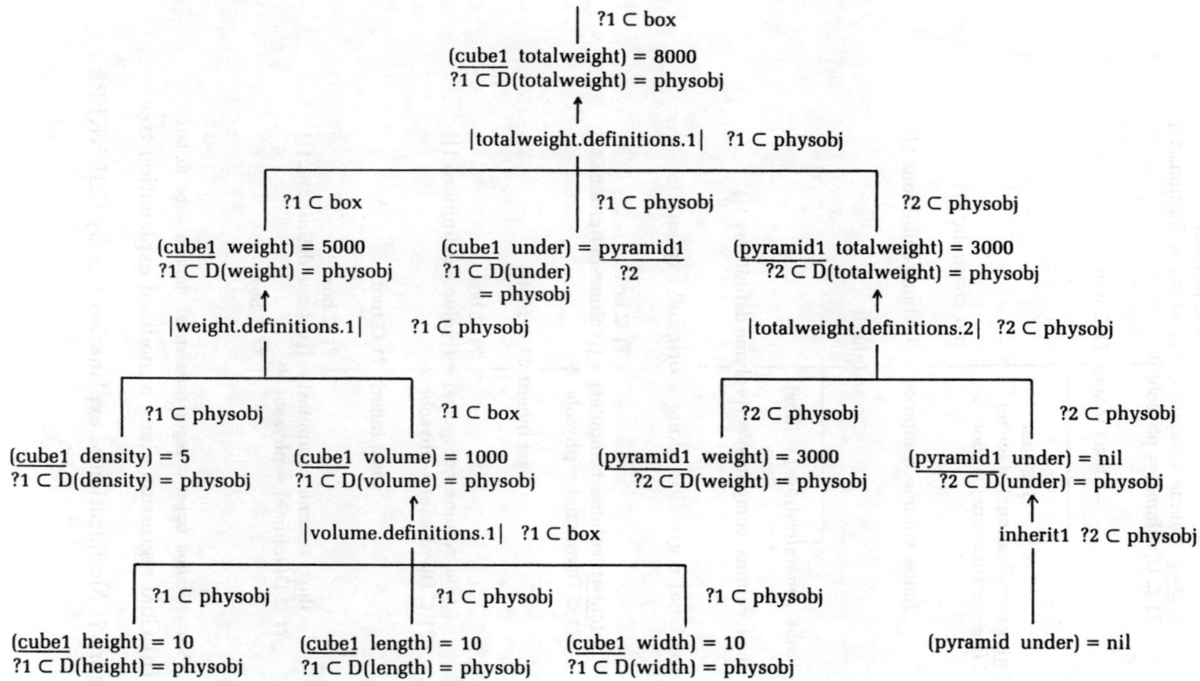

Figure 6: A trace of CHUNKER2

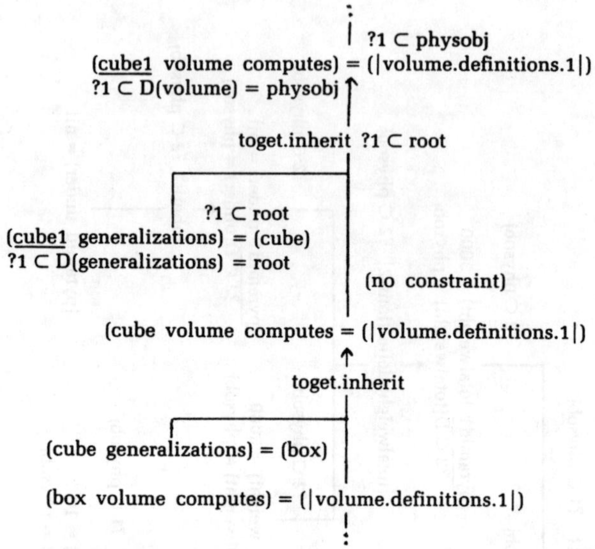

(a) Goal regression along a original explanation tree

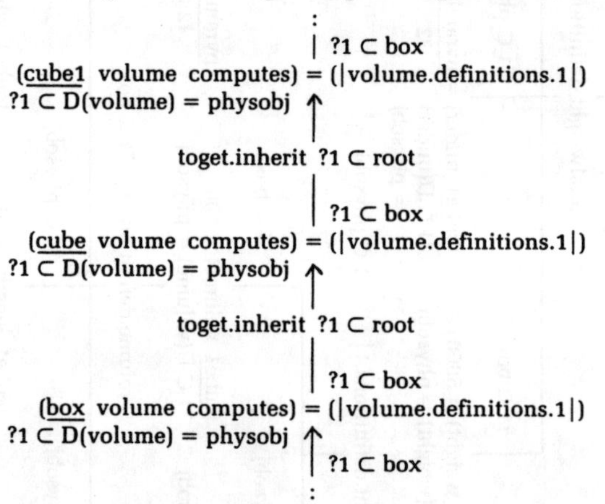

Variablization target changes from cube1 through cube to box.

(b) Goal regression along a modified explanation tree

Figure 7: Modification of explanation tree by CHUNKER2

generalization hierarchy appear in the resulting rule (see Figure 3).

3.4 Storing and using learned rules

A learned rule is stored into the `rules` slot of a problem class for which the rule can be applied. We refer to a set of rules stored in the `rules` slot as *rule set*. Thus, rule sets are indexed by a problem class name, or an address for the problem class. Within a rule set, however, rules are simply stored in a list, and there is no special indexing scheme among these rules. There are, currently, two versions for the rule interpreter: (a) one implemented as a THEO *method*, and (b) one implemented as a Lisp function hard-wired to the THEO architecture.

The `use.rules` method first collects all the available rules for a given problem class, and then tries one rule after another until successful value is returned. If the first rule is inserted into the `slots` slot of a problem class, the `use.rules` method is added to the `available.methods` slot of the problem class so that THEO can use rules to infer a value for problems of the class.

As shown later in the experiment, the method version of interpreter has several disadvantages, and the hard-wired version has been developed as an improvement to it. Unlike the method version, a hard-wired version of the interpreter is invoked before trying the *toget* function associated with the given problem. Like the method version of the interpreter, it first collects all the available rules for the given problem, and then tries one after another until success.

4 Case Study

In order to demonstrate the ability of CHUNKER, consider its application to the problem of learning search control knowledge during state space search. The search procedure (problem solver) developed for this application first chooses a state from the open list where states with untried operators are stored, and then chooses an operator to expand the state. When the search procedure chooses among available operators, it checks the selectability and rejectability of each operator. An operator is defined to be *selectable* if its use is guaranteed to lead to the goal state, and *rejectable* if its use definitely leads to backtracking. First the search program checks whether an operator is selectable or not, and if so, it applies the operator to the current state. Otherwise, it checks whether the operator is rejectable or not, and if so, it skips to the next available operator. Otherwise, it checks the preconditions of the operator so that it can safely apply the operator to the given state.

We consider the Blocks World problem where the agent uses four operators, pickup, putdown, stack, and unstack to rearrange initial configuration of blocks into a configuration which satisfies a given goal. Table 1 shows preconditions and postconditions of the four operators. Figure 8 shows an example of search in the Blocks World

Operator	Preconditions	Postconditions
pickup.x	(S x clear?) = t (S x on-table?) = t (S arm-empty?) = t	(S x clear?) = nil (S x on-table?) = nil (S arm-empty?) = nil (S holding) = x
putdown.x	(S arm-empty?) = nil	(S x clear?) = t (S x on-table?) = t (S arm-empty?) = t (S holding) = nil
unstack.x	(S x clear?) = t (S x on-table?) = nil (S arm-empty?) = t	(S x clear?) = nil (S x on) = nil (S y clear?) = t (S arm-empty?) = nil (S holding) = x
stack.x	(S x clear?) = t (S arm-empty?) = nil	(S x clear?) = nil (S y on) = x (S y clear?) = t (S arm-empty?) = t (S holding) = nil

Table 1: Operators and their Preconditions/Postconditions

from which rules in Figure 9 are learned. In this example, the following user-defined operationality predicates are given:

- Descriptions of the world in the current state or older states are operational.
- Descriptions of the world in the future state are not operational.
- History of operator use is operational.

These operationality predicates are crucial because without them all the beliefs about the world would be traced back to the initial world. The rule shown in Figure 9(a) rejects an operator because it leads to a state loop. Therefore this rule includes features of a previous state for which state loop is checked. Note that since state loop is independent from which goal is given, there is no description about the goal state included. In contrast to the rejection rule, the selection rule in Figure 9(b) does mention the goal state. The selection rule also mentions the description of the current state, but not those of previous states.

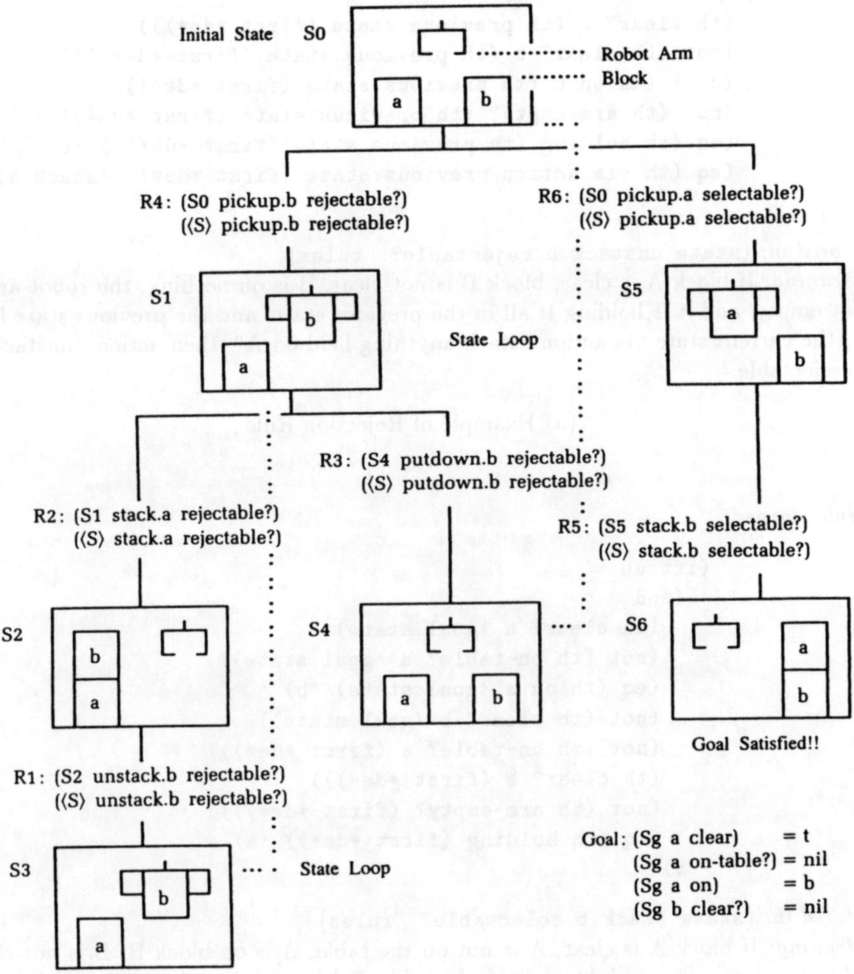

Figure 8: Example of Search in Blocks World

Rule learned:

```
(iftrue
  (and
    (th clear? a (th previous.state (first *de*)))
    (not (th clear? b (th previous.state (first *de*))))
    (null (th on b (th previous.state (first *de*))))
    (not (th arm-empty? (th previous.state (first *de*))))
    (eq (th holding (th previous.state (first *de*))) 'b)
    (eq (th via.action previous.state (first *de*)) 'stack.a))
t)
```

Stored in: `(state unstack.b rejectable? rules)`
Meaning: If block A is clear, block B is not clear, B is on nothing, the robot arm is not empty, and it is holding B all in the previous state, and the previous state leads to the current state via action "stack anything held on A", then action "unstack B" is rejectable.

(a) Example of Rejection Rule

Rule learned:

```
(iftrue
  (and
    (th clear? a 'goal.state)
    (not (th on-table? a 'goal.state))
    (eq (th on a 'goal.state) 'b)
    (not (th clear? b 'goal.state))
    (not (th on-table? a (first *de*)))
    (th clear? b (first *de*)))
    (not (th arm-empty? (first *de*)))
    (eq (th holding (first *de*)) 'a)
t)
```

Stored in: `(state stack.b selectable? rules)`
Meaning: If block A is clear, A is not on the table, A is on block B, B is not clear, all in the goal state, and A is not on the table, B is clear, the robot arm is not empty and it is holding A, then action "stack anything held on B" is selectable.

(b) Example of Selection Rule

Figure 9: Example of Search Control Rules

5 Experimental results

In this section, we present experimental results that illustrate the performance of CHUNKER. These experiments use a knowledge base of physical objects and their properties, part of which is shown in Figure 1. In the following, we describe the purpose of the experiments, details of the experimental setup, and results.

5.1 Basic performance

The first group of experiments examines (a) effect of learning and rule interpreting, (b) interaction between problem solving and learning, and (c) the effect of caching during rule interpretation. For this group of experiments, 2000 new THEO objects representing different cubes were added to the knowledge base, i.e. cubeij, $i = 1, \cdots, 100$, $j = 1, \cdots, 20$. Each cubeij has size $9 + i$ and density $4 + i$. Thus, cubes with the same j subscript have the same size and density.

5.1.1 Basic performance of CHUNKER and its rule interpreter

In this experiment, we compared the problem solving time before and after learning in two cases (see Figure 10(a) and 10(b)). More specifically, the time to solve problems (cube1j weight), $j = 1, \cdots, 20$ before and after learning a rule which can compute the weight of cube1j are measured. This comparison was done in two cases; in one case, the size and density were given as explicit beliefs, and in the other, the volume was given in addition to the size and density.

A detailed description of these experiments is as follows (the two experiments were done independently). First, without any learning or previously learned rules, the time spent to solve each problem (cube1j weight), $j = 1, \cdots, 20$ was measured and recorded (the line connecting boxes in Figure 10(a) and 10(b)). In order to eliminate the effect of caching, the system was exited and started again here. Next, the problem (cube11 weight) was solved, and a rule to compute the weight was learned from this problem solving experience. Lastly, once the rule was learned, problems (cube1j weight), $j = 2, \cdots, 20$ were posed to the system, and the time spent to solve each problem (using the learned rule) was measured and recorded (the line connecting circles in Figure 10(a) and 10(b)).

Figure 10(a) represents a case in which CHUNKER works well. The problem solving time after learning was lower that that before learning. In this case, a sequence of five *method* inference steps was required to compute the weight (see Figure 2). Figure 10(b) illustrates a slightly different situation, for which the learned rules of CHUNKER degrade, rather than improve, performance. In this case, each cube was described by its volume in addition to its size and density . Because the system's initial knowledge base already contains a *definition* for infering weight from volume and density, it could infer the weight in a single inference step even before learning.

Therefore, there is no utility to the additional learned rule, and the overhead of retrieving and matching the learned rule leads only to higher cost.

5.1.2 Interaction between problem solving and learning

In order to clarify the interaction between problem solving and learning, problem solving time (not including learning time) was measured in two extreme cases: (a) learn all rules and then use them, and (b) learn and use one rule after another. Figure 11 shows the result; The line connecting white circles corresponds to the former, *non-interleaving* case, and the line connecting black circles to the latter, *interleaving* case.

A more detailed description of the experiment is as follows. In the *non-interleaving* case, first, for each cube$i$1, $i = 1, \cdots, 100$, query (`cube`i`1 weight`), $i = 1, \cdots, 100$ was given, and rules were learned from these problem solving experience. Next, after all the 100 rules were learned, 100 queries of (`cube`i`2 weight`), $i = 1, \cdots, 100$ were given, and the CPU time was measured for problem solving using the learned rules (shown by the line connecting white circles in Figure 11). In the *interleaving* case, for each $i = 1, \cdots, 100$, the following was done. First, problem (`cube`i`1 weight`) was solved, a rule was learned from this experience, and the rule was used to solve a problem (`cube`i`2 weight`). The time for this problem solving using a rule was measured (shown by a line connecting black circles in Figure 11).

As Figure 11 shows, there is a drastic difference between the curves for the two cases, which is explained as follows. The rule interpreter used in this experiment is implemented as the `use.rules` method, and the method first collects all the rules applicable to a given problem, inheriting down a set of rules from more general problem class of the given problem. Since rule sets associated with more specific classes of problems depend on, or are justified by, the rule sets of more general classes, THEO's truth maintenance system *uncaches* the rule sets associated with more specific problem classes each time a newly learned rule is added to the more general problem class. Therefore, if problem solving and learning are interleaved, THEO must search for and recollect the rule set for each new problem instance following the addition of each newly learned rule.

One way to avoid this degradation is to not maintain dependencies between rule sets and to build rule sets every time they are needed. Another way is to reduce the overhead of collecting rules, by storing them in a less distributed fashion throughout the knowledge base.

5.1.3 Effect of caching during rule interpretation

Since the rule interpreter is implemented as one of the THEO *methods*, it is expected that THEO's other learning mechanisms improve the performance.

In this experiment, 300 queries of (`cube`ij `weight`), $i = 1, \cdots, 100$, $j = 1, 2\, and\, 3$, were given with rules learned for $j = 1$ and used for $j = 2and3$. The following was

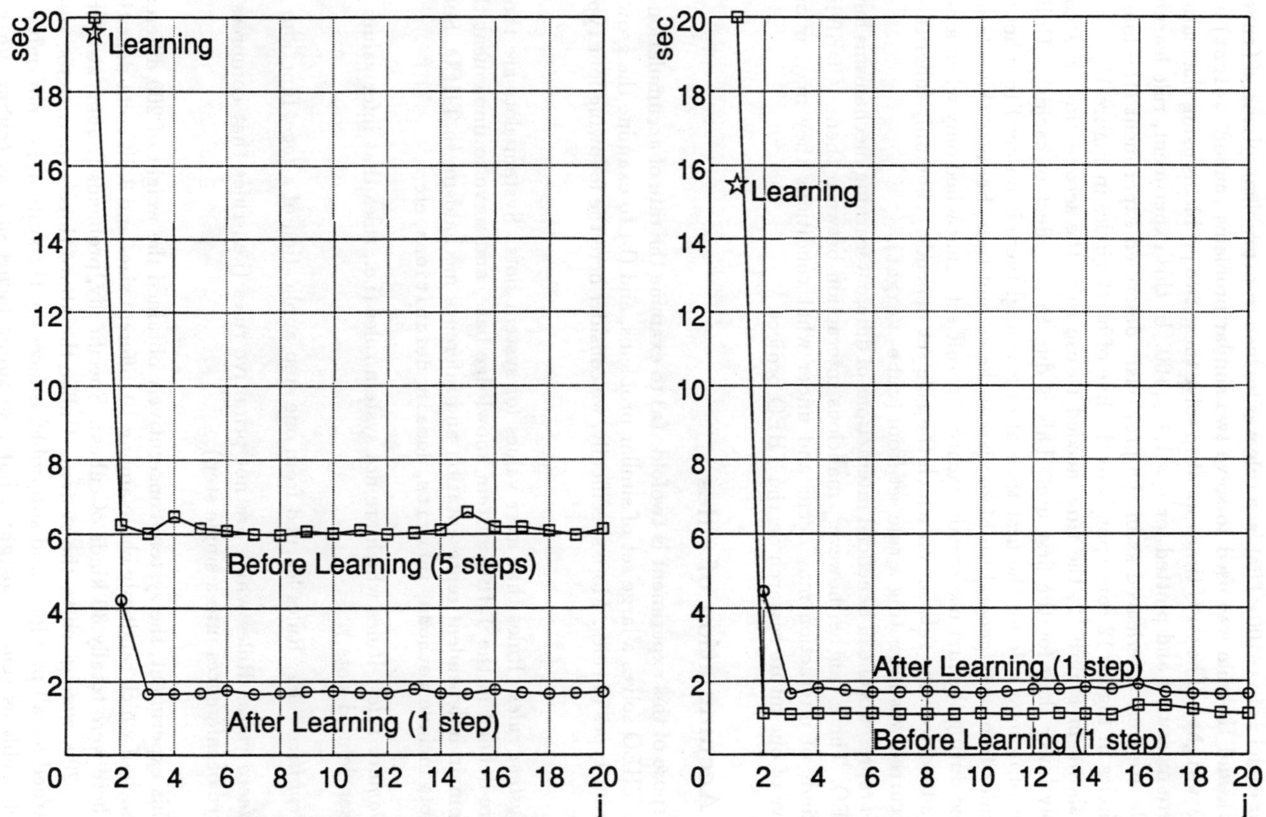

(a) When Chunking Improves Performance

(b) When Chunking Degrades Performance

Figure 10: Effect of Chunking

done for each i = 1,···,100. That is, a rule was learned from problem solving of (**cube**i**1 weight**), and the rule was used to solve two similar problems (**cube**i**2 weight**) and (**cube**i**3 weight**). The CPU times spent to solve these two problems using the learned rule were measured and plotted for $i = 1, \cdots, 100$. In this experiment, rule learning and rule use were interleaved as in the previously described experiment (the line of open circles in Figure 12 corresponds to the line of black circles in Figure 11).

As shown in Figure 12, the time needed to use rules the second time was considerably lower than for the first use. This is due to the effect of caching. Unlike the case of using a rule for the first time after learning, there is no need to inherit a set of rules from very general problem classes the second time, because the rule sets inherited during the first use remain cached. (A rule set is uncached only when a new rule is added to some rule set on which the rule set depends.) The only inheritance that occurred was inheriting a rule set from (**cube weight**).

This represents one of beneficial interactions of different learning mechanisms built on THEO. There can be, however, malicious interaction between them. Clarifying what kinds of interaction may occur and under what conditions they may, is one objective of our future research on the THEO project.

5.2 Accumulation of rules

The purpose of this experiment is twofold: (a) to examine the rate of accumulation of rules as THEO solves a large set of similar problems, and (b) to examine the growth in different types of rules. More specifically, we consider here the following four types of rules:

- *System rules:* Rules that infer values for system slots. System slots are those pre-defined in the THEO system knowledge base, and serve to implement the domain-independent representation and inference mechanisms in THEO. Such slots include **methods**, **inherits**, **domain**, **definitions**, etc.
- *Domain rules:* Rules which are not system rules (i.e., rules that infer values of user-defined slots.).
- *Primitive rules:* Rules learned from one step application of a *toget* function.
- *Macro rules:* Rules which are not primitive rules (i.e., rules that compress a tree of inferences into a single step).

In this experiment, the system consecutively obtained the weight of 200 different cubes, each of which randomly chose among 10 different sizes and 3 different densities (thus, there were totally 30 kinds of cubes). Specifically, problems (**cube**i **weight**), $i = 1, \cdots, 200$, were given to the system. Unlike the above three experiments, rules were learned for all problems and subproblems faced by the system, i.e. not only the top-level problems (**cube**i **weight**), but also subproblems such as (**cube**i **weight**

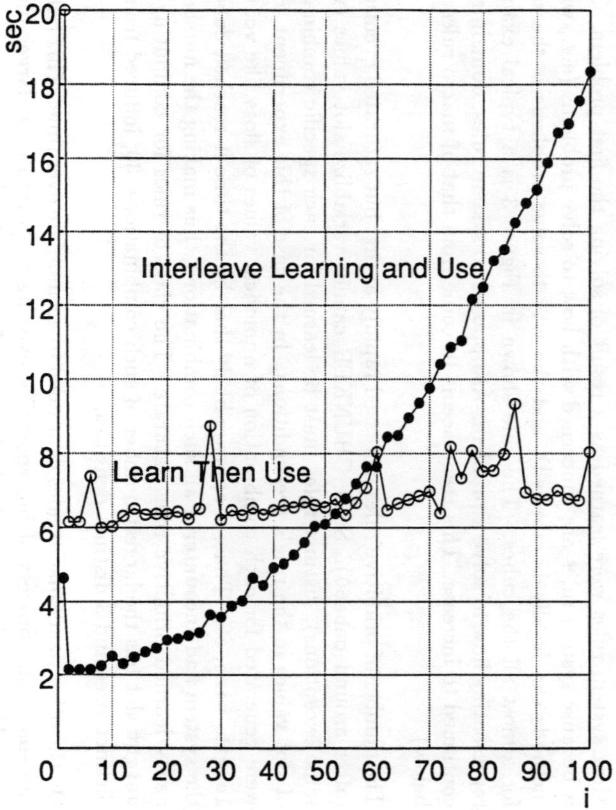

Figure 11: Interleaving Learning and Problem Solving

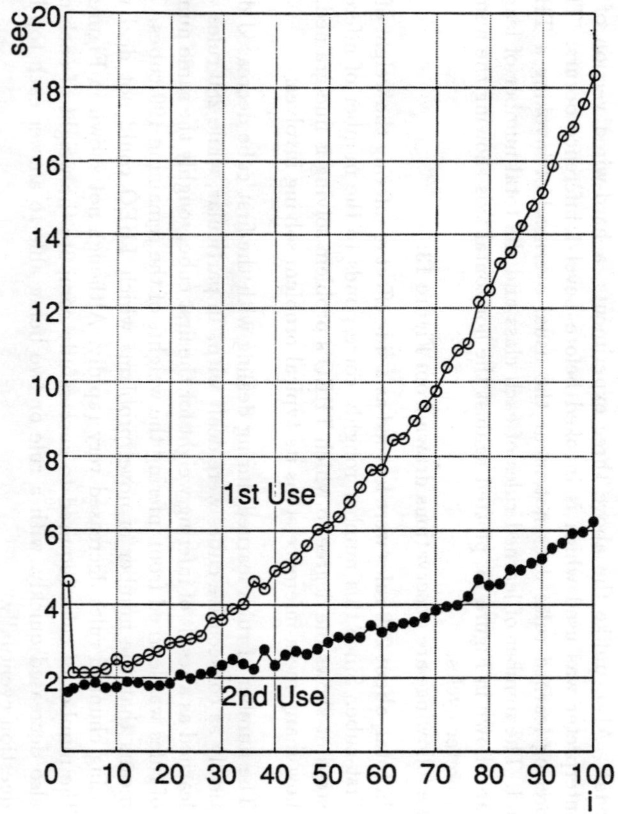

Figure 12: Effect of Caching During Rule Interpretation

methods). Also, unlike the above three experiments, a hard-wired version of the rule interpreter was used which is invoked before Level 1 inference occurs. Thus, interpreting learned rules does not incur the usual overhead of invoking a THEO method. The number of learned rules of each class and the total number of learned rules are shown in Figure 13, plotted against the horizontal axis showing the number of interpreter cycles.

The followings are observations drawn from Figure 13:

- In total, about 280 rules were learned as a side effect of infering the weight of the first cube. Since this number roughly corresponds to the number of inference steps, it shows the degree to which THEO's problem solving is fine-grained, i.e. how many meta-inference steps its typical problem solving involves.

- The number of rules learned during dealing with the first cube decreased drastically as consecutive cubes were dealt with. In particular, while 280 rules were learned as a result of infering weight for the first cube, roughly the same number of rules was learned from infering the weights of the remaining 199 cubes. This means that the number of novel problems which THEO could not deal with using chunked rules decreased very rapidly. Although not shown in Figure 13, the number of rules required to successfully compute the weight of each cube also decreased quickly, with a rule or two being able to answer each top-level question eventually.

- Most system rules were learned as a result of solving the first problem. This is because system rules are concerned with how to solve problems (e.g., which methods can legally be applied), and the way to solve problems is the same for almost all the cubes. The rule shown in Figure 3 is a typical example representing how to solve a problem. As opposed to system rules, domain rules continued to increase. This trend seems to come from that of macro rules (see below).

- The number of primitive rules increased rapidly at first but saturated eventually (after around `cube60`). Since CHUNKER cannot variablize slot values (with some exceptions), distinct rules must be learned for each specific combination of slot values in the rule's preconditions. In the case of this experiment, rules were generated for each combination of numerical values of slots like **volume**, **length**, **size**, etc. Thus, it appears likely that by the time it reached **cube60**, the system had encountered all such combinations, thus making the number of newly learned primitive rules saturate. The class of rules for computing the **weight** slot has the largest number of such combinations, 30, induced from 10 distinct sizes and 3 distinct densities.

- On the other hand, macro rules continued to accumulate over time. This reflects the continuing process of incorporating existing rules into new, larger macro rules. This trend appears also in the curve for domain rules.

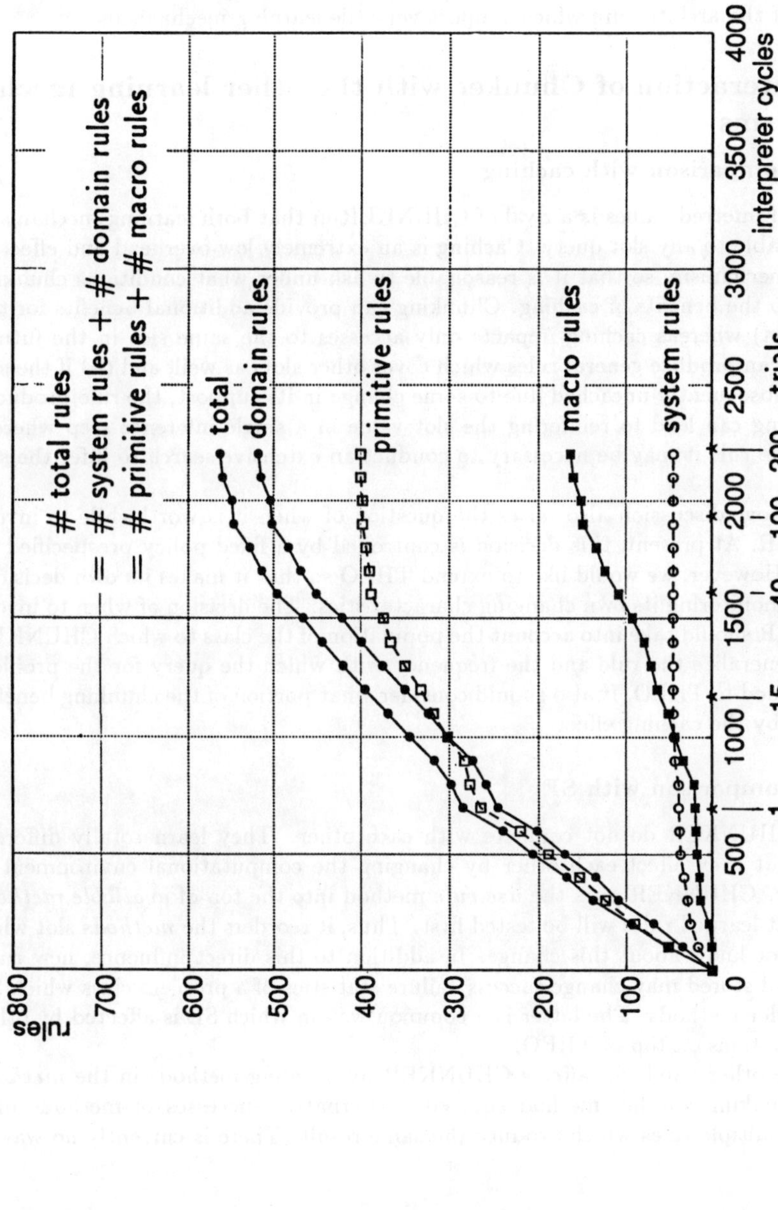

Figure 13: Accumulation of rules

6 Discussion

In this section, we first compare CHUNKER with the other four learning mechanisms developed on top of THEO, exploring the interactions among them. We then discuss features of the architecture which support versatile learning mechanisms.

6.1 Interaction of Chunker with the other learning mechanisms

6.1.1 Comparison with caching

Caching of inferred values is a rival of CHUNKER in that both learning mechanisms are applicable to any slot query. Caching is an extremely low-overhead and effective learning mechanism, so that it is reasonable to ask under what conditions chunking can add to the benefits of caching. Chunking can provide additional benefits for two reasons: (a) whereas caching impacts only accesses to the same slot in the future, chunking can produce general rules which cover other slots as well, and (b) if the slot value is subsequently uncached due to some change in its support, the rule produced by chunking can lead to reinfering the slot value in a single inference step whereas without the rule it may be necessary to conduct an extensive search to infer the slot value.

The above discussion also raises the question of when it is worthwhile to invoke CHUNKER. At present, this decision is controlled by a fixed policy prespecified by the user. However, we would like to extend THEO so that it makes its own decisions based on monitoring its own changing characteristics. The decision of when to invoke CHUNKER should take into account the population of the class to which CHUNKER tries to generalize the rule and the frequency with which the query for the problem class is issued to THEO. It also should consider what portion of the chunking benefits are offset by the caching effect.

6.1.2 Comparison with SE

SE and CHUNKER do not compete with each other. They learn totally different things. But they affect each other by changing the computational environment of each other. CHUNKER puts the *use.rule* method into the top of *available.methods* slot so that learned rules will be tested first. Thus, it reorders the *methods* slot while SE does not know about this change. In addition to this direct influence, new rules learned and stored may change success/failure statistics of a problem class which SE uses to order methods. The latter is a common way in which SE is affected by other learning systems on top of THEO.

On the other hand, SE affects CHUNKER by ordering methods in the *methods* slot and making another method succeed. Alternative successes of methods may result in multiple rules which produce the same result. There is currently no way of choosing among rules, alternative rules for the same problem class become redundant and memory consuming.

6.1.3 Comparison with TMAC

Both CHUNKER and TMAC analyze an explanation tree provided by the TMS to learn compiled knowledge. While CHUNKER learns rules, TMAC creates macro-definitions. While CHUNKER can work in principle on any slot, TMAC only captures *definition* uses. TMAC composes a series of definitions into a macro method and stores it in the `specialized.definitions` slot of a problem class which is a superclass of a given problem. For example, TMAC produces a macro-definition such as the one shown at the bottom of Figure 14, from solving `(cube1 weight)` and stores it in `(cubet weight specialized.definitions)`. Note that even TMAC does not understand that size multiplied by size multiplied by size is equivalent to size cubed. It only captures the chain of definitions used in problem solving and composes them into macro definitions. In contrast, the top of Figure 14 shows the rule formed by CHUNKER from the same training example. Notice in this case the rule produced by CHUNKER is less general than the macro produced by TMAC. This is generally the case: TMAC produces more general results by relying on stronger knowledge associated with level 3 inference, whereas CHUNKER applies to a wider range of slots but uses weaker knowledge to produce weaker generalizations.

Rule learned:

```
(iftrue
   (and (= (th density *de*) 5)
        (= (th size *de*) 10))
   5000)
```

Stored in: `(cube weight rules)`

(a) Rule learned by CHUNKER

Definition learned:

```
(* (th density *de*)
   (* (th size *de*)
      (th size *de*)
      (th size *de*)))
```

Stored in: `(cube weight specialized.definitions)`

(b) Definition learned by TMAC

Figure 14: Comparison between CHUNKER and TMAC

6.2 Requirement for architecture

In this section, we discuss architectural features of THEO which facilitates CHUNKER.

6.2.1 Highly uniform representation

THEO's highly uniform representation is important to CHUNKER for several reasons. The uniformity of the representation and the commitment that all problem solving corresponds to inferring slot values, allows CHUNKER to be applied to all aspects of THEO's reasoning. In addition, the ability to treat problems as entities and hold beliefs about these, leads to a convenient scheme for representing knowledge needed by CHUNKER about methods, their application, and operationality of slots.

CHUNKER can recompile learned rules into another rule. Specifically, this is because a learned rule becomes one of THEO's inference methods, and can then be used by another session of problem solving. Like other methods, rule use is recorded by TMS with supporting beliefs corresponding to conditions of the rule. Therefore, conditions of a rule are combined into a new rule. In this way, a set of rules can evolve. We would like to explore a variety of forgetting mechanisms in order to develop schemes which evolve high-performance sets of rules.

6.2.2 Truth maintenance facility

The explanation structure constructed by the truth maintenance facility is, of course, central to the generalization algorithm of CHUNKER. There are other benefits of the truth maintenance system as well. Since those beliefs collected during search through the explanation tree which are not variablized are stored as assumptions (i.e. as the explanation of the resulting rule), the rule itself is subject to the truth maintenance system. If some assumption ceases to hold, the rule is retracted. Thus, truth maintenance facility can keep track of truth of rules as well as data and other inference method. If the operationality criterion referred to some beliefs existing at the time of rule creation, and if the belief becomes unsatisfied by the current belief, then the rule is retracted. Thus, operationality predicate not only dynamically judges the operationality, but also dynamically maintains the operational set of rules. This mechanism can be used to make a set of rules keep up with the shift of computational environment.

However, THEO's truth maintenance facility is still overly simple in some ways. First, while THEO supports multiple values for one slot, its TMS is not good at separating out the explanations for the individual values. CHUNKER suffers from this shortcoming. This is the case with the *method* slot, and the explanation related to an irrelevant method may constrain the domain of a learned rule. For example, even if inheritance fails, the explanation structure concerning how to inherit is included. Although CHUNKER carefully analyzes explanation structure and eliminates irrelevant substructure, this reasoning would be less complicated with a more sophisticated explanation structure.

Second, while the explanation is intended to provide a logical justification for a slot value, the TMS records all beliefs accessed by the *toget* function during problem solving *regardless* of whether they were used to support the final value. This can lead to extra nodes in the explanation and to rules which are less general than they might be. However, this is part of the price THEO pays for allowing users to attach arbitrary Lisp procedures for inferring slot values. When users program at the more structured inference level 2 or 3, it is possible for the learning mechanism to take advantage of this stronger discipline to provide stronger generalizations.

7 Conclusion

Our initial explorations developing and incorporating the CHUNKER learning mechanism in THEO illustrate the possibility of embedding a general rule-learning ability in a uniform frame-based architecture. While our preliminary results are encouraging, we have yet to extensively test the application of CHUNKER to learning the thousands of rules which in principle are learnable from the thousands of slot accesses in a typical problem solving domain in THEO. No doubt, doing so will lead to new insights and new issues, and to the need for an intelligent selection by THEO of those slots to which it should apply this learning mechanism.

Acknowledgements

We would like to thank the other members of the THEO group for useful discussions, and for their collaboration more generally in developing THEO: Jim Blythe, John Cheng, Lisa Dent, Oren Etzioni, Marc Ringuette, and Jeff Schlimmer. This research has been supported by the National Science Foundation under grant IRI-8740522, and by a grant from Digital Equipment Corporation.

References

[1] Jim Blythe and Tom M. Mitchell. On becoming reactive. In *Proceedings of the Sixth International Workshop on Machine Learning*, pages 255–257, 1989.

[2] Gerald F. DeJong and Raymond Mooney. Explanation-based learning: An alternative view. *Machine Learning*, 1(2):145–176, 1986.

[3] Oren Etzioni. Hypothesis filtering: A practical approach to reliable learnig. In *Proceedings of the Fifth International Conference on Machine Learning*, pages 416–429. Morgan Kaufmann, 1988.

[4] Russell Greiner and Doug B. Lenat. RLL: A representation language language. In *Proceedings of the First National Conference on Artificial Intelligence (AAAI80)*, pages 165–169, 1980.

[5] John E. Laird, Allen Newell, and Paul S. Rosenbloom. Soar: An architecture for general intelligence. *Artificial Intelligence*, 33(1):1–64, 1987.

[6] Steven Minton. Quantitative results concerning the utility of explanation-based lerarning. In *Proceedings of the Seventh National Conference on Artificial Intelligence (AAAI88)*, pages 564–569, 1988.

[7] Tom M. Mitchell. Becoming increasingly reactive. In *Proceedings of the Eighth National Conference on Artificial Intelligence (AAAI90)*, pages 1051–1058, 1990.

[8] Tom M Mitchell, John A. Allen, Prasad Chalasani, John Cheng, Oren Etzioni, Marc Ringuette, and Jeffrey C. Schlimmer. Theo: A framework for a self-improving systems. In *Kurt van Lehn (Eds.), Architectures for Intelligence.* Lawrence Erlbaum, 1990 (to appear).

[9] Tom M. Mitchell, Richard Keller, and Smadar Kedar-Cabelli. Explanation-based generalization: A unifying view. *Machine Learning*, 1(1):47–80, 1986.

Applications of Learning and Planning Methods
edited by N. G. Bourbakis

CHAPTER 2

CONNECTIONIST LEARNING WITH CHEBYCHEV NETWORKS AND ANALYSES OF ITS INTERNAL REPRESENTATION

Akira Namatame

Department of Computer Science
National Defense Academy
Yokosuka, 239 JAPAN

ABSTRACT

We propose a new architecture for supervised neural network learning. We show that the inductive capability or generality of supervised learning depends on how the continuous mappings are representing and obtain the necessary conditions of the networks' architecture for learning the structures of the continuous mappings. We propose a new network architecture, Chebychev networks with some Chebychev polinomials as input units. It is shown that Chebychev networks trained with backpropagation can generalize and infer the highly nonlinear structures of the continuous mappings. The internal representation capability of the high-order functional networks is analyzed. Nonlinear mappings can be characterized by the features of their extremums and curvatures. It is shown that the combination of some nonmonotonic Chebychev polonomials as input units and some monotonic Sigmoid functions as hidden and output units make it possible to realize and learn a proper internal representation of the networks for extracting those features of the continuous mappings. Based on the analyses of the internal representation capabilities, a methodology for determining the size of the network architecture of Chebychev networks is also proposed.

1. Introduction

Connectionist models are known by names including neural network models, artificial neural networks and parallel distributed processing models [14]. Whatever the names, all these models attempt to achieve good performance via dense connections of simple computational elements. Connectionist model research has a long history, and a new resurgence of the field has been led by several new discoveries. This new interest is mainly due to the development of a new learning algorithm, backpropagation [12]. A multi-layer network trained with backpropagation provides a technique that may have the greatest potential in many information processing problems. The potential benefits of connectionist models extend beyond the high computational rates provided by massive parallelism. They also adapt the connection weights in time to improve the performance.

A key facility of intelligence is the ability to infer general regularities or to extract essential features from complex, high-dimensional data. It has been widely recognized that inductive inference capability or generalization is a fundamental issue for machine learning, and there has been a great deal of theoretical and experimental work done on inductive learning in symbolic AI research [10]. Connectionist learning may provide an alternative approach to inductive learning and provide a number of important properties that are not found in symbolic AI. The process of inferring general regularities requires the construction of complex mappings, which utilizes internal representation, from a high-dimensional input space to a lower dimensional output space. The multi-layer neural network may realize a proper internal representation within hidden units [3][12]. Despite its wide popularity, this inference capability of connectionist learning has not been explicitly analyzed [1].

In this paper, the generality of supervised connectionist learning is examined. We choose a set of continuous patterns as a task domain for learning. Depending on the method of encoding a continuous pattern into the networks continuous, some architecture models can be utilized. It is shown that conventional models are inadequate to represent the continuous nonlinear mapping. Although they can represent the nonlinear relation in the training set, they fail to capture structural relation. Therefore, it may require infinite units to represent nonlinear patterns with arbitrary precision and results in the network complexity. We propose a new network architecture with Chebychev polinomials as input units and show that it can generalize a highly nonlinear structural relation within the continuous pattern. We also explore the internal representation power of Chebychev networks and show that the generalization and learning capability of Chebychev networks is only realized by the combination of some nonmonotonic Chebychev polinomials as input units and some monotonic Sigmoid function as hidden and output units.

2. Generalization of Connectionist Learning

The ability of recognizing wholes from partial cues, even for patterns not encountered before is known as generalization or induction [5]. A key facility of intelligence is the ability to infer general regularity or to extract essential features from complex data. The connectionist learning with backpropagation may provide an alternative method that learns from examples, therefore, providing a number of important properties not found in symbolic AI researches. The process of inferring general regularities requires the construction of complex mappings, which utilizes internal representation, from a high-dimensional input space to a lower dimensional space. The internal representation of these mappings are believed to capture high-order correlations which embody the invariant relationships between the input data and the output data. It is widely believed that a multi-layer neural network realizes proper internal representation and poses useful holistic qualities that they may have innate tendencies to recognize wholes from partial examples, even for patterns that they have not encountered before. However, those innate properties can be realized only if the representation model and the corresponding network architecture that fit the task domain for learning are appropriately designed.

We choose a set of continuous patterns as a task domain for learning. Each continuous pattern in 2-dimensional space is defined by a subset $D=\{(x,y);\ x \in K,\ y=F(x)\}$, where K is a compact set of R and F is a continuous function defined on K. The structure of the function F, however, may be difficult to identify. We consider the problem of representing D on a multi-layer neural network utilizing a set of a few training examples $T=\{(x_1,y_1),(x_2,y_2),\ldots,(x_n,y_n)\}\subset D$. The task for learning is then stated as to acquire a functional structure or to generalize a structural relation between an input variable x and an output variable y from a finite training set T.

We assume the function $F(x)$ is a class of C^2 with many extremums. The set of the extremums of $F(x)$ is denoted by G, i.e., $G=\{x_s \in K: F'(x_s)=0,\ s=1,2,\ldots,S\}$. The radius of curvature ρ_s at the extremum x_s is given as

$$1/\rho_s = F''(x_s) \tag{2-1}$$

We consider the following specific functions defined on K=[0, 1] as the test cases.

(Case 1)

$$F(x)= 0.5 + 0.3\sin 8\pi x, \tag{2-2}$$

(Case 2)

$$F(x)= 0.5 + 0.3\sin 12\pi x, \tag{2-3}$$

(Case 3)

$$F(x)=(12+3x-3.5x^2+7.2x^3)(1+\cos 4\pi x)(1+0.8\sin 3\pi x)/44 \tag{2-4}$$

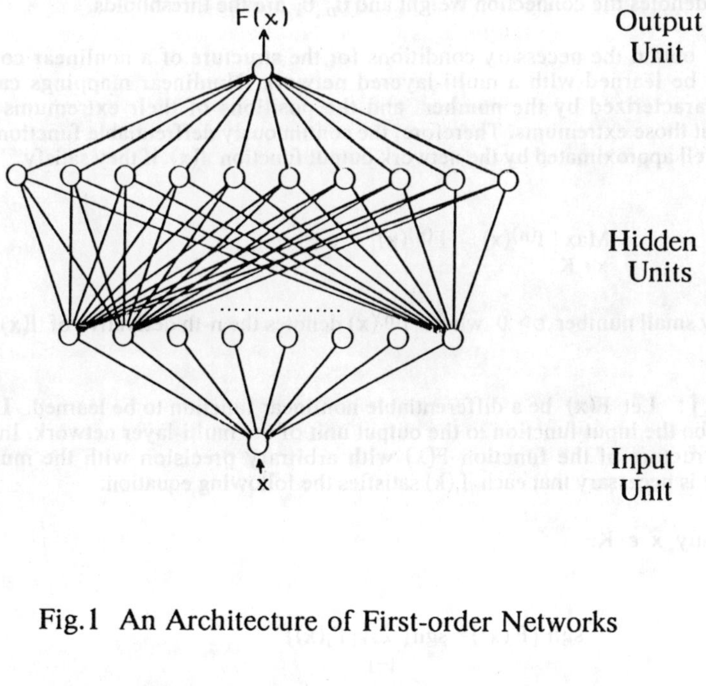

Fig.1 An Architecture of First-order Networks

3. Necessary Conditions of an Network Architecture

It is known that any continuous function can be approximated to arbitrary precision by a three-layer neural network [2][6]. However, there is no systematic method to construct such a network. A basic and commonly used model of representing continuous relation in a neural network is to encode input and output variables by continuous input and output units as shown in Fig.1[3][9]. We term the network architecture in Fig.1 as first-order networks. Learning and representing powers of the multi-layer network is greatly enhanced by nonlinear transformations with sigmoid-type output functions employed with hidden units. However, learning with a first-order network architecture is characterized by the monotonic transformations.

We denote the input function to the output unit of the network from the i-th hidden unit at the final hidden layer by $f_i(x)$ and the sum of the inputs to that unit from the units at the lower layer by $h_i(x)$, i=1,2,...,k. The network output function $f(x)$ is then given as

$$f(x)= 1/[\ 1 + \exp\{- \sum_{i=1}^{k} v_i f_i(x) + \theta_o\}] \qquad (3\text{-}1)$$

and $f_i(x)$ is given as

$$f_i(x)= 1/[\ 1 + \exp\{- h_i(x) + \theta_i\}] \qquad (3\text{-}2)$$

where w_i denotes the connection weight and θ_o, θ_i are the threshholds.

We now obtain the necessary conditions for the structure of a nonlinear continuous function to be learned with a multi-layered network. Nonlinear mappings can be, in general, characterized by the number and the positions of their extremums and the curvatures at those extremums. Therefore, the continuously deffrentiable function $F(x)$ is said to be well approximated by the network output function $f(x)$ if they satisfy

$$\max_{x \varepsilon K} |\ f^{(n)}(x) - F^{(n)}(x)| < \varepsilon \qquad (3\text{-}3)$$

for arbitrary small number $\varepsilon > 0$ where $f^{(n)}(x)$ denotes the n-th derivative of $f(x)$, n=1,2.

Lemma 3.1: Let $F(x)$ be a differentiable nonlinear function to be learned. Let $f_i(x)$, i=1,2,...,k, be the input function to the output unit of the multi-layer network. In order to learn the structure of the function $F(x)$ with arbitrary precision with the multi-layer network, it is necessary that each $f_i(k)$ satisfies the following equation:

(i) For any $x\ \varepsilon\ K$,

$$\text{sgn}\ \{F'(x)\} = \text{sgn}\{\ \sum_{i=1}^{k} v_i f'_i(x)\} \qquad (3\text{-}4)$$

where sgn(x) denotes the sign of x.

(ii) For any extremum $x_s \in G$ of F(x),

$$\left| \sum_{i=1}^{k} v_i f'_i(x_s) \right| < \varepsilon \tag{3-5}$$

Proof: Let f(x) be the output function of the multi-layer network that is given in (3-1). The derivative of f(x) is given as

$$f'(x)=f(x)\{1-f(x)\}\sum_{i=1}^{k} v_i f'_i(x) \tag{3-6}$$

For any $x \in K$, we have $0< f(x)\{1-f(x)\}\leq 1/4$, and therefore,

$$\begin{aligned} \text{sgn}\{F'(x)\} &= \text{sgn}\{f'(x)\} \\ &= \text{sgn}\{\sum_{i=1}^{k} v_i f'_i(x)\} \end{aligned} \tag{3-7}$$

Especially at the extremum x_s of F(x) satisfying $F'(x_s)=0$, we should have

$$\left| \sum_{i=1}^{k} v_i f'_i(x_s) \right| < \varepsilon \tag{3-8}$$

The derivative of $f_i(x)$ is given as $f'_i(x)= f_i(x)\{1- f_i(x)\}h_i(x)$ and $\{f_i(x)\}$, i=1,2,...,k, can be assumed to be linearly independent functionals. Therefore, in order to satisfy the condition (3.8) for all extremum x_s, it is necessary to have

$$f'_i(x_s)=0 \tag{3-9}$$

for any $x_s \in K$ satisfying $F'(x_s)=0$. These results imply that it is necessary for the networks to acquire the proper structures of the input and output functions of the hidden units that satisfy the conditions in (3-9). That is, each output function of the hidden units should be nonmonotonic and the value of its derivative evaluated at each extremum of the learning function should be zero. The limitation of a multi-layered network with the first-order architecture stems from the linear input to each hidden unit and the monotonic transformation with the sigmoid-type output functions at the hidden units and the output unit in the network. Those linear input and monotonic transformations within the multi-layered network do not satisfy the condition of (3-9).

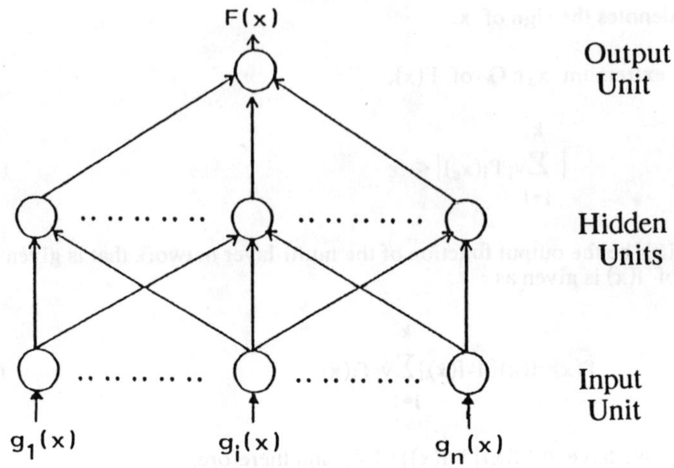

Fig.2 An Architecture of Chebychev Networks

4. Learning with Chebychev Networks' Architecture

In order to learn the structure of a complex continuous mapping with arbitrary precision, it may require a large network with many training examples unless there is a strong generalizing capability with the network and a learning algorithm. We claim that the problem of generalization depends on the manner in which the learning task domain is represented, i.e., the network architecture. In the previous section, we showed that the nonlinear transformation in the network with the monotonic functions such as sigmoid functions are inadequate to learn the complex nonlinear structures of the continuous mappings. The signs of the derivative of the network function are determined by the connection weights of the units in the network and they do not depend on the value of x. This property of the monotonic transformation, therefore, lacks the flexibility of approximating nonlinear structure of the task domain. In order to avoid the limitation of the monotonic transformation of the network, we should implement some nonmonotonic high-order functions as the inputs to the network [7]. We consider three-layer high-order functional networks with some high-order functional units $\{g_1(x), g_2(x), \ldots, g_n(x)\}$ as input units as shown in Fig.2 [11]. The advantage of high-order functional networks only be realized when such high-order functional units are selected that may fit any continuous pattern. If the form of such a continuous patterns is not known as priori, the applicapability of the high-order functional networks may depend on the system designer's ability or a priori knowledge to a task domain for choosing a set of high-order functional units that adequately represent the problem. We term high-order functional networks with Chebychev polynomials as input units as Chebychev networks. While implementing high-order functional that are non-monotonic as input units, the output functions of hidden units after monotonic transformation at the hidden layer become nonmonotonic and it becomes possible to find the connection weights satisfying (3-4).

A Chebychev polynomial $g_i(x)$ defined on $0 \leq x \leq 1$ is given as

$$g_i(x) = [\cos\{i \arccos(2x-1)\} + 1]/2 \qquad i=1,2,3,.... \tag{4-1}$$

Each $g_i(x)$ takes the value of [0,1] and has i-1 extreme points. Some specific Chebychev polynomials are given as $g_1(x)=x$, $g_2(x)=(2x-1)^2$, $g_3(x)=x(4x-3)^2$.

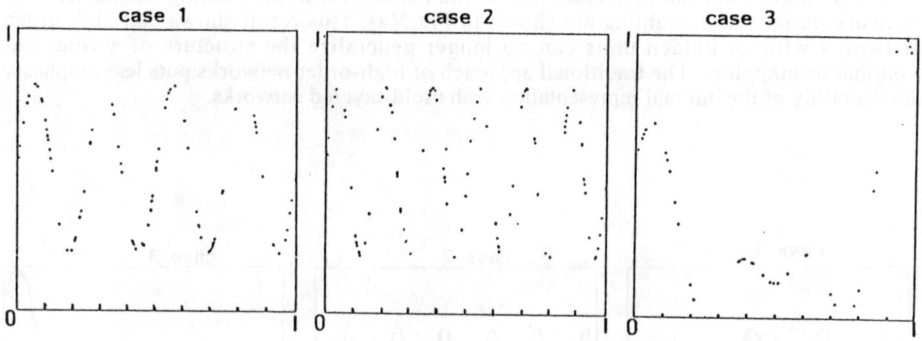

Fig.3 The Depictions of Training Examples of Each Test Case

In order to evaluate the learning capability of Chebychev networks, we constructed the three-layer network with Chebychev polynomials input units up to the 5th order and four hidden units for Case 1 and Case 3 and the up to the 9th order Chebychev polynomials input units and three hidden units for Case 2. We trained each network with the arbitrary chosen training examples (80 training examples for Case 1, 120 training examples for Case 2 and 40 for Case 3). The training examples are depicted in Fig.3. The training cycle for each case was 2,000. The network outputs after training when presented any value x in [0,1] with the interval of 1/500 are shown in Fig.4. These results show that Chebychev networks can learn and infer highly nonlinear structures of continuous mappings via a small set of training examples demonstrating the representation and generalization power of Chebychev networks.

There is a relatively long history of supervised learning of implementing high-order units as input units [4][7] and it is known that the learning power of a network is enhanced by implementing the high-order units. There are essentially two models for implementing high-order units: outer-product and functional link models. In the outer-product model, each component of the input pattern multiplies the input pattern vector and includes the high-order products and powers of input variables. In the functional link model, the input pattern vector is transformed into a larger pattern space through a set of functions such as sine and cosine functions. Those models, however, essentially aim to increase the speed of the learning algorithm by improving the representation power of the input space and deleting the hidden units. The two-layer flat network which uses high-order units can learn polynomial mappings by using a learning algorithm for training. This can eliminate the need for multi-layered networks and provide extremely rapid learning for problems where it is known a priori that a polynomial mapping is appropriate. The derivative of the network output function with a flat network without hidden units is given as

$$f'(x)=f(x)\{1-f(x)\}\sum_{i=1}^{k} v_i\, g'_i(x) \qquad (4\text{-}2)$$

Therefore, each high-order input function $g_i(x)$ should have the same extremums with F(x), the task domain for learning, and this is the unrealistic assumption when we do not have information regarding the structure of F(x), the task domain for learning. We trained two-layer high-order flat networks without hidden units with the training examples. The network outputs after training are shown in Fig.5(a). This result shows that high-order networks without hidden units can no longer generalize the structure of a complex continuous mapping. The traditional approach of high-order networks puts less emphasis on the utility of the internal representation with multi-layered networks.

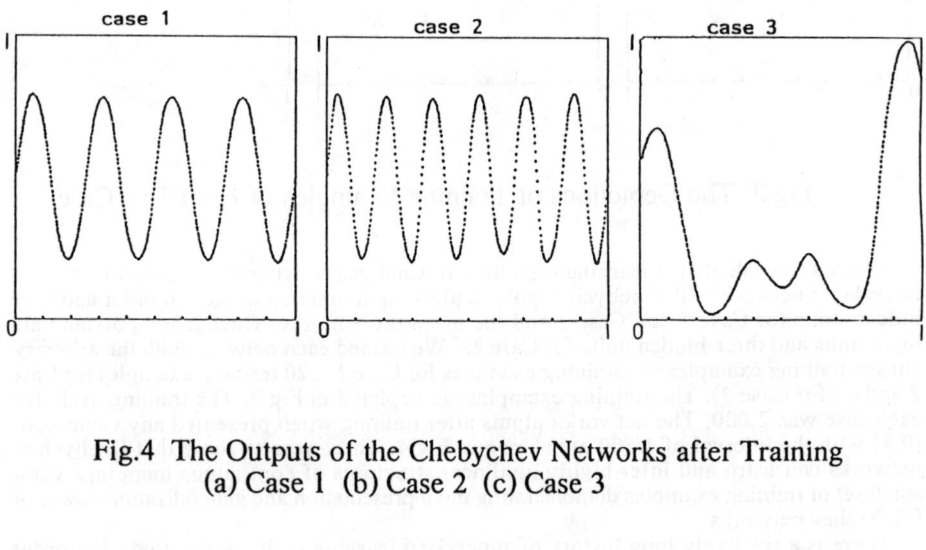

Fig.4 The Outputs of the Chebychev Networks after Training
(a) Case 1 (b) Case 2 (c) Case 3

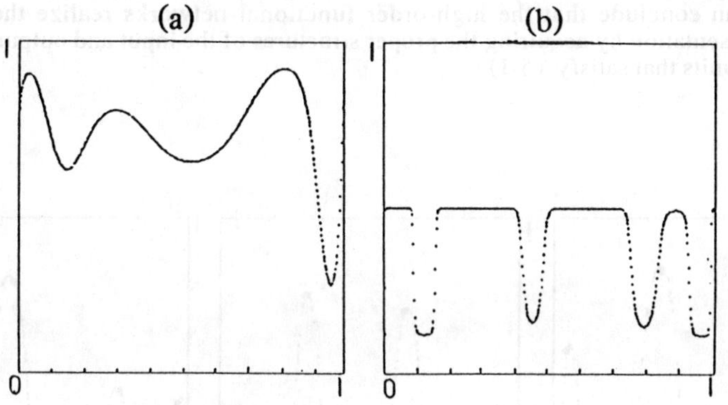

Fig.5 (a)The Outputs of Flat-networks(without hidden unit)afterLearning
(b)The Outputs of Chebychev networks after Learning
for Test Case 2

5. Analyses of Internal Representation of Chebychev Networks

In this section, we show that Chebychev networks can provide excellent generalization by specializing the selection of high-order terms in order that the network output after learning matches any given specific problem. If each output function of the hidden units, $f_i(x)$, i=1,2,...,k, has the form of in (5-1), that is a kind of the set of the step functions, it can be easily shown that they satisfy the conditions of Lemma 3.1:

$$f_i(x) = \begin{cases} \phi_s(x) & \text{for } x \in N(x_s) \\ \{0, 1\} & \text{otherwise} \end{cases} \tag{5-1}$$

where $\phi_s(x)$ is some nonmonotonic function defined over the neighborhood $N(x_s)=(x_s-\delta_s,\ x_s+\delta_s)$ of some extremums x_s of $F(x)$ and satisfying $\phi'_s(x_s)=0$. The output function of each hidden unit of Chebychev networks that were trained with the training examples of Case 2 and 3 are shown in Fig.6 and 7 respectively. From these results, we can conclude that the high-order functional networks realize the internal network representation by acquiring the proper structures of the input and output functions at the hidden units that satisfy (5-1).

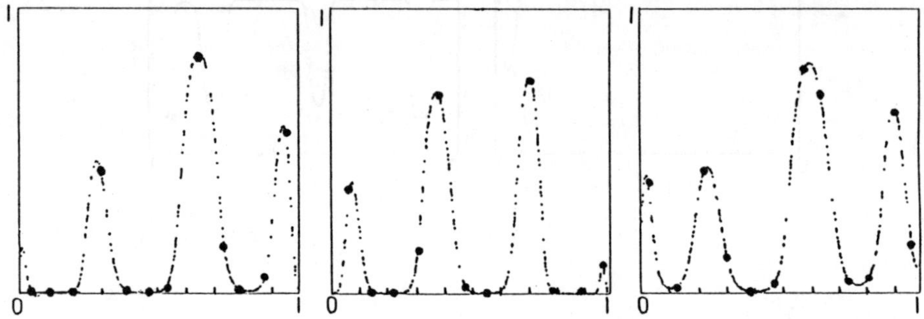

Fig.6 The Output Functions of Hidden Units Acquired after Learning for Test Case 2.
(The dots denote the extremums of the learning function.)

The power of the high-order networks are only realized if some high-order functional input units and hidden units are implemented. The internal representation that may fit to the specific given problem is only realized by the combinations of the high-order input units and hidden units. The roles of the hidden units and the high-order functional input units are characterized to capture the features regarding the number and the positions of the extremums of the learning task function. The positions of the extremums of the output functions from the hidden units should be close to those of the task domain and they should cover all these extremums and, therefore, a sufficient number of the hidden units is required. In Fig.5(b), we show the network output with one hidden unit that trained the Case 2 (2-2) and the features regarding some extremums are not extracted because of the lack of hidden units. The role of the high-order input units is also characterized to capture the curvatures of the task function. Let ρ^* be the minimum absolute value of the radius curvature at the extremum of the task function $F(x)$, i.e.,

$$1/\rho^* = \text{Max } F''(x_s) \tag{5-2}$$

From (3-6)and (3-9), at the extremum x_s of F(x), we have

$$F''(x_s)=f'(x)=f(x_s)\{1-f(x_s)\}\sum_{i=1}^{k} v_i\, g''_i(x_s) \tag{5-3}$$

where $g_i(x)$, i=1,2,...,n, are the Chebychev polynomial with the i-th order. Therefore, the curvature of the network function f(x) at its extremum is essentially determined by the curvature of $g_n(x)$,the high-order functional input unit with the highest order.

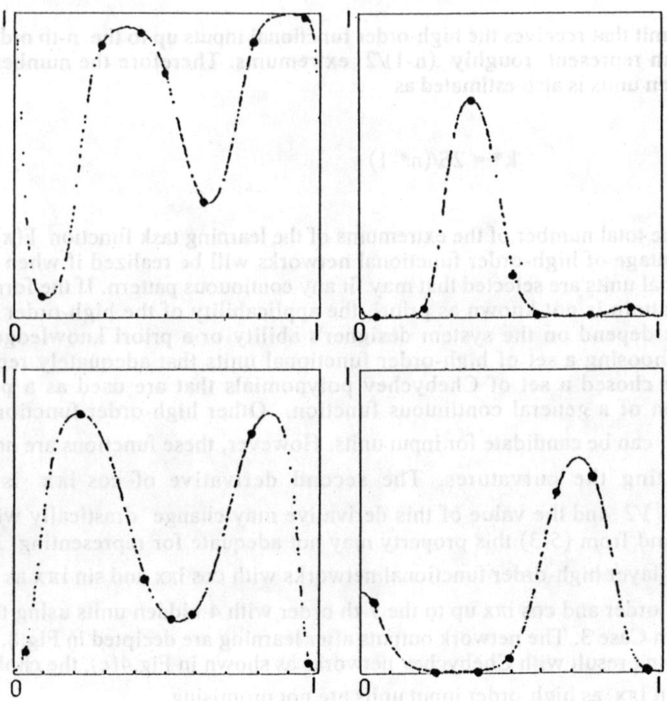

Fig.7 The Output functions of the hidden units acquired after Learning for Test Case 3
(The dots denote the extremums of the learning function.)

Currently, there is no available proven methodology for determining network architecture and parameters. By examining the internal representation capabilities of high-order functional networks, we can gain insights into building neural networks that are applicable to the given task domains. The absolute value of the curvature of the Chebychev polynomial with the n-th order around its extremum is approximated with

$$g''_i(x) = i^2/2x(1-x) \qquad (5\text{-}4)$$

which takes the value of $[2i^2, \ )$ and it can easily represent a small absolute value of the radius of curvature. From (5-2)(5-3) and (5-3) the required order of the Chebychev polinomials as input units n^* can be roughly estimated as

$$n^* = (\rho^*/2)^{1/2} \qquad (5\text{-}5)$$

Each hidden unit that receives the high-order functional inputs up to the n-th order from the input units can represent roughly $(n-1)/2$ extremums. Therefore the number k^* of the required hidden units is also estimated as

$$k^* = 2S/(n^*-1) \qquad (5\text{-}6)$$

where S is the total number of the extremums of the learning task function F(x).

The advantage of high-order functional networks will be realized if when such high-order functional units are selected that may fit any continuous pattern. If the form of such a continuous pattern is not known as priori, the applicability of the high-order functional network may depend on the system designer's ability or a priori knowledge to a task domain for choosing a set of high-order functional units that adequately represent the problem. We chosed a set of Chebychev polynomials that are used as a polynomial approximation of a general continuous function. Other high-order functions such as $g_i(x)=\cos i\pi x$ can be candidate for input units. However, these functions are not adequate for representing the curvatures. The second derivative of $\cos i\pi x$ is given as $-(i\pi)^2(\cos i\pi x)/2$ and the value of this derivative may change drastically with a small change of x and from (5-3) this property may not adequate for representing F"(x). We trained three-layer high-order functional networks with $\cos i\pi x$ and $\sin i\pi x$ as input units up to the 4-th order and $\cos i\pi x$ up to the 9-th order with 4 hidden units using the training examples from Case 3. The network outputs after learning are decipted in Fig.8. Compared with the learning result with Chebychev networks as shown in Fig.4(c), the choice of $\cos i\pi x$ and $\sin i\pi x$ as high-order input units are not promising.

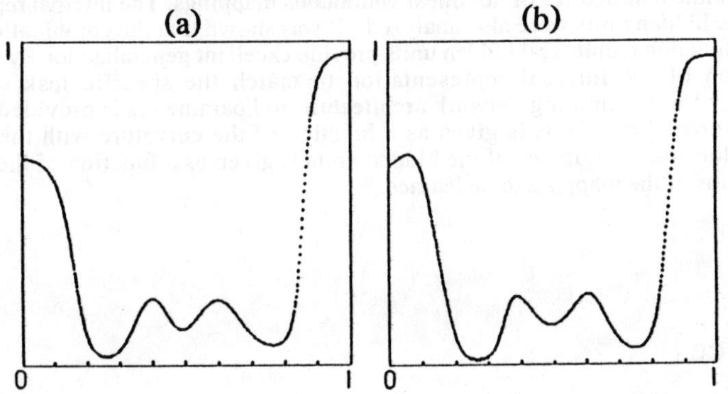

Fig.8 The Outputs of the High-order Functional Networks aftter learning with the Different Types of Input Functions after Learning for Test Case 3 (a) the combination of cos i πx and sin i πx up to the 4-th order (b) cos i πx up to the 9-th order

6. Conclusion

It is often said that distributed representations are inherently posed of useful holistic qualities that they have innate tendencies to recognize wholes from partial examples, even for patterns they have not encountered before [12]. Phenomena of that sort are often described with such words as generalization or induction. We showed that the generalizations made by any particular network are likely to be inappropriate unless there happens to be an appropriate relationship between the network architecture and the manner in which the problem is represented. We considered the set of nonlinear continuous mappings as a task domain for learning that may lay the foundation for many cognitive information processing problems. We showed that the first-order multi-layer networks and the high-order networks without hidden units that are commonly used in the previous researches are inadequate to generalize and to learn the nonlinear structures underlying the continuous mappings. We proposed the new network architecture, high-order functional networks, with some high-order functionals as input units. It was shown that the three-layer networks with some Chebychev polynomials as input units can learn and generalize complex nonlinear structures of nonlinear continuous mappings. The internal representation power of the hidden units were also analyzed. It was shown that the combinations of some high-order functional units and hidden units provide excellent generalization by specializing the selection of the internal representation to match the specific task domain. A methodology for determining network architecture and parameters is provided. The order of the high-order input units is given as a function of the curvature with the maximum absolute value and the number of the hidden units is given as a function of the number of the extremums of the mapping to be learned.

References

1. S.Ahmad, and G.Tesauro.,Scaling and generalization in neural network: A case study, *Proceeding of the 1988 Connectionist Models*, Morgann Kaufman (1988),p.3-10
2. G.Cybenko.,Approximation by superposition of sigmoid function, *Tech.Rep.Dept.of CS*, Tuft University,(1988)
3. J.LElman,and D.Zipser, Learning the hidden structure of speech,*Journal of the Acoustic Society of America*,Vol.183, (1987)p1615-1626
4. C.L.Giles,and T.Maxwell, Learning,invariance and generalization in high-order neural networks, *Applied Optics*,Vol.26, (1987) p4972-4978
5. H.J.Holland,and K.Holyoak, Nisbett,and P.R.Thagard.,Induction, *MIT Press* (1986)
6. K.Funahashi.,On the Approximate Realization of Continuous Mappings by Neural Networks, *Neural Networks*, Vol.2,(1989) p.183-192
7. P.Klassen,Y.H.Pao,and V.Chen, Characteristics of the functional link nets, *IEEE International Conference on Neural Networks*, Vol.1,(1988),p509-513
8. R.P.Lippmann, An Introduction to computing with neural nets, *IEEE ASSP Magazine*, (1987), p.4-22
9. R.P.Lippmann,and P.Beckman, Adaptive neural nets preprocessing for signal detection in non-Gaussian, in *Advances in Neural Information Processing Systems* Vol.1,(Tourretzky,ed), Morgan Kaufmann, (1989)p124-132
10. R.P.Michalski,G.J.Carbonell,and T.M.Mitchell, *Machine Learning*, Vol.1 & 2, Morgann Kaufmann (1983)(1986)
11. A.Namatame,and Y.Kimata, Improving the generalizing capabilities of a back-propagation network, *The International Journal of NeuralNetworks*,Vol.1, (1989)p86-94
12. D.E.Rumelhart,G.E.Hinton,and R.J.Williams, Learning representation by backpropagation, *Nature*, Vol.323, (1986),p533-536
13. D.Waltz,and J.A.Feldman,(Eds), Connectionist Models and Their Implications , *Ablex Publishing*, (1988)
14. P.J.Werbos,Generalization of backpropagation with application to a recurrent gas market model,*Neural Network*,Vol.1, (1988), p339-356

Applications of Learning and Planning Methods
edited by N. G. Bourbakis

CHAPTER 3

Layered Inductive Learning Algorithms and Their Computational Aspects

H Madala

Dept of Math & Comp Sci
Clarkson University
Potsdam, NY 13676, USA

Abstract

The inductive learning algorithms are based on the perceptron theory and on the principle of self organization. These algorithms are widely used for identification and predictions of complex systems. The inductive approach is realized by enumerating a large number of equations to choose the best. The linear combinations of inputs are generated at the simulated units of the network and are evaluated according to the changes in response to comparison with a threshold objective measure. Outputs of some units are discarded as per the measure at various hierarchical levels of the network structure. A parameter estimation technique is adapted in estimating the connection weights by minimizing the output error as a learning law at each unit; the least squares method is demonstrated here. Collective computation of overall process of the self organization helps in obtaining the optimum output response as per the problem objectives. There are different algorithms (multilayer, combinatorial and harmonical) based on the inductive mechanism; they differ in their network structures in generating the unit input-output functions and in using various heuristics. Modular programming is considered in designing the multilayered structure. The algorithms and their computational aspects are described in this chapter.

1 Introduction

Recent years there is rapid development in neural network (NN) based algorithms for various applications. Various scientific journals and international conferences related to neural modeling have revealed various works in this direction. Most of the works depend on the McCulloch and Pitts' concept of brain's functioning [1], which was an important invention used in the basic design of the digital computer to the parallel architectures. The flow and the processing of information in the nervous system is achieved by the parallel distributed activity of a large number of neural cells. Perceptron was described as a theoretical neural structure for information storage and organization in the brain [2]. Concept of the perceptron was developed as it adopts a 'brain metaphor' of information processing, which occurs through interactions involving large number of simulated neurons or units. Each simulated unit consists of four important components: 1.Input connection (synapses), through which unit receives activation from other units; 2.Summation function or input-output relationship that combines various input activations into a single activation; 3.Threshold logic function that makes the unit 'on' or 'off' and coverts the unit activation into output activation (it may be zero output activation, if it falls below some threshold value); and 4.Output connection (axonal paths), by which a unit's output activation arrives as input activation at other units in the system.

During the end of nineteen sixties it was felt that the perceptron theory was applied only to the single layer class [3]. Special networks are obtained by specifying the functions and parameters in the framework either by experimental measurements or by derivation from other specific neural structures and interaction principles [4]. Self organization of powerful internal representations and adaptation of general and powerful training schemes from statistical theory have lead to the multilayer artificial neural network structures [5, 6, 7, 9]. The most essential property of the theory is its suitability for the incorporation of experimental data. They differ in learning mechanisms adapted in them with feedback type of back-coupling. Since from the beginning the studies are concentrated on solving recognition and prediction related problems. There are excellent works describing the former category [10], whereas the latter is done extensively by using inductive learning algorithms, which are emphacized here.

In problem solving, main strategy is to specify a set of proper input-output associations and main goal is to design an efficient learning algorithm, which is regarded as a search procedure that correctly solves the problem. Learning in the networks takes a variety of forms, mainly it discovers statistical features for regularity detections. Self organization is considered while building up the connectionism among the units through a learning mechanism to represent discrete items. The self organizing process which is established through various heuristics in the network structure helps in obtaining the optimum output response. In the network each unit is represented independently as a black box to generate the input-output relationship as a state function and group of units are treated as a layer of certain thresholding stage. The relationships are established through the connecting weights, which are estimated by adapting a parameter estimation technique. The measure of an objective function is used as a threshold value to make the unit 'on' or 'off' and if the unit is 'on' its output is fed forward to the next layer as inputs. The measure is also considered as an objectivity measure of the unit. Overall this works as a searching in the domain of solution space through a sort of competitive learning [8]. Relevance of local minimum depends on complexity of the task on which the system is trained by building up heuristics like design of objective function, design of input-output relationships or summation functions at the units and usage of emperical data for training and testing the network. When one of the units in a particular layer achieves the global minimum on the objective function, the processing is stopped and information about the optimal response is recollected from the associated units in the preceding layers. The global minimum is guaranteed because of steepest descent in the output error with respect to the connection weights in the solution space, in which it is searched as per a specific objective through cross validating the weights.

The inductive learning algorithms are used in the literature with different names as Group Method of Data Handling (GMDH) and self organizing methods [11]. They are based on the layered network structures with a strong learning mechanism built up in them. The inductive approach is realized by sifting a great number of partial functions in a layered network structure to identify the best by an objective function or a selection criterion. Concepts from various disciplines like cybernetics [12], control systems theory, mathematical logic, psychology, regression analysis and so on were taken into the developments of these algorithms. The objective function works as a search rule to obtain the optimum output response in the progressive training of the network. The chapter is meant for describing various inductive learning algorithms and their computational aspects. It is arranged as follows: 'Basic Algorithm', 'Selection Criteria', 'Heuristics in Problem Solving', 'Network Structures and Their Computational Aspects' and 'Conclusions'.

2 Basic Algorithm

Suppose we have a sample of n obseravations, a set of input-output pairs (for simplicity we consider the output as a scalar unit) (I_1, o_1), (I_2, o_2), $\cdots$, $(I_n, o_n) \in O$, where O is a domain of observations corresponding to certain emperical data and we have to train the network using these input-output pairs to solve two types of problems: Identification problem is that for the given input $I_j\,(1 \leq j \leq n)$ of variables x corrupted by some noise expected to reproduce the output o_j and to identify the physical laws, if any, embedded in the system; Prediction problem is that for the given input I_{n+1} expected to predict exactly the output o_{n+1} from a model of the domain it has learned during the training.

In solving these, a general relationship between output and input variables is to be built up in the form of a mathematical description. Usually the description is considered as a discrete form of the Volterra functional series which is also called as Kolmogorov-Gabor polynomial:

$$y = \alpha_0 + \sum_{i=1}^{m} \alpha_i x_i + \sum_{i=1}^{m}\sum_{j=1}^{m} \alpha_{ij} x_i x_j + \sum_{i=1}^{m}\sum_{j=1}^{m}\sum_{k=1}^{m} \alpha_{ijk} x_i x_j x_k + \cdots \tag{1}$$

where the output is designated as y, the external input vector as $x = (x_1, x_2, \cdots)$ and α are the coefficients or weights. This is linear in parameters α and nonlinear in x [9]. Components of the input vector x could be independent variables or functional terms or finite difference terms. That means the function could be either an algebraic equation or a finite difference equation or an equation with mixed terms. The partial form of this function as a state or summation functional is developed at each simulated unit and is activated in parallel to build up the complexity.

2.1 *Unit Level*

Each simulated unit k receives input variables for example, $(x_i;\ x_j) \subset x;\ i \neq j$ and generates a function $f(\,)$, which is a partial form of Eq. 1:

$$f(x_i, x_j) = \nu_0^{(k)} + \nu_1^{(k)} x_i + \nu_2^{(k)} x_j + \nu_3^{(k)} x_i x_j + \nu_4^{(k)} x_i^2 + \nu_5^{(k)} x_j^2 \tag{2}$$

where $\nu^{(k)}$ are the connecting weights. If we denote o as the desired and y as the estimated values of the outputs for the function being considered, the output errors are given by

$$e_p = y_p - o_p; \quad p \epsilon O_A \tag{3}$$

The total squared error for that input vector is

$$E = \sum_{p \epsilon O_A} e_p^2 \tag{4}$$

This corresponds to the minimization of the averaged error E in estimating the weights $\nu^{(k)}$; this is the least squares technique. The weights are computed using a specific training set O_A, which is a part of the whole set O specified for this purpose.

2.2 *Layer Level*

Each layer contains a group of units, which are interconnected to the units in the next layer. The weights at each unit are estimated by minimizing the error E. Measure of an objective function is used as the threshold value to make the unit 'on' or 'off' in comparison with the testing set O_B which is another part of O and at the same time it is considered to obtain the optimum output response, ie. this is used as threshold as well as objectivity measures simultaneously. The outputs of the units which are 'on' as per the threshold values are connected as inputs to the units in the next layer; that means, the output of kth unit if it is in the domain of local threshold measure would become input to some other units in the next level. The process continues layer after layer. The estimated weights of the connected units are memorized in the local memory.

2.3 *Functional Flow of the Algorithm*

The flow of the algorithm can be described as follows. We feed the input data of m input variables of x randomly; for example, if they are fed in pairs at each unit, then a total of C_m^2 $(= m(m-1)/2)$ partial functions of the form below are generated at the first layer:

$$y = f(x_i, x_j); \quad i, j = 1, 2, \cdots, m; \; i \neq j \tag{5}$$

where $f()$ is the partial function as in Eq. 2 and y is its estimated output. Then outputs of $g1$ $(\leq C_m^2)$ functions are selected as per the threshold measure to pass on to the second layer as inputs in pairs. In the second layer we check the functions of the form:

$$z = f(y_i, y_j); \quad i, j = 1, 2, \cdots, g1; \; i \neq j \tag{6}$$

where $f()$ is the same form of the partial function as in Eq. 2 and z is its estimated output. The number of such functions is C_{g1}^2. Outputs of $g2$ $(\leq C_{g1}^2)$ functions are selected to pass on to the third layer. In the third layer we estimate the functions of the form:

$$v = f(z_i, z_j); \quad i, j = 1, 2, \cdots, g2; \; i \neq j \tag{7}$$

where v is the estimated output of the type of function as in Eq. 2. The number of such functions is C_{g2}^2. The process continues and is stopped when the threshold value begins to increase. The parameters of the optimal function are retrieved through the path of the connecting units from the preceding layers.

3 Selection Criteria

Selection criterion is error criterion, the name given to the objective function as it selects the partial functions at each layer. This plays an important role in obtaining the desired model. The threshold measure based on the criteria is computed for each unit by using a test set O_B and is compared with other units in the layer. There are objective functions like regularity, minimum bias, balance of variables, prediction, and so on, which might vary from problem to problem developed as the error criteria. One can also use the standard statistical criteria like Akaike's information criterion (AIC), prediction squared error (PSE) and so on. Keeping in mind the problem objectives, one has to choose the criteria either from existing ones or design a new set of criteria as per the problem and use it in the algorithm.

3.1 *Regularity Criterion*

This consists of a squared error calculated on the basis of testing set O_B.

$$\Delta(B) \triangleq \sum_{p \epsilon O_B} (y_p - o_p)^2 \rightarrow min \tag{8}$$

where $\Delta(B)$ indicates the regularity measure, y_p and o_p are the estimated and the desired outputs respectively.

3.2 *Minimum Bias Criterion*

This consists of a squared error of difference between the outputs of two functions developed on the basis of two distinct sets O_A and O_B.

$$n_{bs} \triangleq \sum_{p \epsilon O} (y_p^A - y_p^B)^2 \rightarrow min \tag{9}$$

where y_p^A is the estimated output of the function, which is obtained on the basis of the set O_A and y_p^B is the estimated output of the function, which is based on the set O_B. Usually the data with higher values of variance are included into the set O_A, while those with the smaller variance are put in the set O_B. When the model is consistent with exact weights on both sets of the data, then the outputs are equal, $y_p^A = y_p^B$ and $n_{bs} = 0$. This criterion enables to obtain consistent models; even it is possible to recover an optimal response which represents a physical law of the system hidden in the noisy experimental data. This criterion has also different versions.

3.3 *Prediction Criterion*

This consists of a squared error calculated on the basis of a separate checking set O_C, which is not used in the training:

$$\Delta(C) \triangleq \sum_{p \epsilon O_C} (y_p - o_p)^2 \rightarrow min \tag{10}$$

In case of finite difference equations, the criterion is evaluated using the testing set O_B by stepwise integration of the function for its predictions.

3.4 *Combined Criterion*

Sometimes we want to choose not only the minimum bias function but also those stable functions which have the best prediction properties. The combined criterion of the form

$$c3 \hat{=} \sqrt{(n_{bs}^2 + \Delta(C)^2)} \rightarrow min \tag{11}$$

provides the most unbiased, stable and accurate predicting models. The data should be divided into three sets O_A, O_B and O_C. $O_A$andO_B are used to calculate the minimum bias measure and O_C for prediction error. There are different types of combined criteria depending upon the combination of criteria used.

4 Heuristics in Problem Solving

Developing a mathematical description as per the input-output characteristics of a system and generating partial functions by linear combinations of the input arguments from the description, splitting of data into number of sets and design of a threshold objective function are noted as common features established in learning mechanism of the inductive algorithms. The output response of the network modeling depends highly on how these features are formed in solving a specific problem. As it depends on the researcher's experience and knowledge about the system, these features are treated as heuristics used in these algorithms.

Mathematical description of the system: One can use the type of mathematical descriptions like algebraic equations, autoregressive or finite-difference equations, single input - single output forms (trend equations), multi input - single output forms (multivariate equations) and multi input - multi output forms (system of equations). Specific terms like moving averages, logarithmic terms, time function, time and harmonic trends and so on can be considered into these descriptions. One can use any parameter estimation technique like maximum likelihood method for estimating the weights at each unit of the network. The generation of partial functions must be thought of how the complexity of state function is built up in the structure layer after layer. Higher ordered functions could be used for checking up nonlinearity of the given data.

Splitting data into training and testing sets: There are various techniques adopted in splitting the data; using dispersion analysis, splitting into two halves, splitting as odd and even numbers of observations and using any other statistical evaluation techniques.

Objective functions: Thinking of objectives in mathematical form is one of the difficult tasks in these algorithms. There is extensive work done in this direction, but the existing error criteria as objective functions are limited in number which are related to the standard mathematical modeling objectives like regularization, prediction, unbiasedness and so on. One can choose these criteria or use the standard statistical criteria or design a set of criteria as per the specific objectives.

In the following section, algorithms based on single and multilayer network structures and their computational aspects are explained.

5 Network Structures and Computational Aspects

The network structures differ as per the interconnections among the units and their hierarchical levels. There are three main inductive learning networks: multilayer, combinatorial and harmonical. There are other networks based on these three using the concepts from the objective systems analysis. Multilayer algorithm uses a multilayered network structure with linearized input arguments and generates simple partial functionals. Combinatorial algorithm uses a single layered structure with all combinations of input arguments including the full description [13]. This could be realized at each layer of multilayer structure. Harmonical algorithm follows the multilayered structure in obtaining the optimal harmonic trend with non-multiple frequencies for oscillatory processes.

5.1 *Multilayer Algorithm*

Multilayer network is a parallel bounded structure built up based on the connectionistic approach and information flows forward only. Each layer is having number of simulated units depending upon the number of input variables. Two input variables are passed on through each unit. For example, x_i and x_j pass on through kth unit and builds a summation function. Weights are estimated using the training set. At the threshold level, error criterion is used to evaluate this function using the test set. If there are m input variables, first layer generates $r\,(= C_m^2)$ functions. Usually 15 - 20 ($\leq r$) units as per the threshold values are made 'on' to the next layer. Outputs of these functions would become inputs to the second layer and same procedure would be taken place in the second layer. Further it is repeated in successive layers until a global minimum on the error criterion is achieved. If it is not achieved, that means the heuristic specifications must be considered for alteration. The partial function which achieves the global minimum is treated as an optimal model under the given specifications.

The mathematical description of a system can be considered as a nonlinear function in its arguments which may include delayed values of the input variables.

$$y = f(x_1, x_2, \cdots, x_1^2, x_2^2, \cdots, x_1x_2, x_1x_3, \cdots, x_{1(-1)}, \cdots, x_{1(-2)}, \cdots) \tag{12}$$

where $f()$ is a function of higher degree and y is its estimated output. This can be renotated as a linearized function by calculating all arguments x in the following form of full description:

$$\begin{aligned} y &= f(u_1, u_2, \cdots, u_m) \\ &= \alpha_0 + \alpha_1 u_1 + \alpha_2 u_2 + \cdots + \alpha_m u_m \end{aligned} \tag{13}$$

where u_i, $i = 1, 2, \cdots, m$ are the renotated terms of the arguments x; α_k, $k = 0, 1, \cdots, m$ are the coefficients and m is total number of arguments. These m input variables become inputs to the first layer. The partial functions generated at this layer are:

$$\begin{aligned} y_1 &= \nu_{01}^{(1)} + \nu_{11}^{(1)} u_1 + \nu_{21}^{(1)} u_2 \\ y_2 &= \nu_{01}^{(2)} + \nu_{11}^{(2)} u_1 + \nu_{21}^{(2)} u_3 \\ &\cdots \\ y_r &= \nu_{01}^{(r)} + \nu_{11}^{(r)} u_{m-1} + \nu_{21}^{(r)} u_m \end{aligned} \tag{14}$$

where r $(= C_m^2)$ is number of partial functions generated at the first layer, y_j and $\nu_{i1}^{(j)}$, $j = 1, 2, \cdots, r$; $i = 0, 1, 2$ are the estimated outputs and corresponding weights of the functions. Let us assume that $g1$ functions are selected for the second layer and there are g $(= C_{g1}^2)$ partial functions generated at the second layer.

$$\begin{aligned} z_1 &= \nu_{02}^{(1)} + \nu_{12}^{(1)} y_1 + \nu_{22}^{(1)} y_2 \\ z_2 &= \nu_{02}^{(2)} + \nu_{12}^{(2)} y_1 + \nu_{22}^{(2)} y_3 \\ &\cdots \\ z_g &= \nu_{02}^{(g)} + \nu_{12}^{(g)} y_{g1-1} + \nu_{22}^{(g)} y_{g1} \end{aligned} \tag{15}$$

where z_j and $\nu_{i2}^{(j)}$, $j = 1, 2, \cdots, g$; $i = 0, 1, 2$ are the estimated outputs and corresponding coefficients of the functions. In the same way, assume that $g2$ functions are passed on to the third layer; that means there are l $(= C_{g2}^2)$ partial functions generated in this layer.

$$\begin{aligned} v_1 &= \nu_{03}^{(1)} + \nu_{13}^{(1)} z_1 + \nu_{23}^{(1)} z_2 \\ v_2 &= \nu_{03}^{(2)} + \nu_{13}^{(2)} z_1 + \nu_{23}^{(2)} z_3 \\ &\cdots \\ v_l &= \nu_{03}^{(l)} + \nu_{13}^{(l)} z_{g2-1} + \nu_{23}^{(l)} z_{g2} \end{aligned} \tag{16}$$

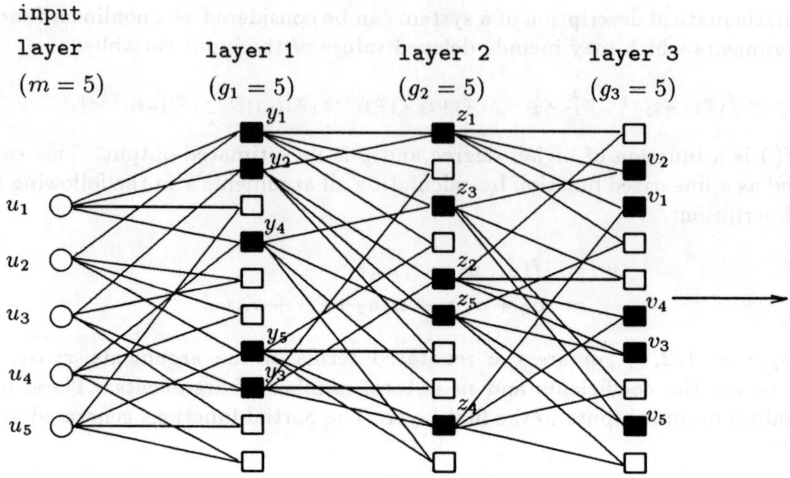

Figure 1: Multilayer network structure with 5 input arguments and selected nodes.

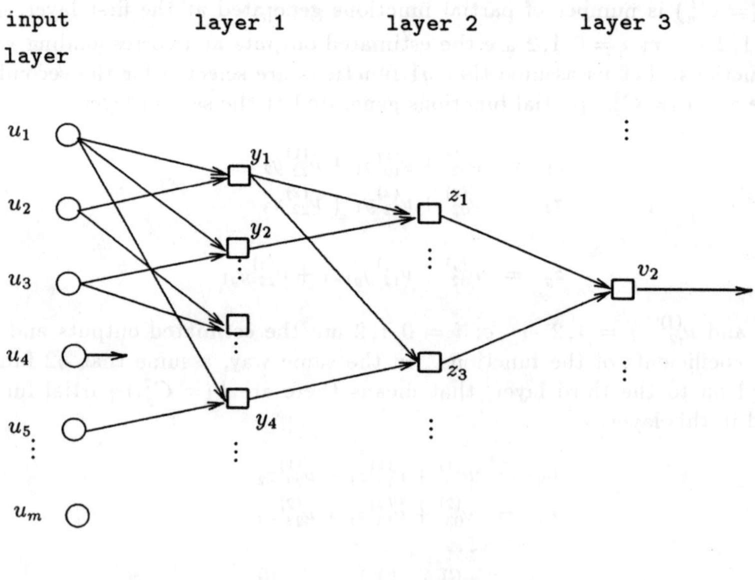

Figure 2: Schematic output flow to unit 2 of layer 3 in multilayer algorithm.

where v_j and $\nu_{i3}^{(j)}$, $j = 1, 2, \cdots, l$; $i = 0, 1, 2$ are the estimated outputs and corresponding coefficients of the functions. The process is repeated by imposing threshold levels of $m \geq g1 \geq g2 \geq g3 \geq \cdots \geq go$ so that finally an unique function is selected at one of the layers. The multilayer network structure with five input arguments and with five selected nodes at each layer is exhibited in fig. 1. For example, if the function $v_2 = \nu_{03}^{(2)} + \nu_{13}^{(2)} z_1 + \nu_{23}^{(2)} z_3$ in Eq. 16 achieves the global minimum, then it traces back to the preceding layers to recollect the functional values of all connecting weights from the associated units. Finally we get the optimal function in terms of the input arguments as shown below:

$$\begin{aligned} v_2 &= f(z_1, z_3) \\ &\equiv f(\ f(y_1, y_2), f(y_1, y_4)\ ) \\ &\equiv f(u_1, u_2, u_3, u_5) \ = \ f(X) \end{aligned} \tag{17}$$

This is demonstrated in fig. 2. One could obtain more functions, which are nearer to global measure for further evluation.

5.1.1 *Computational Aspects*

As the multilayer network procedure is having more repetitive in nature, it is important to consider the algorithm in modules and facilitate repetitive characteristics. The most economical way of constructing the algorithm is to provide three main modules: i) First module is for computations of common terms in the symmetric matrix (XH) of the normal equations for all input variables. This must be done at the beginning of each layer with all fresh input variables entering into the layer using the observations from 1 to NA of training set (O_A), ii) Second module is for generating the partial functions by forming symmetric matrices (HM) of the normal equations for all pairs of input variables, estimation of their weights (CO), evaluation of their threshold values (C) on testing set (O_B) from NA+1 to N1 and memorizing the information of coefficients and input variables of the best functions in COE and NK respectively; this must be done for each layer, and iii) third module is for computing the coefficients (SK) of optimal model and its input variables (ST) by recollecting the information from the associated units. The modules could be modified easily for any variations.

Here it is explained for a simple case in use of the regularity criterion. To initiate the program one has to specify control parameters like number of total data points (N1), number of training set points (NA), number of input arguments (P2 = M), number of layers (PM), number of functions to be selected as per the local minimum at each layer (P1 = CHO( )) and data table of input arguments (XX) and output variable (Y), which are in discrete analog form. The input data might be normalized within the range of {-1,+1} or {0,1}. FORTRAN language is used to describe the modules.

In the first module, control parameters are given; some common terms in the matrix XH is computed for P2 input variables using the training set points NA and the output variable Y. P1 and PU indicate number of functions to be selected at the first layer and number of the layer correspondingly.

```
       ...
      PU = 1
      P2 = M
      P1 = CHO(1)
 10   DO 12 J = 1,P2
      XH(J,1) = 0.0
      XH(J,2) = 0.0
      XH(J,3) = 0.0
      DO 11 K = 1,NA
      XH(J,1) = XH(J,1)+XX(J,K)
      XH(J,2) = XH(J,2)+XX(J,K)**2
      XH(J,3) = XH(J,3)+XX(J,K)*Y(K)
 11   CONTINUE
 12   CONTINUE
      XM = 0.0
      DO 13 K = 1,NA
      XM = XM+Y(K)
 13   CONTINUE
       ...
```

In the second module, it forms the matrix (HM) of normal equations for each pair of input variables J and I and uses the subroutine GAUSS for estimating the weights or coefficients (CO). All partial functions are evaluated by the regularity criterion using data from NA+1 to N1. Threshold values (C) are computed. Subroutine RANG arranges all values in ascending order. It stores the information on coefficients (COE) and input variables (NK) of the best P1 functions. Standard subroutines GAUSS and RANG are used.

```
       ...
      SH = 1
      J = 0
 21   J = J+1
      I = J+1
 22   HM(1,1) = NA
      HM(1,4) = XM
      H21 = 0.0
      DO 23 K = 1,NA
 23   H21=H21+XX(J,K)*XX(I,K)
```

```
      CM(SH) = C
      DO 25 K = 1,3
 25   COE(SH,K,PU) = CO(K)
      NK(SH,2,PU) = J
      NK(SH,3,PU) = I
      IF(SH.EQ.P1)CALL RANG(CM,NC,P1)
      SH = SH+1
      GOTO 28
 26   IF(C.GT.CM(NC(P1)))GOTO 28
```

```
      HM(2,3) = H21
      HM(3,2) = H21
      CALL NM(HM,XH,J,I)
      CALL GAUSS(HM,3,4,CO)
      C = 0.0
      DO 24 S = NA+1,N1
 24   C = C+(Y(S)-(CO(1)+CO(2)*
     $ XX(J,S)+CO(3)*XX(I,S)))**2
      IF(SH.GT.P1)GOTO 26
      CM(NC(P1)) = C
      DO 27 K = 1,3
 27   COE(NC(P1),K,PU) = CO(K)
      NK(NC(P1),2,PU) = J
      NK(NC(P1),3,PU) = I
      CALL RANG(CM,NC,P1)
 28   I = I+1
      IF(I.LE.P2)GOTO 22
      IF(J.LT.P2-1)GOTO 21
      ...
```

In continuation of this, the outputs (YY) of P1 functions are calculated to send it to the next layer. To repeat the above two modules we have to convert the outputs (YY) as inputs (XX) and initialize with fresh control parameters of the layer as below: number of the layer PU is updated as PU+1, number of input arguments P2 is equated to P1 and number of functions to be selected is taken from CHO(PU) as it is specified at the beginning.

```
        ...
      DO 20 I = 1,P1
      DO 20 I = 1,N1
 20   XX(J,I) = YY(J,I)
      IF(PU.EQ.PM)GOTO 29
      PU = PU+1
      P2 = P1
      P1 = CHO(PU)
      GOTO 10
 29     ...
```

The above procedure is repeated until PU becomes the number of specified layers (PM). Then it comes to the third module to recollect the information for the function which has achieved global minimum. The parameter PDM is calculated in advance as an indicator of number of original input arguments u activating in the function at a particular layer. In the first layer PDM = 2 and in consecutive layers PDM = PDM*2. The coefficients and number of input arguments of the optimal function are computed as in Eq. 17 using the stored information from COE and NK.

```
        ...
      DO 30 I = 0,M
      CTRO(I) = 0
 30   CT(I) = 0
      DO 31 I = 1,PDM
      A(I) = 0.0
 34   CONTINUE
      SH = SH+3
      ENDIF
      ENDIF
 35   CONTINUE
      DO 36 I = 2,PDM
```

```
      AD(I) = 0.0
      AN(I) = 0
31    AND(I) = 0
      DO 32 I = 1,3
      A(I) = COE(NC(1),I,PU)
32    AN(I) = NK(NC(1),I,PU)
      AD(1) = A(1)
33    SH = 1
      DO 35 I = 2,PDM
      IF(A(I).NE.0)THEN
      IF(AN(I).EQ.0)THEN
      SH = SH+1
      AD(SH) = A(I)
      AND(SH) = 0
      ELSE
      DO 34 S = 1,3
      AD(SH+S)=A(I)*COE(AN(I),S,PU-1)
      AND(SH+S) = NK(AN(I),S,PU-1)
      A(I) = AD(I)
36    AN(I) = AND(I)
      PU = PU-1
      IF(PU.GT.1)GOTO 33
      DO 37 I = 1,PDM
      S = AN(I)
37    CT(S) = CT(S)+A(I)
      DO 38 I = 0,M
38    CTRO(I) = CT(I)
      DO 39 J = 1,M
      IF(CTRO(J).NE.0.0)THEN
      K = K+1
      ST(K) = J
      SK(K) = CTRO(J)
      WRITE(*,*)ST(K),SK(K)
      ENDIF
39    CONTINUE
      END
```

Modules 1 and 2 along with the subroutine NM help in forming normal equations for each pair in more economical way of utilizing the computer time.

```
      SUBROUTINE NM(HM,XH,J,I)
      INTEGER S,R
      DIMENSION XH(15,3), HM(15,16)
      S = 2
      R = J
1     HM(1,S) = XH(R,1)
      HM(S,1) = HM(1,S)
      HM(S,S) = XH(R,2)
      HM(S,4) = XH(R,3)
      S = S+1
      R = I
      IF(S.EQ.3) GOTO 1
      RETURN
      END
```

5.2 *Combinatorial Algorithm*

This uses a single layered structure. Summation functions are generated for all combinations of input variables; ie. this is like 'all types of regressions' in the regression analysis. Let us describe the mathematical description of a system as shown below with three input arguments:

$$y = \alpha_0 + \alpha_1 u_1 + \alpha_2 u_2 + \alpha_3 u_3 \tag{18}$$

where y is the estimated output; u_1, u_2, and u_3 are the input arguments; and α are the weights. The algorithm as it is used in a single layered structure and it could be used at each layer of a multilayered structure to increase complexity of the model. The algorithm is given as below.

5.2.1 *Layer Level*

1. Summation functions for all combinations of the input arguments, u_1, u_2, u_3 (in this case there are 7 units) are generated:

$$\begin{aligned}
y_1 &= \nu_0^{(1)} + \nu_1^{(1)} u_1 \\
y_2 &= \nu_0^{(2)} + \nu_2^{(2)} u_2 \\
y_3 &= \nu_0^{(3)} + \nu_1^{(3)} u_1 + \nu_2^{(3)} u_2 \\
y_4 &= \nu_0^{(4)} + \nu_3^{(4)} u_3 \\
y_5 &= \nu_0^{(5)} + \nu_1^{(5)} u_1 + \nu_3^{(5)} u_3 \\
y_6 &= \nu_0^{(6)} + \nu_2^{(6)} u_2 + \nu_3^{(6)} u_3 \\
y_7 &= \nu_0^{(7)} + \nu_1^{(7)} u_1 + \nu_2^{(7)} u_2 + \nu_3^{(7)} u_3
\end{aligned} \tag{19}$$

where y_k is the estimated output of kth unit, $k = 1, 2, \cdots, 7$; and $\nu_i^{(k)}$, $i = 0, 1, 2, 3$ are their connecting weights.
2. The weights are estimated by using the least squares technique with a training set at each unit.
3. Then the unit errors are compared as per the threshold objective function using a testing set, and
4. Units with selected output responses are made 'on' and evaluated further.

The combinatorial algorithm could be realized at each layer of the multilayer network structure by keeping the limit on the input arguments at each layer. The unit outputs are fed forward layer by layer as per the threshold measure to obtain the global output response by increasing complexity in the partial models. This structure is exhibited in fig. 3 with three input arguments and with three selected nodes at each layer.

5.2.2 *Computational Aspects*

If there is increase in the input arguments, there is corresponding increase in the combinations of them. Suppose there are m variables, then the total combinations are $(2^m - 1)$. This is the major difference in this in comparison with multilayer algorithm.

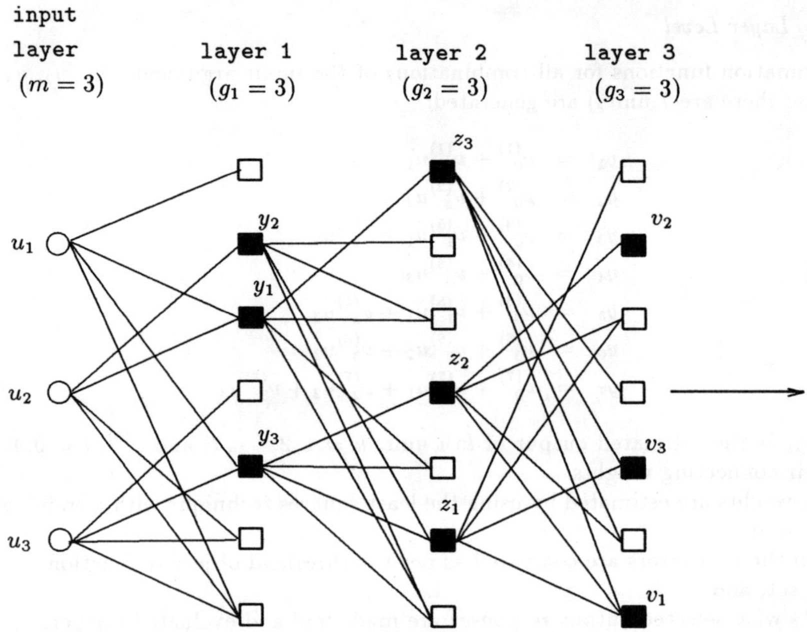

Figure 3: Multilayer network structure with combinatorial set up at each layer.

There are restrictions to use this algorithm as per the capacity of the computer. Efforts are given to build up the algorithm more economical way in generating the functions with all combinations. It constructs a matrix structure of '0's and '1's, where each row indicates a partial function with its parameters represented by '1', number of rows indicates total number of units in the layer, and number of columns indicates total number of parameters in the full description. Terms in the matrix are made equal to '0' if the parameters are not present in the function. This is called as 'structure of functions', which includes the full description, the function with all arguments. The 'structure of functions' is as shown below, where ν_0 is the constant term present in all functions:

$$\begin{array}{cccc} \nu_0 & \nu_3 & \nu_2 & \nu_1 \\ 1 & 0 & 0 & 1 \\ 1 & 0 & 1 & 0 \\ 1 & 0 & 1 & 1 \\ 1 & 1 & 0 & 0 \\ 1 & 1 & 0 & 1 \\ 1 & 1 & 1 & 0 \\ 1 & 1 & 1 & 1 \end{array} \qquad (20)$$

This is referred further in forming the normal equations for each function. Connection weights of each unit are estimated using a training set and evaluated for its threshold measure in comparison with a testing set. Finally it gives out the selected output responses as per the threshold values. One can improve the speed of the algorithm in case of finite-difference input arguments.

5.3 *Harmonical Algorithm*

The principal aim of this algorithm is to extend the use of inductive self organization principle to identify the oscillatory processes. It is assumed that the effective reference functions of such processes are in the form of a sum of harmonics with non-multiple frequencies. The harmonical function is formed of several sinusoids with arbitrary frequencies which are not necessarily related [15]. This type of function produces a multifrequency resultant and exhibits similar spectral characteristics [14]. A balance relation plays an important role in obtaining the frequencies of the process and as an objective function in selecting the optimal trend in the multilayered structure. The algorithm is explained below with the derivation of the balance relation:

Suppose that a function $f(t)$ is a process having sum of m harmonic components with pairwise discrete frequencies $w_1, w_2, \cdots, w_m$.

$$f(t) = \sum_{k=1}^{m} [\mathcal{A}_k \, sin\,(w_k t) + \mathcal{B}_k \, cos\,(w_k t)]$$

$$\equiv \sum_{k=1}^{m} \Phi(w_k, t) \tag{21}$$

where $\mathcal{A}_k$ and $\mathcal{B}_k$ are the coefficients; and $w_i \neq w_j,\ i \neq j,\ 0 < w_i < \pi,\ i = 1, 2, \cdots, m$. The process has total interval length of $N\ (1 \leq t \leq N)$ and points at discrete intervals of time δt.

For a fixed point i and any p of Eq. 21, one can obtain the formula:

$$f(i+p) + f(i-p) = 2 \sum_{k=1}^{m} cos\,(pw_k)\ \Phi(w_k, i) \tag{22}$$

Summing the Eq. 22 for p from 0 to $m-1$, with weighing coefficients $\mu_0, \mu_1, \cdots, \mu_{m-1}$, we derive a relation:

$$\begin{aligned} \sum_{p=0}^{m-1} \mu_p\,[f(i+p) + f(i-p)] &= 2 \sum_{k=1}^{m} \Phi(w_k, i)[\mu_0 + \sum_{p=1}^{m-1} \mu_p cos\,(pw_k)] \\ &= 2 \sum_{k=1}^{m} \Phi(w_k, i)\ cos\,(mw_k) \\ &= f(i+m) + f(i-m) \end{aligned} \tag{23}$$

This is considered as a balance relation of the process.

$$b_i = [f(i+m) + f(i-m)] - \sum_{p=0}^{m-1} \mu_p [f(i+p) + f(i-p)] \tag{24}$$

If the process is expressed exactly in terms of a given sum of harmonic components, then $b_i = 0$, ie. the discrete values of $f(t)$ which are symmetric with respect to a point $i\,(m+1 \leq i \leq N-m)$, satisfy the balance relation. The coefficients μ_p are independent of i.

The trignometric relation used in the calculations of Eq. 23 is:

$$\mu_0 + \sum_{k=1}^{m-1} \mu_p\, cos\,(pw_k) = cos\,(mw_k) \tag{25}$$

This could be formed as mth degree algebraic equation in $cos\,w$ by using the recursive trignometric relations:

$$\mathcal{D}_m(cos\,w)^m + \mathcal{D}_{m-1}(cos\,w)^{m-1} + \cdots + \mathcal{D}_1(cos\,w) + \mathcal{D}_0 = 0 \tag{26}$$

where $\mathcal{D}_i,\ i = 0, 1, \cdots, m$ are the functions of μ_p.

It is possible to determine uniquely the coefficients μ_p, $p = 0, 1, \cdots, m-1$ from the relation (Eq. 23) for $i = m+1, \cdots, N-m$. $(N-m) - (m+1) \geq m-1$, ie. $N \geq 3m$. Substituting the values of μ_p in Eq. 26, it can be solved for m frequencies w_k of harmonics by using standard numerical techniques. These frequencies are passed through a multilayer network to form various combinations of the harmonic components. The parameters of the harmonic components $\mathcal{A}_k$ and $\mathcal{B}_k$ are estimated at each unit by using the least squares technique. The optimal trend is obtained at one of the units by considering the balance relation as a threshold objective function.

5.3.1 *Computational Aspects*

Suppose $y(t)$ is the given discrete time series data and is to be identified for its harmonic trend $f(t)$. The data is to be separated into two data sets as training set points N_A and testing set points N_B. We can allot some more points N_C as checking set for checking the trend, ie. $N = N_A + N_B + N_C$. Maximum number of harmonics is chosen as M_{max} $(< N/3)$. The coefficients μ_p are estimated by using the least squares technique by forming the balance equations with the training set. The system of equations has the form:

$$\sum_{p=0}^{m-1} \mu_p[y(i+p) + y(i-p)] = y(i+m) + y(i-m);$$
$$i = m+1, \cdots, N_A - m \tag{27}$$

By substituting the values of μ_p, the frequencies of the harmonics are determined by solving the mth degree Eq. 26. This has m roots that uniquely determines m frequencies w_k. These frequencies are fed through the input layer of multilayer structure as components of the harmonic terms. The procedure of complete sifting of trends would take place by multilayer selection of trends as per the inductive principle of self organization. The linear normal equations are constructed in the first layer for any $1 \leq m \leq M_{max}$ number of harmonics and the coefficients $\mathcal{A}_k$ and $\mathcal{B}_k$ are estimated for all combinations of M_{max} harmonics based on the training set using the least squares technique. All harmonic trends are evaluated for their threshold values in comparison with the testing set and the outputs of the best trends from F units are fed forward as inputs to the second layer. This procedure is repeated in all subsequent layers. The complexity of the model increases layer by layer as long as the value of the 'imbalance' decreases on the testing set points N_B. The balance criterion B (Eq. 28) is used as the objective function, which takes into account the balance relation, b_i (Eq. 24).

$$B \triangleq \sum_{i=m+1}^{N_B-m} b_i^2 \;\rightarrow\; min \tag{28}$$

Unique solution is guaranteed as it is in the multilayer algorithm; the performance of the optimal trend can be tested further using the checking set N_C.

6 Conclusions

The description of the inductive learning algorithms and their computational aspects are given. The layered networks used in these algorithms are the bounded parallel structures which could be modified as per the problem requirements. Obtaining the best model depends on how best the heuristics are designed: the mathematical description of the system, threshold criteria and splitting of data. The best description could be developed by proper study of the system and its dynamics. The data could be analyzed and normalized using the statistical analysis techniques. Multi-objective functions as threshold criteria could be used in obtaining the optimum output responses.

These algorithms are used in complex systems modeling like weather forecasting, pollution problems (air and water), agricultural systems, economical systems, time series evaluations, digital signal processing, river flow predictions, fish productions and so on. There are various experimental studies in evaluating these algorithms without splitting the emperical data into training and testing sets. One can also find in the literature application of this modeling in solving mathematical programming and vector optimization problems. This could be extended further to a specific problem solving by modifying the algorithms: i) without discarding the unit outputs by using sigmoid functions, ii) with 'back propagation' type of learning mechanism, iii) with and without a large volume of emperical data, iv) with orthogonalization of the data and so on.

Acknowledgements

The author is very much grateful to Dr A S Fokas for his encouragement and support during the work.

References

1. W. S. McCulloch and W. Pitts, 'A Logical Calculus of the Ideas Immanent in Nervous Activity', *Bull. Math. Biophys.* **5** (1943) 115-133.
2. F. Rosenblatt, 'Perceptron - A Probabilistic Model for Information Storage and Organization in the Brain', *Phys. Rev.* **65**, N6 (1958) 386-408.
3. M. Minsky and S. Papert, *Perceptrons: an Introduction to Computational Geometry*, (MIT press, Boston, 1969).
4. A. G. Ivakhnenko, 'Polynomial Theory of Complex Systems', *IEEE Trans. Systems, Man and Cybernetics* **SMC 1**, N4 (1971) 364-378.
5. J. J. Hopfield, 'Neural Networks and Physical Systems with Emergent collective Computational Abilities', *Proc. Natl. Acad. Sci. USA* **79** (1982) 2554-2558.

6. T. Kohonen, *Self Organization and Associative Memory*, (Springer verlag, Berlin, 2nd ed., 1988), p. 312.
7. D. E. Rumelhart, G. E. Hinton and R. J. Williams, 'Learning Representations by Back Propagation Errors', *Nature* **323** (1986) 533-536.
8. D. E. Rumelhart and J. L. McClelland, *Parallel Distributed Processing*, (MIT press, Vols 1 & 2, Boston, 1986).
9. B. Widrow, R. G. Winter and R. A. Baxter, 'Layered Neural Nets for Pattern Recognition', *IEEE Trans. ASSP* **36**, N7 (1988) 1109-1118.
10. R. P. Lippmann, 'An Introduction to Computing with Neural Nets', *IEEE ASSP Magazine* **4** (1987) 4-22.
11. *Self Organizing Methods in Modeling - GMDH Algorithms*, ed. S. J. Farlow, (Marcel Dekker Inc., New York, 1984), p. 350.
12. N. Wiener, *Cybernetics*, (MIT press, Boston, 1947).
13. H. Madala, 'Self Organization GMDH Computer Aided Design for Modeling of Cyclic Processes', *Proc. IFAC Computer Aided Design*, Indiana, USA (1982) 611-618.
14. H. Madala, 'A New Harmonical Algorithm for Digital Signal Processing', *Proc. IEEE ASSP*, San Diego, USA (1984) 671-674.
15. F. B. Hildebrand, *Introduction to Numerical Analysis*, (McGraw-Hill, New York, 1956).

Applications of Learning and Planning Methods
edited by N. G. Bourbakis

CHAPTER 4

An Approach to Combining Explanation-based and Neural Learning Algorithms

Jude W. Shavlik
Geoffrey G. Towell
Computer Sciences Department
University of Wisconsin – Madison
1210 West Dayton St.
Madison, WI 53706
USA

ABSTRACT

Machine learning is an area where both symbolic and neural approaches to artificial intelligence have been heavily investigated. However, there has been little research into the synergies achievable by combining these two learning paradigms. A hybrid system that combines the symbolically-oriented explanation-based learning paradigm with the neural back-propagation algorithm is described. In the presented EBL-ANN algorithm, the initial neural net work configuration is determined by the generalized explanation of the solution to a specific classification task. This approach overcomes problems that arise when using imperfect theories to build explanations and addresses the problem of choosing a good initial neural network configuration. Empirical results show that the hybrid system more accurately learns a concept than the explanation-based system by itself and learns faster and generalizes better than the neural learning system by itself.

1 Introduction

Explanation-based learning (EBL) and artificial neural networks (ANNs) are two actively pursued approaches to machine learning. These two styles approach the learning problem from very different viewpoints. EBL falls within the traditional symbolically-oriented approach to artificial intelligence. It has shown promise in high-level tasks such as planning,[26] natural language processing,[29] and diagnosis.[15] Neural (or connectionist) networks are largely numerically-oriented. Large numbers of processing units, each with minimal capability, operate without global control and together produce intelligent behavior. ANNs have proven successful on low-level tasks, such as perception[22] and signal processing,[1] as well as on classification and diagnostic tasks.[32] Each approach has its strengths and weaknesses. A method for combining the two approaches, building on the strengths and overcoming the weaknesses of each, is described and a successful implementation reported. The method is not intended to produce a detailed model at the neurophysiological level. Rather, it investigates whether symbolic and neural learning techniques can be profitably combined to produce more powerful, general-purpose machine learning algorithms.

Explanation-based learning is a recently developed and actively pursued approach to machine learning.[9, 28] In this type of learning, the solution to a sample problem is generalized into a form that can later be used to solve conceptually similar problems. The generalization process is driven by the explanation of why the solution worked. Knowledge about the domain allows the explanation to be developed, and then generalized, thereby producing a general concept from the solution to a specific problem. EBL systems can be used to produce general classification rules by analyzing the classification of a specific object (for example, a rule for recognizing cups[28]) or to produce general plans by observing the achievement of a goal in a specific situation (for example, a plan for building towers[47]).

One advantage of EBL systems is that they only require a small number of examples to learn a concept, unlike empirically-based learning methods[36, 24, 27, 41] which require large numbers of examples. Only a few examples are needed because EBL systems possess the ability to explain what is relevant in these examples. Other approaches, including connectionist ones, rely on statistics to determine relevance and hence need many examples. They may be mislead by spurious correlations unless someone initially selects a small set of relevant features and uses them to describe the training examples.

The strength of EBL arises largely from it use of prior knowledge to guide its learning. However, dependence on prior knowledge is also the major source of EBL's weaknesses. The basic algorithms[30, 46, 17, 19] assume a complete and correct "domain theory".[28, 38] This domain theory, which is usually a collection of inference rules describing a given domain, is used to build the explanations that are then generalized. If the domain theory is not complete, some examples cannot be explained and, hence, nothing is learned. Incorrect domain theories lead to incorrectly learned results. A final problem is that, even if correct and complete, a domain theory may not be usable because its complexity makes it intractable.

In any real-world domain, a computer system's model can only approximate reality. Needed inference rules may be missing, while many of those possessed will lead to incorrect results in some situations. Furthermore, the complexity of problem solving prohibits any semblance of completeness and tractability. Finally, some domains are inherently uncertain. The actions of other agents, both competitors and cooperators, cannot be fully determined, nor can the physical world be exactly predicted. In these circumstances a domain theory is necessarily approximate.[4] To be truly useful, EBL systems must be able to work in uncertain, intractable, incompletely-specified, and changing environments, something they currently are incapable of doing.

Artificial neural networks have also generated much interest recently. ANNs consist of a collection of simple, neuron-like processing units. These units are connected by weighted links, where the weights indicate the degree of influence one unit has on another. The activation of a unit is determined by a weighted function of the activations of all connecting units. Computation proceeds by activating *input* units and waiting until the activation of the *output* units stabilizes. During this process, activation may traverse through intermediary, or *hidden*, units.

One area in which ANNs have appeal is that of inductively acquiring concepts from examples. A set of examples, each labeled as belonging to a particular class, is provided to the ANN. Its task is to determine a procedure for classifying future examples. Impressive performance has been achieved on problems such as converting text to speech,[43] evaluating moves in backgammon,[50] and diagnosing diseases.[32]

Several algorithms have been developed for training an ANN to classify new examples.[39, 41, 16] Training is accomplished by analyzing a set of provided, pre-classified examples to determine how to appropriately modify the connection weights so that these training examples are more accurately classified.

While the performance of ANNs that adjust their connection weights has been impressive, there are some problems with this form of machine learning:

1. *Training times are lengthy.* Recent experiments[32, 48, 12, 53] indicate that while ANNs classify new examples slightly better than do symbolic learning algorithms (such as ID3 [36]), they require 100 to 1000 times as much training time.

2. *The initial weights can greatly effect how well the concept is learned.* Because training ANNs usually involves a hill-climbing algorithm, local minima may be a problem. Occasionally the learning algorithms cannot find local weight adjustments that improve performance. As a result. classification correctness on new examples can depend significantly on the initial set of weights of the ANN. For example, in some experiments[48] involving the conversion of text to speech, correctness on a large dictionary of examples ranged from 0.1% to 19%. The only difference between the runs was the randomly-chosen initial set of weights. Others have also reported that performance can be highly dependent on the initial set of weights.[2] One way to overcome this problem is to train several networks, keeping the one that performs best on the training examples. For example, Ahmad[2] usually performs 20 runs. However, this substantially increases the training time.

3. *There is no known problem-independent way to choose a good network topology.* The performance of an ANN can be greatly effected by the layout and number of hidden units, as well as the specification of which units are connected to which other units. Again, runs with several different topologies can be performed, keeping the best.

4. *Many training examples are needed.* As is standard in learning systems that inductively acquire concepts from examples, ANNs, unlike EBL systems, need many training examples to acquire a concept.[3]

5. *There is no way to focus attention on significant features.* ANNs have traditionally been trained in environments where the set of features to be considered is carefully selected by the experimenter. Without this pre-selection, the ANN would be forced to consider thousands of features normally dismissed as irrelevant by the human user of the system.

6. *The learned network is difficult to interpret.* While the meanings of the ANN's input and output units are easily understood because they are fixed by the environment, the meaning of individual hidden units frequently cannot be determined. Hence, it is usually hard to ascertain reasons for the answers provided by the ANN.

The next section describes the hybrid approach that overcomes the weaknesses of both EBL and ANNs. Following that, the algorithm is presented in detail and empirically applied to two sample problems. Related research and some open issues are then described. An appendix reports data from the two experiments.

2 An Integrated Approach

EBL systems can take advantage of existing knowledge and, hence, need few examples to learn a new concept. However, they require a well-formed base with which to build explanations. Incorrect concepts may be learned if a faulty domain theory is used. Conversely, ANNs provide a way to incrementally adjust the representation of a concept on the basis of additional examples. Their weakness is that they require a large number of examples and many training epochs to converge on an acceptable concept, especially if they start from a randomly chosen initial state.

An approach that integrates these two styles of learning may accrue the benefits of each while avoiding their shortcomings. The hybrid EBL-ANN system described below has been developed with this goal in mind. Basically, the EBL system produces categorization rules rapidly from a small number of examples. These rules are then refined by an ANN, using some additional examples. In this hybrid system, a complete and correct domain theory is not necessary, as the ANN refines learned concepts. Also, fewer training examples are needed than in a pure ANN, as the EBL portion of the system produces an initial network that is close to the correct concept.

In the hybrid EBL-ANN approach, an approximate domain theory may be used. With this domain theory, the EBL portion of the system can build only roughly-correct explanations. These explanations need not be proofs, but only reasonable arguments as to why the current item belongs to a specific category or why the current plan meets the given goal. That is, the domain theories can be incorrect. For domain theories that are intractable or incomplete, simplifying or default assumptions can be made in order to produce these roughly-correct explanations.[7, 28]

Many EBL systems have domain theories of the desired form already, in particular those involving natural language.[31] However, they treat their domain theories as being as logically correct as theories representing mathematical calculations.[45] The hard part is not writing roughly-correct domain theories. The hard part is developing a learning system that properly treats the domain theory as roughly correct.

In the hybrid EBL-ANN system, the EBL portion of the system produces a rough explanation and then generalizes it using a standard explanation-based generalization algorithm. The final, generalized result is then converted into a neural network. The antecedents of each classification rule determine the weights running from the input units to a hidden unit, while the consequents determine the weights from a hidden unit to the output unit. Features that are believed to be important, according to the result produced by the EBL portion of the system, are highly weighted. Features that the rule says should be absent receive large negative weights. Finally, features that appear to be irrelevant receive weights near zero. These apparently irrelevant features must appear in the ANN because the roughly-correct EBL system may have mistakenly thought they were unimportant.

Once the ANN is produced, additional training examples are provided to refine the network, using standard neural network learning techniques. The final result can be some function of the inputs that is not expressible using the symbols and vocabulary in the domain theory. Hence, the hybrid system can increase the expressiveness of the basic EBL system, thereby increasing accuracy. Also, since the approach is naturally incremental, if the concept being learned is slowly varying over time, the ANN can capture these changes if it periodically receives feedback. Using only an EBL system would require rewriting portions of the domain theory whenever a concept changed. Finally, ANNs can store exceptions along with general concepts,[23] another way to correct an EBL domain theory.

The major assumption underlying the EBL-ANN approach is that the space of possible concepts is reasonably "smooth". When this assumption holds, it is likely that starting at a nearly-correct point in the space will facilitate reaching the correct concept. The validity of this assumption in real-world domains must be tested empirically. The assumption about the smoothness of concept space is also made by *case-based reasoning* researchers.[20] In this paradigm, previous solutions are slightly modified "tweaked" so that they apply to the current problem.[42]

Ahmad[2] reports on an experiment that supports the hypothesis underlying the EBL-ANN approach. In this experiment, a neural network is configured with the "correct" set of weights. Next, the weights are perturbed. After some retraining, the network performs substantially better than if it had started off with random weights, even when the perturbations are large. This indicates the value of starting off with a set of weights that is close to a correct solution. Sejnowski and Rosenberg[43] report similar experimental results.

3 The EBL-ANN Algorithm

Table 1 lists the hybrid EBL-ANN learning algorithm. A roughly-correct domain theory involving binary features[a] is assumed, as are a collection of training examples and criteria for termination of learning.

Table 1: The EBL-ANN Algorithm.

Given:

1) A list of binary features used to describe examples.
2) A roughly-correct domain theory that can explain why an example is a member of the concept being learned.
3) A collection of training examples, classified as to whether of not they are members of the concept being learned.
4) A specification of the desired classification accuracy on the training examples.
5) A bound on the maximum number of training epochs

Do:

1) Use the roughly-correct domain theory to explain one of the positive examples of the concept to be learned.
2) Generalize the explanation using the EGGS algorithm.
3) Map the generalized explanation into an ANN (see text).
4) Use the classified examples to train the ANN using the backpropagation algorithm.
5) If the training examples are correctly classified to within the specified accuracy or if the maximum number of epochs is reached, *STOP*. Otherwise, randomly permute the training examples and go to step 4.

In the hybrid algorithm, the EBL portion first uses the domain theory to classify one of the positive examples.[b] An explanation can be viewed as a deductive proof

[a]Being binary is not a restriction, in that multiple-valued features can easily be converted into binary features. For example, assume *color* is a feature that can take on the values *red, blue,* or *yellow* . This can be converted to the following binary features: *color-is-red; color-is-blue; color-is-yellow* . Note, though, that this conversion leads to some loss of information; namely, that *color* can take on only one value. In the binary representation, *color-is-red* and *color-is-blue* are no more related than are *color-is-red* and *size-is-small.*

[b]If the domain theory can also explain why something is *not* a member of the concept being learned, two separate ANNs can be produced. Alternately, a single ANN with two output nodes could be used. However, the integration of multiple explanations into a single ANN is an open research issue.

that the training example is a member of the concept being learned. The resulting explanation is then generalized using the EGGS algorithm.[30] Currently, the algorithm uses the explanation generated for the first example encountered which can be explained. (An open research issue is how to best integrate the explanations of several examples or several different explanations of the same example into a single ANN.) The EGGS algorithm assumes that, in the course of solving a problem, a collection of pieces of general knowledge (e.g., inference rules, rewrite rules, or plan schemata) are interconnected, using unification to insure compatibility. The resulting explanation is generalized by determining the most general unifier that allows the general pieces of knowledge to be connected in the same way. This involves replacing the constants in the specific explanation with constrained variables. The result is a new composite knowledge structure that contains the unifications that must hold in order for the knowledge pieces to be combined in the given way.

To visualize the EBL process, assume a tree-structured explanation such as the one in Figure 1. In this figure, all conclusions are conjunctive in that all antecedents must be satisfied in order for a node[c] to be true. If the leaf nodes can be consistently satisfied, the root node, which concludes the example is a member of the concept, will also be satisfied. Internal nodes can be ignored. There is no need to again reason about combining the pieces of knowledge together to make the classification. Since a substantial amount of work can be expended constructing the original solution, the new knowledge structure can lead more rapidly to future solutions. A generalized version of the solution is saved and can be efficiently applied to future problems. Also notice that the explanation selects which features are relevant to classification.

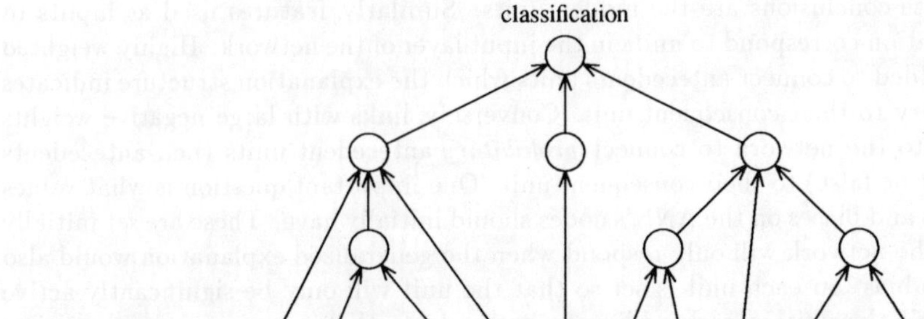

Figure 1: A sample explanation for a classification task.

Notice, though, that in this standard style of EBL, no *knowledge level learning*[11] occurs. That is, only concepts in the deductive closure of a system's existing knowl-

[c]For clarity, *node* will always be used to refer to components of explanations while *unit* will always be used to refer to processing elements in neural networks.

edge can be learned. Basic EBL systems speed up problem solvers but do not change what a system could solve if given enough time. The ANN portion of the hybrid system addresses the issue of knowledge level learning by refining the rules encapsulated in the ANN.

Since dealing with variables in neural networks is an open research area, the EBL-ANN algorithm currently assumes that domain theories are purely propositional (i.e., they do not contain any variables). This means that finding the most general unifier is trivial. In this case, the primary role of the EBL system is to focus attention on the features most likely relevant to classifying an example. The relevant features are those appearing in the leaf nodes of the explanation.

Following the application of the EGGS algorithm, the generalized explanation is converted into an ANN, as explained below. From this point, training continues following a standard neural network training procedure called *back-propagation*.[41] Back-propagation adjusts inter-unit weights to reduce the difference between the ANN's actual and desired output. Training by repeated presentation of the set of pre-classified examples in random order[d] continues until enough training examples are properly classified or until a training limit is reached. An example is considered correctly classified if the ANN's output is within 0.25 of the correct output. (The setting of this parameter is discussed further later.)

Translation of the classification rule learned by the EBL component into an ANN network is guided by the structure of the explanation. The core layout of the ANN is isomorphic to the explanation structure; Table 2 lists the correspondences. Each conclusion, final or intermediate, in the explanation structure has a corresponding unit in the ANN. The final conclusion becomes the output unit for the network, while intermediate conclusions are the hidden units. Similarly, features used as inputs to the explanation correspond to units in the input layer of the network. Highly weighted links are added to connect antecedents units which the explanation structure indicates are necessary to their consequent unit. Conversely, links with large negative weights are added to the network to connect *prohibitory* antecedent units (i.e., antecedents which must be false) to their consequent unit. One important question is what values the weights and biases on the ANN's nodes should initially have. These are set initially such that the network will only respond when the generalized explanation would also do so. The bias[e] on each unit is set so that the unit will only be significantly active when: 1) all the units corresponding to antecedents that are necessary are active; 2) none of the units corresponding to antecedents that are prohibitory are active.

To understand how weight and bias setting works, consider a unit with $n+m$ links, n from necessary antecedents, and m from prohibitory antecedents. The weights on

---

[d] A single exposure to each of the training examples constitutes a *training epoch*.

[e] The activation of a unit i is defined to be $\frac{1}{1+e^{-(\sum_j w_{ij} * activation_j - \theta_i)}}$, where j ranges over the units that connect to unit i. θ_i is often called the *bias* of unit i. Roughly speaking, unless the sum of weighted activations on connected units exceeds this bias, unit i is not active. That is, the bias indicates how much weighted activation a unit needs to receive from its neighbors before it "fires".

Table 2: Correspondences between explanations and neural network structures.

Explanation	ANN
conclusion (root node)	output unit
supporting facts (leaf nodes)	input units
intermediate conclusions (internal nodes)	hidden units
dependencies (arcs)	weighted links

the n links from the necessary antecedents are set to ω while the weights on the m links from the prohibitory antecedents are set to $-\omega$. The bias for the unit is set to $n\omega - \omega/2$. When the bias is set to this level, all of the necessary antecedents must be present and none of the prohibitory antecedents can be present for the unit to have a significant activation. To prevent symmetries that are hard to break,[41] all weights and biases are adjusted by a random amount between $-\epsilon$ and $+\epsilon$. Currently, we use $\omega = 5$ and $\epsilon = 0.1$, values empirically found to work well.

Recall that the generalized explanation may be incorrect. The neural learning algorithm will refine the learned concept, but it must have the nodes and links that allow it to do so. At a minimum, this means additional input units must be added and connected to the network for every possible binary feature not mentioned in the explanation. Additional connections are made by first grouping the units according to the maximum length path from some input unit to each unit. New links, with a randomly-chosen weight within ϵ of zero, are made between units in successive groups. New connections are not made if there already is a link between a pair units or a loop would be formed. The addition of these low-weight connections allows the insertion into the ANN network of features which were thought to be irrelevant by the roughly-correct domain theory.

After the ANN is initialized in a nearly correct configuration, training using back-propagation and the complete set of pre-classified examples completes the learning process.

3.1 A Sample Translation

Consider the artificial domain theory presented in Table 3 which defines when an object is a member of category A. Figure 2a represents the only explanation possible given the rules in Table 3. The solid and dotted lines in Figure 2a respectively represent necessary and prohibitory dependencies.

Figure 2b represents the network which results from the translation into a neural network of the explanation in Figure 2a. The thick lines in Figure 2b represent the links in the neural network which correspond to dependencies in the explanation. Thus, with $\omega = 5$, the weight on the thick solid lines is 5 while the weight on the thick dotted line is -5. The lighter solid and dotted lines represent the links added to

Table 3: An artificial domain theory.

A	:—	B, E.
B	:—	C, D, H.
C	:—	not F, G.
D	:—	H.
E	:—	I, J.

the network to allow refinement of the domain theory; all these lines have a randomly-chosen weight within ϵ of zero. The number in parentheses next to each hidden and output unit is the unperturbed bias of that unit (i.e., its bias before adding a random number within ϵ of zero).

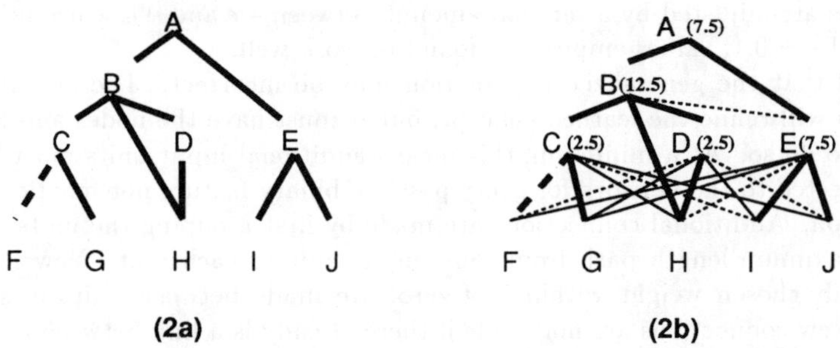

Figure 2: Translation of an explanation into a network.

4 Two Experiments

The implementation of the EBL-ANN hybrid algorithm has been tested using two simple domain theories. The first involves recognizing cups, and the second involves recognizing chairs. Using the same parameter settings, the hybrid system EBL-ANN outperforms the EBL-ALONE (i.e., stand-alone EBL) and ANN-ALONE (i.e., stand-alone back-propagation) systems in both domains.

4.1 Learning to Recognize Cups

The first experimental domain is an adaptation of the traditional cup recognition problem.[28] Table 4 lists a set of PROLOG-like rules which define a roughly-correct theory for the recognition of cups. The first rule can be read *"X is a cup if it is stable, liftable, and an open vessel."* Using these rules something analogous to a coffee mug can be recognized as a cup. However, cups without handles (e.g. Styrofoam cups) cannot be recognized. Also, a pail, where drinking is prevented by the handle across the mouth of the cavity[9] is falsely classified as a cup.

Table 4: A roughly-correct domain theory for recognizing cups.

cup	:—	stable, liftable, open-vessel.
stable	:—	bottom-is-flat.
liftable	:—	graspable, light.
graspable	:—	has-handle.
open-vessel	:—	has-concavity, concavity-points-up.

Table 5: The training examples for learning to recognize cups.

	Cups					*Non-Cups*				
Features	c1	c2	c3	c4	c5	n1	n2	n3	n4	n5
Has Handle	√			√	√		√			
Handle on Top							√			
Handle on Side	√			√	√					
Bottom is Flat	√	√	√	√	√	√	√		√	√
Has Concavity	√	√	√	√	√		√	√	√	√
Concavity Points Up	√	√	√	√	√		√	√	√	
Light	√	√	√	√	√	√	√		√	
Made of Ceramic	√							√		
Made of Styrofoam		√	√			√				√
Made of Paper				√	√		√		√	
Expensive	√		√		√		√		√	
Fragile	√	√			√	√		√		√

The hybrid algorithm is run using the training set of five cups and five non-cups listed in Table 5. Notice that Styrofoam cups, which cannot be explained by the domain theory, are included (Cups *c2* and *c3*), as are pails (Non-cup *n2*). Also notice that irrelevant features (e.g., *expensive* and *fragile*) are included.

Figure 3 depicts the ANN as initially constructed from the explanation of the first positive training example, Cup *c1*.[f] For clarity, units are labeled by the name of the concept they represent. While only the links and units mentioned in the explanation are shown, but recall that additional low-weighted links are also in the ANN. Figure 4

[f] As only one explanation is possible given the cup domain theory of Table 4, the network of Figure 3 could also have been based upon explanations of Cups *c4* or *c5*.

presents EBL-ANN's performance on the ten training examples averaged over 50 trials, where each trial is initialized using a different random perturbation of the weights.

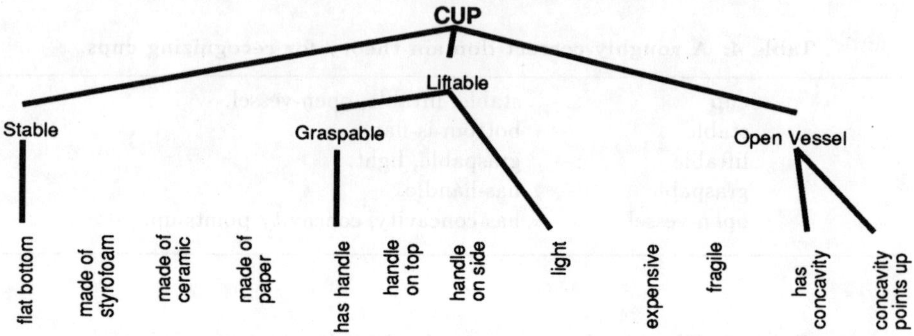

Figure 3: Initial network for recognizing cups (low weighted links not shown).

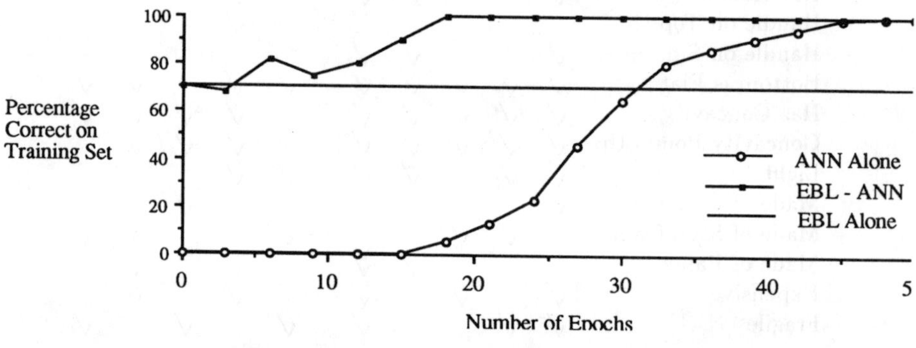

Figure 4: Classification accuracy for cups as a function of training amount.

For comparison, the performances of EBL-ALONE and ANN-ALONE also averaged over 50 trials, are included. As EBL-ALONE can use only the rules appearing in Table 4, and those rules will correctly classify seven of the ten examples, it can never do better or worse than 70% correct. Thus, in Figure 4 the line representing the performance of EBL-ALONE is horizontal at 70% correct. For ANN-ALONE, the conventions used by Mooney, Shavlik, Towell and Gove[32] are followed: there is one-layer of hidden units; the number of hidden units is set at 10% of the sum of the number of input and output units (with a minimum of two hidden units); each input unit is connected to each hidden unit; each hidden unit is connected to the output unit; there are no direct connections between the input and output units. Having one layer of hidden units and full connectivity between layers is a common topology in neural network experiments.

The performance of the hybrid algorithm is initially the same as EBL-ALONE because the procedure which translates explanations into networks generates networks which emulate the EBL rules. After a brief period of floundering, the hybrid quickly achieves 100% correctness on the training examples. Averaged over the 50 trials, the hybrid system takes 16 epochs and has a standard deviation of 2.6 epochs. ANN-ALONE takes an average of 37 epochs and has a standard deviation of 4.8 epochs.

Figure 4 indicates that the hybrid algorithm achieves the goals which motivated its creation. First, it is able to correctly classify the training set using one-half the number of training epochs required by stand-alone back-propagation. Second, 100% correctness is achieved, even though an imperfect domain theory is used.

Furthermore, the results listed in Table 6 indicate that EBL-ANN is better able to generalize its learning than either system alone. The numbers in this table represent the average performance after the 50 learning trials used in Figure 4. For each trial, a different collection of 50 positive and 50 negative cups was used to test each system's ability to generalize to novel examples. The *test set* collections were generated by randomly choosing values for the binary features to form an object, then testing the created object with an oracle domain theory[g] to determine if it should be considered a cup or not. (This procedure allows duplicates and members of the training set in the testing set. However, the space is large enough that this is not a significant issue.) On these test set collections of cups and non-cups, EBL-ANN performed better than either ANN-ALONE or EBL-ALONE. An analysis of variance[h] on these results suggests that the difference in correctness between EBL-ANN and ANN-ALONE is significant with 99.9% confidence, as is the difference in correctness between EBL-ANN and EBL-ALONE.

The results for EBL-ALONE in Table 6 also show that EBL-ANN has to both generalize and specialize the initial cup domain theory. EBL-ANN generalized the domain theory at a small cost of incorrectness on the non-cups, with the result that cups and non-cups were correctly classified at about the same rate. By contrast, ANN-ALONE learned an overly general definition of cups; it was correct on 99% of the cups but only 68% of the non-cups.

---

[g]The oracle for cups expands the definition contained in Table 4 to include cups without handles so long as they are made of Styrofoam. Also, the obstacle classifies objects with handles on the top as non-cups.

[h]Analysis of variance is a standard statistical test. It compares the variance within a group to the variance between groups, in order to determine if the population means of two or more groups are different. The results of an analysis of variance are expressed in terms of two quantities, F and p. F is the ratio of the variations explained by the grouping to variations left unexplained by the grouping. So, large values of F indicate that the groups are likely to be meaningful, while p is based upon the F statistic, the number of groups, and the total number of observations. It is the probability that the population means of the groups are the same. Therefore, $p < .001$ means that it is possible to conclude with 99.9% certainty that the population means of the groups being considered are different.

Table 6: Correctness of randomly chosen sets of cups.

System	Percentage Correct Chairs	Non-chairs	Overall	Standard Deviation	ANOVA Results
ANN-ALONE	99.2	68.1	83.7	2.9	
					$F(1,98)=16, p < 0.001$
EBL-ANN	88.3	84.1	86.2	2.9	
					$F(1,98) = 35, p < 0.001$
EBL-ALONE	71.9	91.1	81.5	3.4	

Figure 5 presents the average of 50 ANNs following learning by the EBL-ANN system. Although it is not explicit in the figure, the weight on *made of styrofoam* is such that either *made of styrofoam* or *has handle* is sufficient to activate *graspable.* Strongly negative links connect *handle on top* to both *stable* and *open vessel.* Thus, the presence of *handle on top* precludes activation of either of these units. Also, a strongly negative link connects *made of paper* to *graspable.* Its presence is due to spurious correlations in the training data and is one of the principle reasons that EBL-ANN correctly classified only 88% of the randomly generated cups. The thin solid and dotted lines respectively represent negative and positively weighted links with a weight which is significantly different from zero but not sufficient to play a major role in the calculations of the network. The exact weights and biases, averaged over 50 trials, are reported in the appendix.

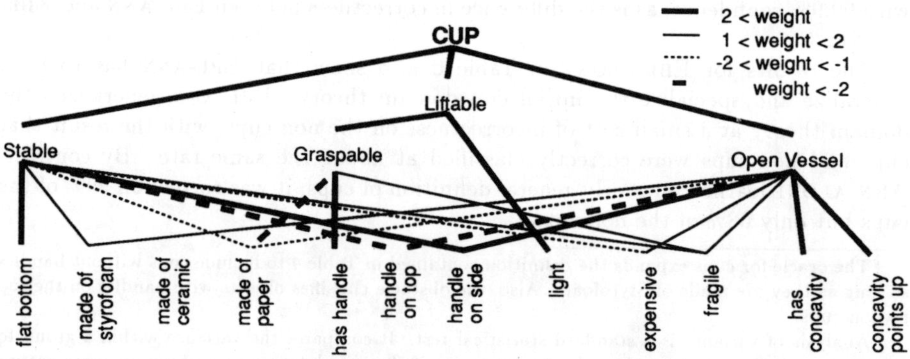

Figure 5: Final network for recognizing cups (low weighted links not shown).

The result of neural network learning in EBL-ANN is that both false negatives and false positives in the training set, with respect to the original domain theory, are now correctly classified. This correction resulted from the network learning the importance of two features, *made of styrofoam* and *handle on top*, which had previously been ignored. The network also regarded the feature *made of paper* as significant due to correlations in the training data. A method to eliminate the class of problems exemplified by the connection from *made of paper* to *graspable*, such as using using EBL failure to direct back propagation, is an open research issue (see Section 6).

The hybrid system needed about half as many training epochs to learn the training set. Speedup achieved by using the hybrid heavily depends on the setting of the parameter that determines how close the output must be in order to be considered correct. If this parameter is set too small (e.g., 0.05), many epochs are needed to perform the final refinements. This reduces the relative advantage of starting with a roughly-correct ANN. The trade-off is that if this parameter is set too large (e.g., 0.5), the final ANN is less likely to appropriately correct the domain theory. Spurious changes to inter-unit connections can do a good enough job to properly classify the training examples, thereby prematurely terminating training. (With more training examples, a higher setting may be appropriate. A good setting for this parameter is an open research issue.)

While this study, as well as the next one involving chairs, indicates the promise of the EBL-ANN approach, it should be remembered that these are two rather artificial domains with very small training sets. Before firm conclusions are drawn, similar experiments must be run on "real-world" domains, using large training and test sets containing actual examples.

4.2 Learning to Recognize Chairs

The second example, which replicates the methodology of the first example, involves learning how to recognize chairs. A roughly-correct set of rules appear in Table 7. Among other things, this theory requires that chairs have four legs, a padded seat, and back support.

Table 7: A roughly-correct domain theory for recognizing chairs.

chair	:—	can-suppoort, comfortable, stable.
can-support	:—	supports-bottom, supports-back.
supports-bottom	:—	horizontal-seat, seat-at-medium-height, medium- sized-seat.
comfortable	:—	smooth-seat, padded-seat.
stable	:—	four-legs, equal-length-legs.

The EBL-ANN algorithm is run using the training set of five chairs and five non-chairs listed in Table 8. Only the first chair is recognized by the domain theory, so the explanation of Chair *c1* is always used to establish the network for the EBL-ANN system. Chairs *c2*, *c3* and *c4* are not recognized as chairs because they are three-legged while Chairs *c2*, *c4* and *c5* lack back support. Conversely, Non-chairs *n2* and *n5*, which cannot support weight, are classified as chairs by this domain theory. Hence, the domain theory properly classifies only four of the sample chairs.

Table 8: The training examples for learning to recognize chairs.

	Chairs					*Non-Chairs*				
Features	c1	c2	c3	c4	c5	n1	n2	n3	n4	n5
Vertical Back	√		√				√	√	√	√
Smooth Seat	√	√	√	√	√		√			√
Padded Seat	√	√	√	√	√	√	√	√		√
Horizontal Seat	√	√	√	√	√		√	√	√	√
Medium Sized Seat	√	√	√	√	√		√	√	√	√
Seat at Medium Height	√	√	√	√	√	√	√	√	√	√
Four Legs	√				√	√	√			√
Three Legs		√	√	√				√	√	
Equal Length Legs	√	√	√	√	√	√	√		√	√
Supports Medium Weight	√	√	√	√	√	√		√		
Has Arms	√			√		√		√	√	
Has Wheels	√	√			√	√	√	√	√	

Figure 6 contains the ANN as initially constructed from the explanation of the first training example. As before, units are labeled by the name of the concept they represent, and only the links and units mentioned in the explanation are shown. The performance of the three systems appears in Figure 7. The curves are the averages over 50 trials, where each trial involves different random perturbations of the initial weights. The average number of epochs needed by the EBL-ANN hybrid is 20 with a standard deviation of 5.7, while for ANN-ALONE it is 34 with a standard deviation of 4.4. As happened when learning to recognize cups, the hybrid system is more accurate than EBL-ALONE and learns faster than ANN-ALONE.

Testing the classification accuracy of the three systems on novel examples was performed.[i] The results listed in Table 9 show that EBL-ANN achieves a better representation of chairs than either ANN alone or the initial domain theory, outperforming both on sets of randomly chosen test examples. An analysis of variance indicates that the superiority of EBL-ANN is significant at the 99.9% confidence level both by comparison to EBL-ALONE and by comparison to ANN-ALONE. As also seen in the cup example, ANN-ALONE over generalizes, with the result that it correctly classifies all the chairs but incorrectly classifies almost 70% of the non-chairs. EBL-ANN also learns an overly general definition of chairs; however, it only misclassifies about 50% of the non-chairs.

[i]The oracle for chairs constrains Table 7's domain theory by requiring that chains be able to support a load. This oracle also classifies as chairs objects that the original domain theory rejects for having three legs (instead of four) or for lacking back support.

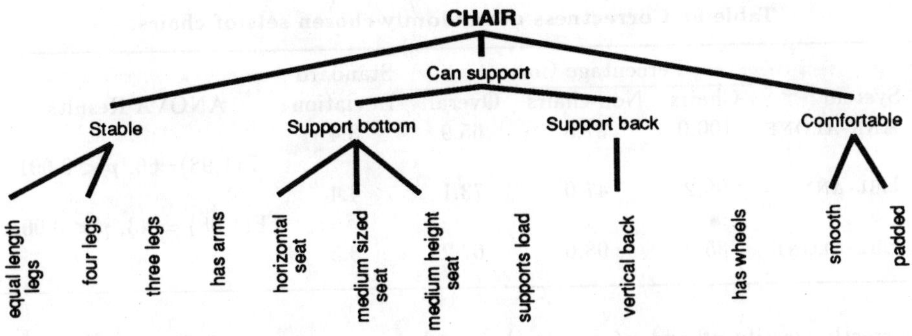

Figure 6: Initial network for recognizing chairs (low weighted links not shown).

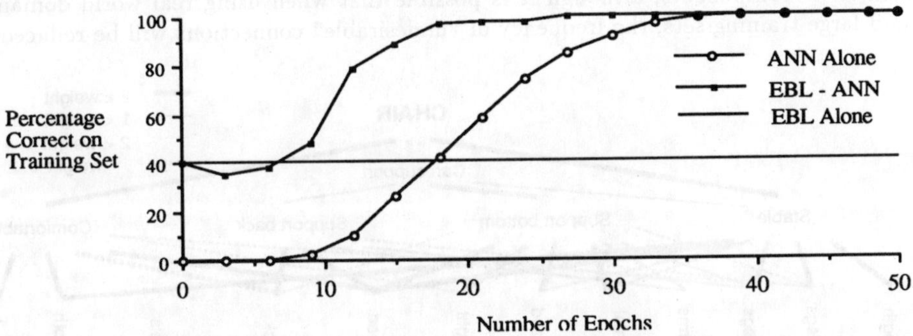

Figure 7: Classification accuracy for chairs as a function of training amount.

Figure 8 shows the average final network for recognizing chairs. (The exact weights and biases are reported in the appendix.) Comparing Figure 8 with Figure 6, it is clear that results of learning by the EBL-ANN system are less obvious for chairs than for cups. However, close scrutiny of Figure 8 reveals that back-propagation training corrects the results of the EBL system in several ways. First, the requirement for back support is dropped allowing Chairs *c2*, *c4* and *c5*, which do not have back support, to be accepted as chairs. Recall that in the cup domain theory, links are added but not dropped. The dropping of the requirement for back support shows that EBL-ANN has the ability to generalize rules by dropping antecedents in addition to its already demonstrated ability to specialize. Second, three-legged chairs are recognized (Chairs *c2*, *c3*, and *c4*). The weights are such that the *stable* is active when a chair has either three or four legs, so long as they are equal length. Third, to be a chair, an object must be able to support a sizable load. This feature, whose inclusion is required to

Table 9: **Correctness of randomly chosen sets of chairs.**

System	Percentage Correct: Chairs	Non-chairs	Overall	Standard Deviation	ANOVA Results
ANN-ALONE	100.0	31.8	65.9	3.4	
					F(1,98)=66, $p < 0.001$
EBL-ANN	99.2	47.0	73.1	4.4	
					F(1,98) = 43, $p < 0.001$
EBL-ALONE	35.9	98.6	67.2	3.5	

correctly classify *n2* and *n5* as non-chairs, plays a significant role in the calculation of every hidden unit to which it is attached. This significant linkage of *supports load* to several hidden units again points to the need for a tighter integration of the EBL and ANN components, although it is possible that when using real world domains and large training sets, the frequency of "undesirable" connections will be reduced.

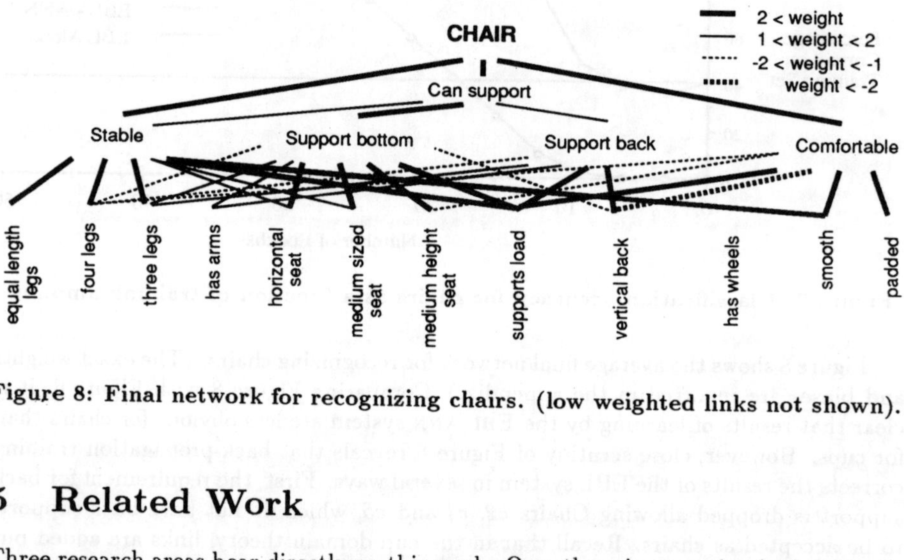

Figure 8: Final network for recognizing chairs. (low weighted links not shown).

5 Related Work

Three research areas bear directly on this work. One, there is a growing body of work on combining explanation-based and empirical approaches to learning. Two, there is additional research on using symbolic rules to guide neural network learning. Three, work on representing symbolic data structures in neural networks is relevant to the issue of mapping from the EBL subsystem into ANNs.

Others have investigated the combination of explanation-based and empirical learning methods.[21, 34, 13, 5, 8] Some systems use empirical learning to provide input to an EBL system. For example, the UNIMEM system[21] analyzes a collection of examples and finds similarities. These similarities are then explained using EBL. Approaches more closely related to the EBL-ANN approach are those that use EBL to produce training examples for an empirical learning system. For instance, the IOE system[13] uses a conventional inductive learning algorithm to empirically analyze a collection of explanations, thereby refining the domain theory. However, the bulk of this research on hybrid systems has focused on combining systems that perform conventional, "symbol-based" computation.

Two other projects which explore the possible synergies between explanation-based learning and neural networks were reported at the 1989 Machine Learning Workshop. Katz[18] describes an approach which, like EBL-ANN, maps explanations into neural networks. However, the focus of his work is on improving the time required to classify an object rather than on revising the rules used to create the explanation. As a result, the initial resemblance of a network to an explanation structures is rapidly lost during learning. Thus, although the network may correct errors in an explanation structure, it is impossible to interpret those corrections as symbolic rules which could be used in the construction of future networks. Work by Diederich[10] also integrates knowledge into a neural network. But the knowledge in Diederich's networks consists of partial ISA hierarchies and the information learned by the network is the presented training instances and more complete ISA hierarchies. While this approach may be applied to classification tasks, it does not address the problems of incorrect or incomplete domain theories. In addition, Oliver and Schneider[33] are investigating using rules to decompose problems into subproblems. Their results show that decomposing problems significantly reduces the learning time of neural networks. Unlike the EBL-ANN approach, they do not address the problem of dealing with incorrect rules.

Many researchers have investigated the representation of symbolic data structures in ANNs.[40, 44, 51, 54] This work is relevant to the further development of techniques for mapping an EBL system's symbolic data structures into neural networks. Rumelhart, Smolensky[40] discuss the representation of schemata in neural networks. The results of EBL systems, which learn useful compositions of other concepts, can be viewed as schemata. Mappings from rules to neural networks have been proposed.[52, 14] These approaches, though, do not deal with mapping useful combinations of rules into ANNs, nor do they address the issue of correcting rules. In case-based reasoning systems, concrete episodes are stored; a neural approach for representing and selecting cases is described by Thrift.[51] EBL systems produce generalized cases, and the issue of selecting the appropriate generalization to apply to a new problem is important,[25] but is currently ignored in the EBL-ANN system.

6 Open Research Issues

While the current implementation of the EBL-ANN system has produced promising results, several aspects of it require refinement and extension. In addition, the approach must be tested in several real-world domains, in order to get a better estimate of its value.

6.1 Adding Additional Hidden Units

The current algorithm adds input units in addition to those involved in the explanation, but does not add any hidden units. Incorporating more hidden units may increase the power of the algorithm. Rules of thumb for the amount and connectivity of additional hidden units are needed.

6.2 Allowing Variables

Another extension of the initial algorithm involves the use of variables. The current algorithm assumes explanations are *propositional* and do not involve variables. A major strength of EBL algorithms is that they determine the most general constraints on the objects in a problem, a task usually accomplished by reasoning about which variables have to be equated with constants or other variables. Techniques for variable binding in ANNs are being investigated.[52, 49] These techniques will be extended so that variables initially believed to be identical by the EBL subsystem can become separate variables following further training. Conversely, initially distinct variables may need to become unified.

6.3 Allowing Real-Valued and Hierarchical Features

A major shortcoming of the initial implementation is that all features must be *nominal.* That is, the possible values of a feature are described in an unordered set. The algorithm will be extended so that features that are *linear* (e.g., volume as a real number) or *hierarchical* (e.g., shapes where squares are polygons) can be handled. These three are the standard types of features used in inductive learning.[24]

One way of using real-valued features that nicely falls within the EBL-ANN approach is to have the EBL subsystem reason using qualitative values, while having the ANN subsystem use quantitative values. For example, in order to simplify the building of explanations, the EBL system could use *small*, *medium*, and *large* as possible sizes. After the initial ANN is constructed, actual measurements could be used.

6.4 Providing Feedback from the ANN Subsystem to the EBL Subsystem

The research undertaken so far investigates having an EBL system provide initial guidance to an ANN. It is also desirable to have the results produced by the ANN affect the EBL portion of the system. Two approaches for doing so are under investigation. The first is to map the refined ANN back into the vocabulary of the EBL system so that the acquired concept can be symbolically analyzed and used by the EBL subsystem to acquire additional concepts. The second is to analyze what has changed in the ANN in order to locate the most uncertain portions of the concept, which can then be focussed upon. These two approaches are elaborated below.

After the ANN refines a rule acquired by the EBL portion of the system, the rule cannot be used by the EBL subsystem to produced additional explanations because it is not in the vocabulary of the EBL subsystem. Intuitively, the hybrid approach nicely matches the phenomenon where after extensive experience with a task, one can no longer articulate how decisions are made. However, techniques are being developed for mapping the ANN back into the EBL subsystem's vocabulary. The re-represented rules can then be used to build new explanations. If additional hidden units appear in the ANN, some undefined symbols may appear in the rules, thereby extending the vocabulary of the EBL subsystem. The techniques could also be used to make the knowledge encoded in an ANN human-comprehensible and to refine the basic rules of the EBL system. For example, after training the domain theory could be re-expressed and presented to the system's user.

By having a symbolic system analyze which weights in the ANN are fluctuating the most and by considering the magnitude of the back-propagated error signals, the most uncertain aspects can be localized. This information can be used in several different ways. For example, it can help produce focused questions of a human advisor, e. g., "Is feature x important?". Also, if the system designs its own experiments,[6, 37] this information can suggest which experiments will prove informative.

6.5 Using EBL Failure to Direct Back-Propagation

An issue previously mentioned is that the EBL-ANN system does not always add new links in the most appropriate place. For example, the hybrid system learned that a strong positive weight from *supports load* to any of the four intermediate conclusions to which it is connected produces the correct classifications of the chairs and non-chairs in the training set. This problem is reduced by having a low setting for the parameter that determines when the output is close enough to being correct, because low settings lead to additional back-propagation training but at the cost of reducing the training time advantage of the hybrid. Methods of localizing back-propagation by examining explanation failure are being investigated in order to reduce the chance of inappropriate connections.

This method would only operate when back-propagation produced a false negative response. Rather than simply propagating error backward through the entire network, the EBL subsystem would be called, and an attempt to explain how the example fits the concept would be made. If the EBL subsystem is unable to complete the explanation, the areas of the explanation which are incomplete could be identified. Those identified areas would be the only areas of the ANN to receive back-propagated error. For example, in attempting to explain that a Styrofoam cup is a cup, both the stable and open-vessel consequents can be fully explained by the provided rules. However, liftable cannot be explained because graspable cannot be explained. Thus, the EBL component would indicate that back-propagation should focus its efforts on determining some other way to suggest that an object is graspable.

6.6 Applying EBL-ANN to Planning

Classification tasks provide an attractive test bed in which to initially develop the EBL-ANN approach. Inductive learning systems, such as ANNs, have long done well on this type of task. Planning is a more challenging task with which to further develop the approach. There has been much less progress on using inductive learning to produce plans. Largely, this is due to the interdependencies among the steps in a plan. A learner must be able to reason about the causal relationships among the entities in the plan in order to decide what is and what is not relevant. EBL systems are well-suited to learning about plans, as they are capable of explaining the interdependencies among the steps in a plan.[47] However, as previously mentioned, EBL systems are limited by the accuracy and coverage of their domain theories. Applying the EBL-ANN approach to planning problems promises to eliminate this limitation.

Techniques for representing and executing stored plans in neural networks are reported by Whitehead and Ballard.[54] The applicability of these techniques to the EBL-ANN approach are being investigated.

6.7 Acquiring Recursive Concepts

Many, if not most, concepts and plans have a recursive or iterative nature. That is, a concept can involve any number of repeated components and some actions can be performed as many times as necessary in order to achieve the goal of a plan. For example, a planner may need to make a number of adjustments in order to complete a task. Sometimes four adjustments may be needed, sometimes six, other times none. The exact number will depend on the details of the current situation (presence of other agents, weather conditions, etc). Nevertheless, underlying this variability is essentially a single plan. It would be foolish for a performance system to have a different plan for each possible number of adjustments.

The recently-developed BAGGER2 algorithm[46] recognizes the presence of recursive processes in the explanations of specific problems. When any such traces are encountered, a recursive concept or plan is produced. When recursive processes are not detected, the result is the same as that produced by EGGS. The BAGGER2 algorithm greatly increases the generality of acquired concepts and plans, thereby reducing training time and improving the effectiveness of the system that uses the acquired results. This ability to acquire a recursive concept from a single training example is especially important in domains where only a small number of training scenarios are available.

Within neural network research, *recurrent* networks are used to represent recursive processes.[35] EBL-ANN is being extended so that recursive concepts can be acquired. This is being accomplished by using the BAGGER2 algorithm to produce roughly-correct concepts, which are then mapped into recurrent ANNs and refined.

6.8 Integrating Multiple Explanations into a Single ANN

A major issue in EBL is that a single training example may not suffice to fully learn the complete concept. For example, a general plan may involve several different ways to accomplish a subgoal while the training example may only use one of these. One way this problem can be addressed is by merging the results of multiple explanations into a single learned concept. Techniques are being developed for combining multiple explanations into a single ANN. A complicated concept, possibly containing several special cases, will be acquired by generalizing the explanations of several training examples. This means that the individual training examples can be simpler and, thus, more easily explained.

7 Conclusion

Using an explanation-based learning system to guide the construction of the initial neural network is an approach that overcomes the shortcomings of both explanation-based and neural learning. In the presented EBL-ANN approach, the initial neural network configuration is determined by the generalized explanation of the solution to a specific classification task. Since the construction of the initial network is motivated, fewer training examples are needed than if the initial network configuration is randomly chosen. The presence of irrelevant features is also less of a problem because the EBL subsystem focuses attention on the features most likely to be relevant. In addition, this hybrid system overcomes a major shortcoming of explanation-based learning, namely, the problem of incorrect domain theories.

In the EBL-ANN system, a "roughly-correct" domain theory leads to the acquisition of a classification rule that is nearly correct. Subsequent refinement by the neural network improves it. Preliminary empirical results indicate that the hybrid system more accurately learns concepts than just the explanation-based system by itself, that the hybrid learns concepts much faster than the neural learning system by itself, and that the results of its learning generalize to unseen examples better than either the neural learning system by itself or the explanation-based system by itself.

Most realistic problems can never be formalized exactly. However, there is much to be gained by utilizing the capability to reason approximately correctly. EBL provides a way to profitably use causal models of the world, while ANNs provide a way to refine roughly-correct concepts. By combining symbolic and neural learning algorithms, the presented research promises to broaden the applicability of machine learning techniques, producing learning algorithms that are not brittle and which can produce concepts whose accuracy improves through experience.

Acknowledgements

We would like to thank James Hendler and the reviewers for their detailed and valuable comments and suggestions. This research was partially supported by the University of Wisconsin Graduate School.

This paper originally appeared in Connection Science, **1**, (1989), 231–252.

References

[1] S. C. Ahalt, F. D. Garber, I. Jouny, and A. K. Krishnamurthy, Performance of Synthetic Neural Network Classification of Noisy Radar Signals, *Advances in Neural Information Processing Systems, Volume 1*, (Denver, CO, 1989), pp. 281–288.

[2] S. Ahmad, A Study of Scaling and Generalization in Neural Networks, Technical Report CCSR-88-13, (University of Illinois, Center for Complex Systems Research, 1988).

[3] E. B. Baum and D. Haussler, What Size Net Gives Valid Generalization?, *Neural Computation* **1** (1989) 151–160.

[4] S. Bennett, Real World EBL: Learning Error Tolerant Plans in the Robotics Domain, *Proceedings of the AAAI Explanation-Based Learning Symposium*, (Stanford, CA, 1988), pp. 122–126.

[5] F. Bergadano and A. Giordana, A Knowledge Intensive Approach to Concept Induction, *Proceedings of the Fifth International Conference on Machine Learning*, (Ann Arbor, MI, 1988), pp. 291–297.

[6] J. G. Carbonell and Y. Gil, Learning by Experimentation, *Proceedings of the Fourth International Conference on Machine Learning*, (Irvine, CA, 1987), pp. 256–266.

[7] S. A. Chien, Using and Refining Simplifications: Explanation-Based Learning of Plans in Intractable Domains, *Proceedings of the Eleventh International Joint Conference on Artificial Intelligence*, (Detroit, MI, 1989), pp. 590–595.

[8] A. P. Danyluk, The Use of Explanations for Similarity-Based Learning, *Proceedings of the Tenth International Joint Conference on Artificial Intelligence*, (Milan, Italy, 1987), pp. 274–276.

[9] G. F. DeJong and R. F. Mooney, Explanation-Based Learning: An Alternative View, *Machine Learning* **1** (1986) 145–176.

[10] J. Diederich, Knowledge-Intensive Recruitment Learning, Technical Report ICSI-TR-88-010, (International Computer Science Institute, Berkeley, CA, 1988).

[11] T. G. Dietterich, Learning at the Knowledge Level, *Machine Learning* **1** (1986) 287–316.

[12] D. H. Fisher and K. B. McKusick, An Empirical Comparison of ID3 and Back-Propagation, *Proceedings of the Eleventh International Joint Conference on Artificial Intelligence*, (Detroit, MI, 1989), pp. 788–793.

[13] N. S. Flann and T. G. Dietterich, A Study of Explanation-Based Methods for Inductive Learning, *Machine Learning* **4** (1989) 187–226.

[14] S. I. Gallant, Connectionist Expert Systems, *Communications of the ACM* **31** (1988) 152–169.

[15] K. J. Hammond and N. Hurwitz, Extracting Diagnostic Features from Explanations, *Proceedings of the AAAI Explanation-Based Learning Symposium*, (Stanford, CA, 1988), pp. 31–35.

[16] G. E. Hinton and T. J. Sejnowski, Learning and Relearning in Boltzmann Machines, eds. D. E. Rumelhart and J. L. McClelland, *Parallel Distributed Processing: Explorations in the Microstructure of Cognition. Volume 1: Foundations.*, (MIT Press, Cambridge, MA, 1986).

[17] H. Hirsh, Explanation-Based Generalization in a Logic-Programming Environment, *Proceedings of the Tenth International Joint Conference on Artificial Intelligence*, (Milan, Italy, 1987), pp. 221–227.

[18] B. F. Katz, EBL and SBL: A Neural Network Synthesis, *Proceedings of the Eleventh Annual Conference of the Cognitive Science Society*, (Ann Arbor, MI, 1989), pp. 683–689.

[19] S. T. Kedar-Cabelli and L. T. McCarty, Explanation-Based Generalization as Resolution Theorem Proving, *Proceedings of the Fourth International Conference on Machine Learning*, (Irvine, CA, 1987), pp. 383–389.

[20] P. Koton, Evaluating Case-Based Problem Solving, *Proceedings of the Case-Based Reasoning Workshop*, (Pensacola, FL, 1989), pp. 173–175.

[21] M. Lebowitz, Integrated Learning: Controlling Explanation, *Cognitive Science* **10** (1986) 219–240.

[22] J. L. McClelland and D. E. Rumelhart, An Interactive Activation Model of Context Effects in Letter Perception: Part 1, *Psychological Review* **88** (1985) 375–407.

[23] J. L. McClelland and D. E. Rumelhart, A Distributed Model of Human Learning and Memory, eds. D. E. Rumelhart and J. L. McClelland, *Parallel Distributed Processing, Vol. II*, (MIT Press, Cambridge, MA, 1986), pp. 170–215.

[24] R. S. Michalski, A Theory and Methodology of Inductive Learning, eds. R. S. Michalski, J. G. Carbonell, and T. M. Mitchell, *Machine Learning: An Artificial Intelligence Approach*, (Tioga, Palo Alto, CA, 1983), pp. 83–134.

[25] S. N. Minton, Quantitative Results Concerning the Utility of Explanation-Based Learning, *Proceedings of the National Conference of Artificial Intelligence*, (St. Paul, MN, 1988), pp. 564–569.

[26] S. N. Minton, *Learning Search Control Knowledge: An Explanation-Based Approach*, (Kluwer, Hingham, MA), 1989.

[27] T. M. Mitchell, Generalization as Search, *Artificial Intelligence* **18** (1982) 203–226.

[28] T. M. Mitchell, R. Keller, and S. Kedar-Cabelli, Explanation-Based Generalization: A Unifying View, *Machine Learning* **1** (1986) 47–80.

[29] R. J. Mooney, *A General Explanation-Based Learning Mechanism and its Application to Narrative Understanding*, (Pitman, London, 1989).

[30] R. J. Mooney and S. W. Bennett, A Domain Independent Explanation-Based Generalizer, *Proceedings of the National Conference of Artificial Intelligence*, (Philadelphia, PA, 1986), pp. 551–555.

[31] R. J. Mooney and G. F. DeJong, Learning Schemata for Natural Language Processing, *Proceedings of the Ninth International Joint Conference on Artificial Intelligence*, (Los Angeles, CA, 1985), pp. 681–687.

[32] R. J. Mooney, J. W. Shavlik, G. G. Towell, and A. Gove, An Experimental Comparison of Symbolic and Connectionist Learning Algorithms, *Proceedings of the Eleventh International Joint Conference on Artificial Intelligence*, (Detroit, MI, 1989), pp. 775–780.

[33] W. L. Oliver and W. Schneider, Using Rules and Task Division to Augment Connectionist Learning, *Proceedings of the Tenth Annual Conference of the Cognitive Science Society*, (Montreal, Canada, 1988), pp. 55–61.

[34] M. J. Pazzani, Integrated Learning with Incorrect and Incomplete Theories, *Proceedings of the Fifth International Conference on Machine Learning*, (Ann Arbor, MI, 1988), pp. 291–297.

[35] F. J. Pineda, Generalization of back-propagation to recurrent neural networks, *Physics Review Letters* **59** (1987) 2229–2232.

[36] J. R. Quinlan, Induction of Decision Trees, *Machine Learning* **1** (1986) 81–106.

[37] S. A. Rajamoney, *Explanation-Based Theory Revision: An Approach to the Problems of Incomplete and Incorrect Theories*, PhD thesis, (Department of Computer Science, University of Illinois, Urbana, 1989).

[38] S. A. Rajamoney and G. F. DeJong, The Classification, Detection and Handling of Imperfect Theory Problems, *Proceedings of the Tenth International Joint Conference on Artificial Intelligence*, (Milan, Italy, 1987), pp. 205–207.

[39] F. Rosenblatt, *Principles of Neurodynamics: Perceptrons and the Theory of Brain Mechanisms*, (Spartan, New York, 1962).

[40] D. Rumelhart, P. Smolensky, J. McClelland, and G. Hinton, Schemata and Sequential Thought Processes in PDP Models, eds. J. L. McClelland and D. E. Rumelhart, *Parallel Distributed Processing: Explorations in the microstructure of cognition; vol. 2: Psychological and Biological Models*, (MIT Press, Cambridge, MA, 1986), pp. 7–57.

[41] D. E. Rumelhart, G. E. Hinton, and R. J. Williams, Learning Internal Representations by Error Propagation, eds. D. E. Rumelhart and J. L. McClelland, *Parallel Distributed Processing: explorations in the microstructure of cognition. Volume 1: Foundations*, (MIT Press, Cambridge, MA, 1986).

[42] R. C. Schank and D. B. Leake, Creativity and Learning in a Case-Based Explainer, *Artificial Intelligence* **40** (1989) 353–385.

[43] T. J. Sejnowski and C. Rosenberg, Parallel Networks that Learn to Pronounce English Text, *Complex Systems* **1** (1987) 145–168.

[44] L. Shastri, A Connectionist Approach to Knowledge Representation and Limited Inference, *Cognitive Science* **12** (1988) 331–392.

[45] J. W. Shavlik, Learning about Momentum Conservation, *Proceedings of the Ninth International Joint Conference on Artificial Intelligence*, (Los Angeles, CA, 1985), pp. 667–669.

[46] J. W. Shavlik, Acquiring recursive and iterative concepts with explanation-based learning, *Machine Learning* **5** (1989) 39–70.

[47] J. W. Shavlik, *Extending Explanation-Based Learning by Generalizing Explanation Structures*, (Pitman, London, 1990).

[48] J. W. Shavlik, R. J. Mooney, and G. G. Towell, Symbolic and Neural Net Learning Algorithms: An Empirical Comparison, To appear in *Machine Learning*, in press.

[49] P. Smolensky, On Variable Binding and the Representation of Symbolic Structures in Connectionist Systems, Technical Report CU-CS-355-87, (University of Colorado - Boulder, 1987).

[50] G. Tesauro and T. J. Sejnowski, A Parallel Network that Learns to Play Backgammon, *Artificial Intelligence* **39** (1989) 357–390.

[51] P. Thrift, A Neural Network Model for Case-Based Reasoning, *Proceedings of the Second Case-Based Reasoning Workshop*, (Pensacola Beach, FL, 1989), pp. 334–337.

[52] D. S. Touretsky and G. E. Hinton, A Distributed Connectionist Production System, *Cognitive Science* **12** (1988) 423–466.

[53] S. M. Weiss and I. Kapouleas, An Empirical Comparison of Pattern Recognition, Neural Nets, and Machine Learning Classification Methods, *Proceedings of the Eleventh International Joint Conference on Artificial Intelligence*, (Detroit, MI, 1989), pp. 688–693.

[54] S. D. Whitehead and D. H. Ballard, Connectionist Designs on Planning, eds. D. Touretsky, G. Hinton, and T. Sejnowski, *Proceedings of the 1988 Connectionist Models Summer School*, (Morgan Kaufmann, San Mateo, CA, 1988), pp. 357–370.

Appendix

The weights and biases for the network in Figures 5 and 8 appear in Tables 10 and 11, respectively. These numbers are the means of 50 experimental runs. Blanks entries indicate the absence of a link between the two units.

Table 10: Average weights and biases of EBL-ANN for cups.

Feature	Open Vessel	Graspable	Stable	Liftable	Cup
BIAS	7.63	1.97	2.57	7.19	11.52
liftable					4.02
open vessel				0.45	6.58
graspable				4.71	
stable				0.15	5.58
flat bottom	-0.28	0.52	5.11		
made of Styrofoam	1.21	3.01	1.68		
made of ceramic	0.21	0.05	0.12		
made of paper	-1.74	-2.55	-1.79		
has handle	0.24	4.68	0.15		
handle on top	-2.24	-0.47	-2.25		
handle on side	2.46	0.34	2.41		
light	0.29	0.57	0.15	5.20	
expensive	-1.42	-0.69	-1.55		
fragile	1.26	1.22	1.69		
has concavity	4.80	0.64	0.08		
concavity points up	5.36	0.71	0.11		

Table 11: Average weights and biases of EBL-ANN for chairs.

Feature	Can Support	Stable	Support Bottom	Support Back	Comfortable	Chair
BIAS	7.87	5.65	13.44	3.21	8.70	9.16
can support						4.05
stable	-1.57					5.05
support bottom	5.23					
support back	3.36					
comfortable	0.74					9.47
equal length legs		7.43	-0.24	-0.22	0.42	
four legs		4.22	-1.68	-1.81	-1.64	
three legs		2.45	0.97	1.20	0.25	
has arms		0.87	1.15	1.28	1.02	
horizontal seat		1.73	4.31	-0.58	0.10	
med. sized seat		1.72	4.30	-0.60	-1.40	
med. height seat		1.68	4.30	-0.60	-1.40	
supports load		3.88	2.85	2.77	4.30	
vertical back		-0.83	-1.69	3.43	-2.39	
has wheels		0.48	-0.15	-0.25	0.18	
smooth		2.51	-0.22	-0.19	6.94	
padded		1.68	-0.68	-0.57	3.60	

Applications of Learning and Planning Methods
edited by N. G. Bourbakis
© 1991 World Scientific Publishing Co., pp. 99–127

CHAPTER 5

THE APPLICATION OF SYMBOLIC INDUCTIVE LEARNING TO THE ACQUISITION AND RECOGNITION OF NOISY TEXTURE CONCEPTS

Peter W. Pachowicz

Center for Artificial Intelligence, George Mason University
4400 University Drive, Fairfax, VA 22030
ppach@gmuvax2.gmu.edu

ABSTRACT

The article presents and tests a methodology developed to acquire and recognize noisy texture concepts. The introduced methodology distinguishes the following three complementary research issues that are finally integrated within a system. The first issue deals with the acquisition of a texture concept incorporating symbolic inductive learning from provided examples. These examples consist of low-level numerical attribute values extracted from texture images incorporating two well-known methods, Laws' masks and co-occurrence matrices. The learned description of a texture class is a rule where a conditional part is a disjunctive normal form. The second issue deals with the manipulation of acquired concept descriptions in order to remove their noisy components. This manipulation optimizes the class description truncating its less significant components and preserving the most typical components of the concept. We show that the application of such optimized descriptions to the recognition of test datasets increases the system recognition effectiveness. The third issue deals with the application of concept optmization performed within a closed-loop system. This closed-loop system integrates the concept optimization, the recognition, control, and final decision making, and it works for gradually increasing optimization degree. This iterative optimization builds the characteristics of the recognition curves taken versus increasing optimization degree. The system makes decisions based on the dynamic pattern rather than static performance of such recognition curves. The proposed methodologies are tested in three separate experiments the results of which are presented in this paper.

1. Introduction and Motivation

One of the features of intelligent systems is their ability to acquire a concept of object such as a description of its characteristics or properties and to apply this concept to recognize a particular object among many other objects. Since our environment contains textured objects and the texture data contains noise and other irregularities, the acquisition and robust recognition of such noisy and imperfect concepts is crucial for many vision tasks; e.g., inspection, object recognition, object localization for navigation, remote tracking of a moving object, and intelligent surveillance. The success of these tasks depends both on the effectiveness of image understanding processes and the quality of data which a system has to deal with during both the learning phase and the recognition phase.

Once the system is trained and tuned to a given type of data, it is difficult to assume without testing that the system will be working with the same data but effected by a different noise. Moreover, a given object can be perceived differently when perceptual conditions are changing. Variability of object characteristics (particularly a texture) requires the development of capabilities that will allow the system to reconfigurate, tune, and update its model (knowledge) regarding object or sensor movement or the variation of sensing conditions (e.g., changing resolution, lighting, overlapped noise).

This paper presents our approach and methodology (along with tested results) to the following three complementary parts of the texture recognition problem:

1. the application of inductive learning methodology to the acquisition of texture concept from low-level numeric texture features,

2. the manipulation of acquired concepts in order to optimize them and reduce some of the noisy components from a concept description,

3. the recognition of texture datasets through flexible matching and newly developed multiple-matching methodology.[1]

Traditional adaptability and learning in engineering systems apply control engineering[23] and/or pattern recognition[10] techniques. These approaches work well using quantitative information. But typical processes of high-level vision are based on symbolic computation. The border line between numeric and symbolic computation has not been established. There are methods, which successfully apply the iconic approach alone to invariant recognition and data fusion.[33] However, the application of such methods to the object inspection task is difficult. For example, the inspection task uses object models or object variances that are acquired through explanation, technical diagrams, and tolerances, rather than from presentation of various real objects during a teaching phase.

Our approach to computer vision incorporates machine learning methodologies applied to object location, recognition, inspection, and scene understanding tasks, and it is based on the creation of system adaptability functions.[27] The learning capabilities of the system use context dependence to adapt to the dynamic environment. In the proposed approach, models used to represent data and system knowledge must possess a dynamic structure that can be modified by the system itself based on its own experience. Incorporating machine learning and cognitive processes such as induction, deduction, discovery, and analogy, we plan to combine elements of visual recognition and visual imagery[28] within a unified system.

Currently, we are investigating vision adaptation mechanisms that allow a system to recognize objects in new external situations. Such situations include 3-D rotation, changed resolution, quantitative and qualitative noise, changing lighting conditions, and mutual positional relations between objects such as occlusion. The system has to perform the following vision tasks: object localization, recognition, and inspection. We assume that prior to the system work, it either possesses an initial model of objects, or it acquires the optimal model from the environment based on background meta-knowledge. The adaptive system must be able to manage its internal structure and data (dynamic memory) in order to increase/decrease the number of distinctive features, add new objects, and modify process flow through the network. One of the most challenging issues that is currently under investigation is learning object description from variable object characteristics. This problem attacks an issue of texture model evolution performed in time (i.e., for a sequence of texture images) and for moving objects. Model evolution through machine learning technique incorporates an incremental mode of learning that is typical for a human being.

A typical task that depends on the system adaptation capability is texture-based object recognition for robot navigation in a natural terrain. The difficulties of robust recognition of 3-D objects is caused by a great variability of images projected onto the sensory array. This variability is caused not only by the 3-D rotation of an object, but also by changes of resolution, positioning of light sources (e.g., light reflection, shadows), mutual positional relations with other objects at the scene (e.g., touching, occlusion), etc. In the traditional approach to object recognition, the system is associated with a set of characteristic object features, which are used for recognizing particular objects. Such a set of features is constant[4] in that the perceptual system is unable to verify the usefulness of the object model. Thus, the classification is correct as far as the accuracy of the off-line created recognition model is concerned.

Recent progress in visual recognition of objects has generally been based on the extension of traditional visual recognition processes. Marr's[17] 2.5-D sketch theory introduces directional information to object shapes reconstructed from contour, color, texture, stereo, motion, etc. Later, this data is integrated into a perceptual system. Currently, this concept is being applied to the creation of the MIT Vision Machine.[29] The recognition of an object and its position, which is insensitive to 3-D rotation, has been proposed by Ikeuchi and Kanade.[13] In the teaching phase, the system creates internal multi-view object representations based on the CAD model. The recognition phase matches currently visible features of an object (where features can possess a 2.5-D description) with predicted features of the object's model. In these approaches, the recognition systems have been created without adaptive capabilities. One can say that these systems are closed and are unable to verify and modify the object model. These typical limitations of object recognition systems have been discussed by Bhanu[3] within the context of automated target recognition. As a consequence, a new approach has been suggested that integrates machine learning with target recognition in order to create an adaptive target recognition system.[2] This new approach integrates two machine learning techniques with a knowledge-based reasoning system. The first technique, explanation-based learning, provides the ability to build generalized descriptions of object classes. The second technique, conceptual clustering, allows the system to make decisions based on simple symbolic descriptions rather than numerical similarity measures. When created, the system will be capable of modifying its internal data concerning objects on the basis of its experience.

Searching the area of integrating vision and symbolic machine learning for texture recognition task, we conclude that such an approach has not been explored before. It was, however, initiated and illustrated by a simple example by Michalski.[19] Most works on adaptive vision systems are limited to the improvement of image segmentation[12], and the application of expert systems to the automatic configuration of vision systems.[15,18,24] Most of this effort, however, is conducted in the area of shape recognition and relatively less effort is paid to the adaptability of texture recognition system dealing with great variability of texture occurrence. Researchers try generally to avoid the problem of texture variability by the application of more powerful methods for the acquisition of texture characteristics; e.g., Markov random fields[6], Gibbs random fields[8], and the spatial frequency spectrum.[16,34]

On the other hand, the traditional object recognition applies classifiers in such a way that they are adapted to the feature set to take advantage of the extracted information. Such an approach belongs to a class of feature extraction oriented methods, where an extraction of relevant features plays a very important role.[9] The main problem with traditional approach to the recognition task is that we do not have a universal feature extraction method that works effectively with noisy and imperfect data. When one considers that the training data can be noisy and imperfect, one has to agree that the derived descriptions of texture classes consequently contain noisy components. Therefore, the recognition based on the direct matching of such noisy and imperfect descriptions with testing data follows all above mentioned disadvantages; i.e., the results are affected by the noise and imperfectness of learning and testing data. This is one of the reasons why most efforts to the solution of the recognition problem focus on the improvement of the feature extraction phase to reduce the influence of noise and data imperfectness on the system performance. Contrary to the traditional approach, we modify concept descriptions to reduce hypothetical noisy componets and apply such description to the recognition phase.

This paper presents methodology of the integration of machine vision with machine learning in the area of texture recognition and it summarizes our first introductory experiments. Section 2 of this paper presents the preparation of training and testing data along with their conversion from numeric to symbolic representations. Section 3 overviews applied methodology to acquire concept descriptions, while the principles of matching such concepts with a testing dataset are presented in Section 4. The comparison of proposed machine learning approach and traditional pattern recognition approach to the problem of texture recognition is demonstrated in Section 5. Section 6 presents the idea of optimization of concept descriptions and its influence on the recognition performance. The integration of such optimization processes with the recognition processes performed iteratively within a closed-loop system architecture is presented in Section 7. Finally, Section 8 summarizes our work and presents problems for the future research.

2. Selection of Learning and Testing Data

This section presents the process of extracting texture attributes and examples (where a single example is a vector of attributes) obtained from input texture images. We begin with the presentation of experimental images of texture classes used in three experiments discussed in this article. Next, the process of extracting numeric attributes and their conversion to their symbolic representation (suitable for the applied machine learning methodology) is presented.

2.1. Preparation of Texture Images

All texture images have been obtained from Brodatz album[5] and they have been coded as black-and-white images of 512x512 pixels. Each pixel was coded onto 256 grey levels. Image data was grouped for three experiments.

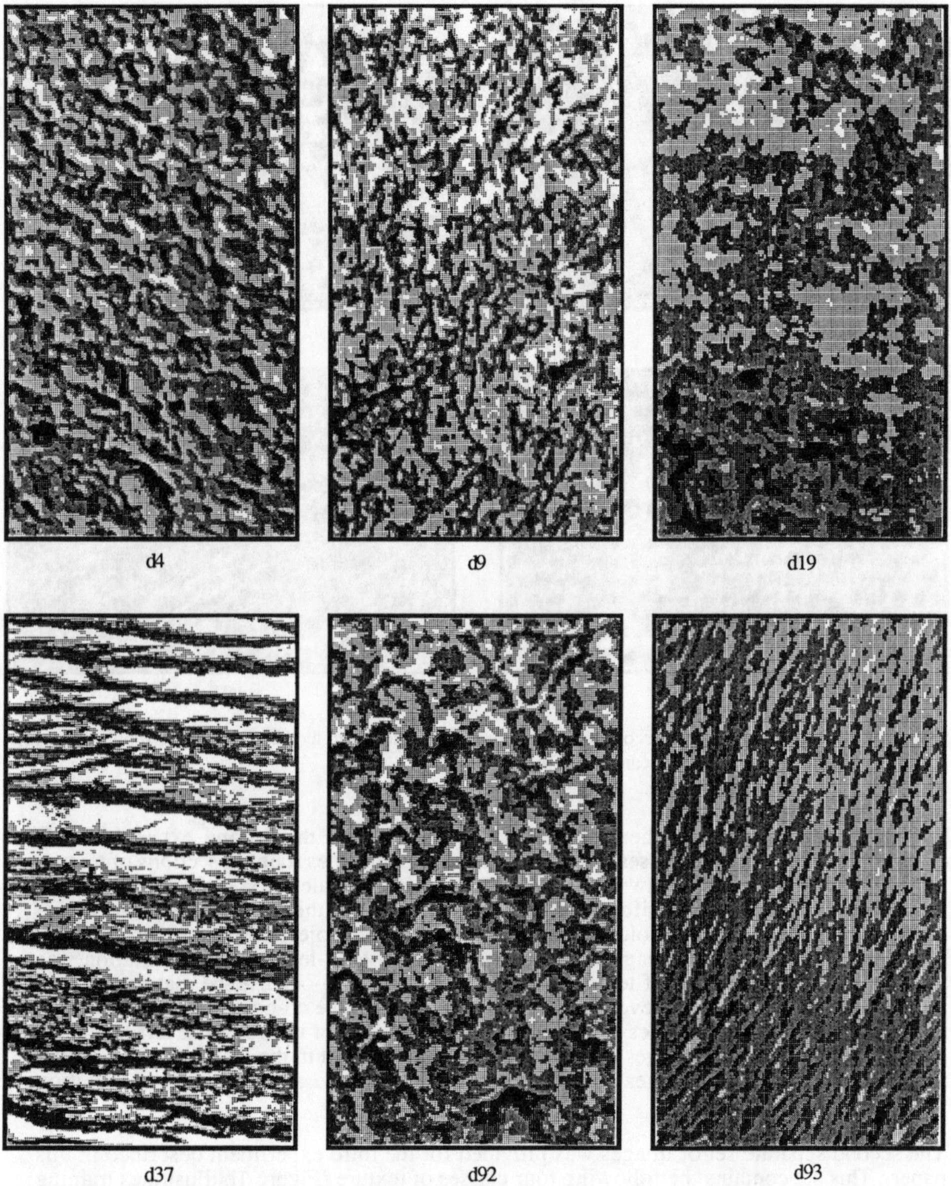

Fig. 1a Training sections of the fist set of texture images taken from the Brodatz album: pressed cork (d4), lawn (d9), woolen cloth (d19), water (d37), pigskin (d92), and fur (d93)

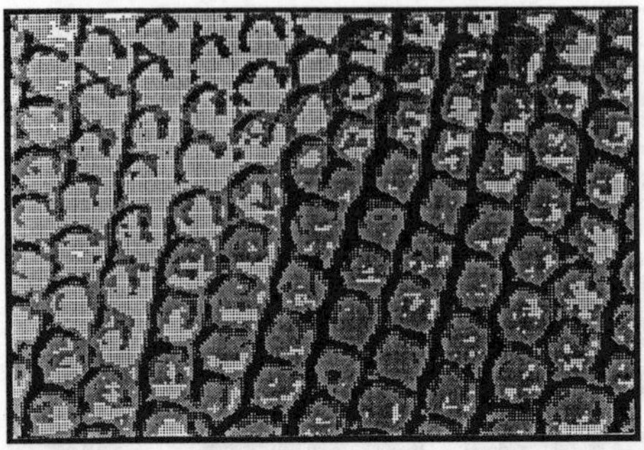

d3

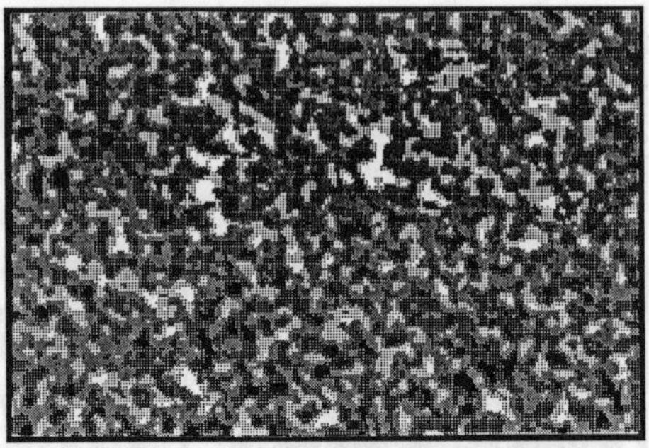

d29

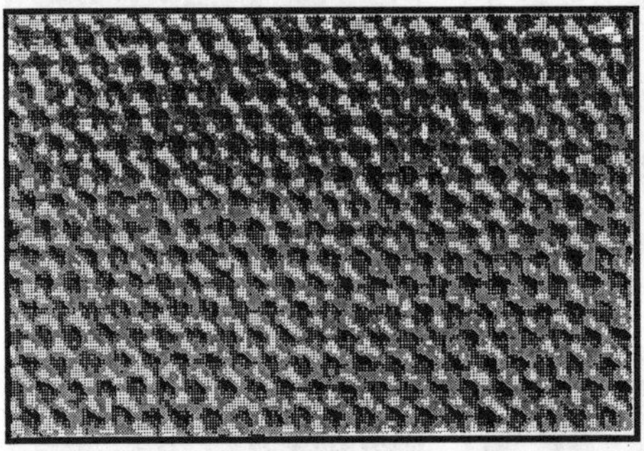

d77

d93

Fig. 1b Training sections of the second set of texture images taken from the Brodatz album: reptile skin (d3), beach sand (d29), cotton canvas (d77), and fur (d93)

The first set of images has been provided for the first and the second experiment. It contains the following six classes of texture (Figure 1a illustrates training sections of texture images): pressed cork (d4), lawn (d9), woolen cloth (d19), water (d37), pigskin (d92) and fur (d93). They were non-uniformly illuminated and two of them (lawn and water) have smoothly changed texture resolution caused by the angled projection of a texture surface. Next, the intensities of all images were decreased to 64 gray-levels. This limitation was imposed by the application of texture-feature detection by co-occurrence matrices (i.e., a large number of image gray-levels causes the dimension of the co-occurrence matrix to be very large). These input images were used to create two files of images. The left-hand side of the input images (sub-images of 512x256 pixels) were used in the learning phase, while the right-hand side (sub-images of 512x256 pixels) were used as data for the recognition phase.

The second separate set of images was provided for the third experiment described in this paper. This set contains the following four classes of texture (Figure 1b illustrates training sections of texture images): reptile skin (d3), beach sand (d29), cotton canvas (d77), and fur (d93). A set of relatively small sections of these images was shown to the system during the learning phase, while the other set of image sections was shown during the recognition phase. Before the feature extraction was executed (for the learning and recognition phases), all textural images were preprocessed. The number of pixel grey-levels was changed from 256 to a number of levels below 55.

2.2. *Extraction of Texture Numeric Attributes*

The extraction of attributes is the first phase of the transformation of raw image into its more specific characteristics. The extraction of texture numeric attributes has been executed for each image of the first group of six texture images presented in Figure 1a. A single attribute determines local specific characteristics of a given number of pixels. For a given pixel, we can extract a certain number of different attributes that form an event such as a vector of these attributes.

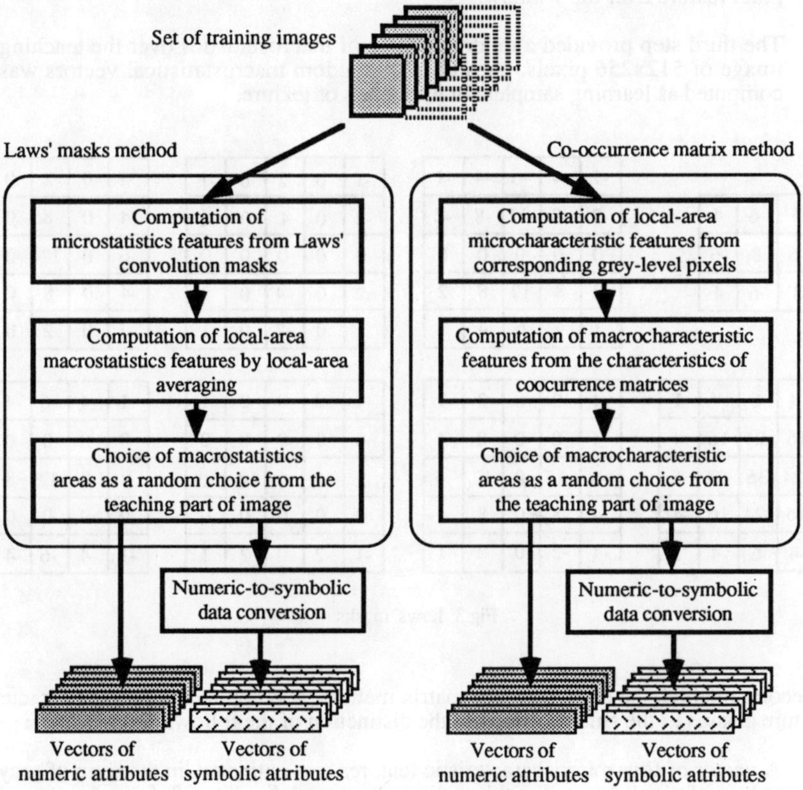

Fig. 2 Extraction of texture attributes

We applied two methods to extract local numeric attributes. The first method filters an input image incorporating Laws' energy masks[14], while the second method computes the characteristics of co-occurrence matrices extracted for a given texture area.[35] The following steps of the extraction of texture numeric attributes are presented in Figure 2.

The extraction of texture numeric attributes for the Laws' energy masks method was performed in the following three steps:

1. The computation of a single vector of microstatistical features was executed for a central pixel using one 3x3 and seven 5x5 convolution masks shown in Figure 3.
2. Based on the vectors of microstatistical features, the system computed local-area macrostatistics over a larger window of 20x20 pixels, where a vector of macrostatistical properties was computed as the average absolute value of each pixel feature over the window.
3. The third step provided a random choice of macrostatistics over the teaching image of 512x256 pixels. A set of 200 random macrostatistical vectors was computed as learning samples for each class of texture.

4	-6	4
-6	8	-6
4	-6	4

-1	-4	-6	-4	-1
-2	-8	-12	-8	-2
0	0	0	0	0
2	8	12	8	2
1	4	6	4	1

-1	0	2	0	-1
-2	0	4	0	-2
0	0	0	0	0
2	0	-4	0	2
1	0	-2	0	1

-1	0	2	0	-1
-4	0	8	0	-4
-6	0	12	0	-6
-4	0	8	0	-4
-1	0	2	0	-1

1	-4	6	-4	1
-4	16	-24	16	-4
6	-24	36	-24	6
-4	16	-24	16	-4
1	-4	6	-4	1

-1	-2	0	2	1
-4	-8	0	8	4
-6	-12	0	12	6
-4	-8	0	8	4
-1	-2	0	2	1

-1	-2	0	2	1
0	0	0	0	0
2	4	0	-4	-2
0	0	0	0	0
-1	-2	0	2	1

-1	-4	-6	-4	-1
0	0	0	0	0
2	8	12	8	2
0	0	0	0	0
-1	-4	-6	-4	-1

Fig.3 Laws' masks

The second method, the co-occurrence matrix method, was applied to build a characteristics of texture classes in the similar way with the distinction of the following three steps:

1. A vector of four microcharacteristic features was extracted, in the form of gray values of pixels located within a distance d, and direction ß, from the central pixel (see Figure 4), and associated with the gray value of the corresponding central pixel.

2. For each microcharacteristic feature and a class of texture, the system computed local-area diagonal co-occurrence matrices over a larger window (20x20 pixels). Each co-occurrence matrix was then used to calculate both the angular second momentum (also called the matrix uniformity) and the contrast as a measure of the spread of values away from the main diagonal.[31] Finally, we obtained a vector of numeric parameters, where each position within the vector represents local-area macrocharacteristics for a given distance d, and direction ß.

3. This step of the extraction of texture attributes was the same as for the Laws' masks method and it provided a random choice of macrostatistic vectors over the teaching image, where a set of 200 learning examples was selected randomly for each class of texture.

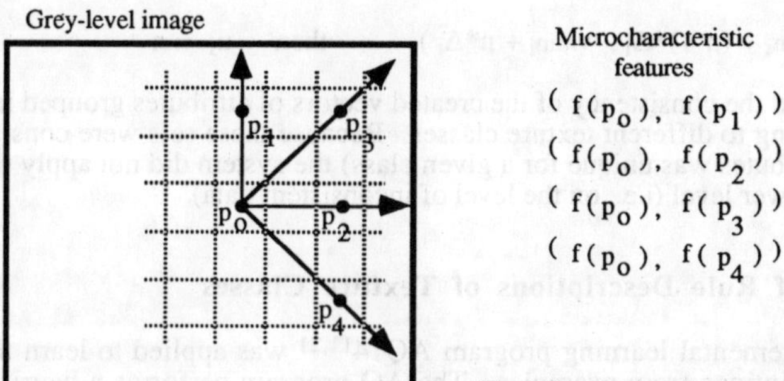

Fig.4 Local microcharacteristic features for a given direction and distance

2.3. Conversion of Numeric Attributes into Their Symbolic Representation

Symbolic inductive learning, as applied to acquire texture rule descriptions, requires to code the data as symbolic values. To fulfill this requirement, we converted numerical attributes into their symbolic intervals incorporating a special interface. As a consequence, this conversion called as scaling process is an early generalization of examples of numeric form into their more general symbolic representation as an elementary cubic-cell in the symbolic attribute space. This conversion is applied in case of a constant number of texture classes, as well as the constant number of attributes. The scaling was determined by an a-priori given set of events, which cannot be changed or extended by adding data from the environment, such as that obtained during on-line system experience. On the other hand, future use of dynamic scaling assumes that a system will be able to extend the number of texture classes, to add new events that are characteristic of a single class, and to change the number of attributes or modify them.

Let us assume that $\mathbf{V} = \{ \mathbf{V}^1 , \dots , \mathbf{V}^j , \dots, \mathbf{V}^k \}$ is a set of numerical events for a single class of texture, and $\mathbf{V} \ni \mathbf{v} = (v_1 , \dots , v_i , \dots, v_m)$ is a learning event expressed as a vector of m numerical attributes. Then, for each i-th attribute, we compute the following scaling parameters: $vmin_i$, $vmax_i$, and Δ_i, which are the minimum value, maximum value, and data interval of each i-th attribute respectively. These parameters were found for all elements of the set **V**:

$$vmin_i = \min_{j \in [1,k]} \{ v_i \in V^j \} \qquad vmax = \max_{j \in [1,k]} \{ v_i \in V^j \} \tag{1}$$

$$\Delta_i = (vmax_i - vmin_i)/(nint - 1) \tag{2}$$

where nint is the number of intervals accepted by an inductive learning algorithm (in our case nint was equal to 50).

We applied the conversion of numeric attributes to their symbolic representation by the computation of a new attribute vector $\mathbf{u} = (u_1 , \ldots , u_i , \ldots , u_m)$ with the following formula:

$$\text{if} \quad v_i \in [\, vmin_i + (n-1)*\Delta_i \,;\; vmin_i + n*\Delta_i \,) \qquad \text{then} \quad u_i = n \tag{3}$$

The system checked the consistency of the created vectors of attributes grouped into sets of events corresponding to different texture classes. Because these sets were consistent (i.e., each vector of attributes was unique for a given class) the system did not apply the scaling processes on the lower level (i.e., on the level of inconsistent data).

3. Acquisition of Rule Descriptions of Texture Classes

The inductive incremental learning program AQ14[11,21] was applied to learn the texture attributional descriptions from examples. The AQ program performs a heuristic search through a space of symbolic expressions, and its goal is to find the most preferred expression according to a specified criterion. The input to the AQ program consists of a string of learning events, and each event is a vector of attribute values. The set of events obtained for one class is called a set of positive examples. With respect to this particular class, all other events are negative examples. The program finds an optimal cover of all the positive examples. This cover cannot include any negative examples. The process repeated for each class of learning events produces decision rules to discriminate all classes of texture images amongst themselves.

The conditional part of a rule is defined as a cover and coded as a disjunctive normal form. A cover is a disjunction of complexes (using the OR operator), where a complex is decomposed into selectors (using the AND operator). A selector is a form [L # R], where L is called the referee which is an attribute, R is called the referent which is a set of values in the domain of the attribute in L, # is one of the following relational symbols: =, <, >. An example of a rule, complex and selectors can then be illustrated as below, however, the domain of texture attributes operates on symbolic values coded as numbers in range from 0 to nint:

rule:	[Transport=car] <== [Weather=bad] v [Temp<60]
complex:	[Weather=bad] <--- [Weather_type=cloudy] & [Temp>60] & [Wind_dir=South v West]
selector:	[Weather_type=cloudy] (4)

Each generated complex is associated with a pair of weights; i.e., t-weight and u-weight. The t-weight of a complex is the number of positive examples covered by the complex, and the u-weight is the number of the positive examples uniquely covered by the complex. The following is the example of a complex for texture attribute domain:

[x1=5..10 v 15][x2>34][x6=5..34][x8<45]   (t-weigth: 18, u-weight: 9)   (5)

where x-i is an attribute. The complexes of a rule are then ordered according to decreasing values of the t-weight.

The AQ14 inductive incremental learning program can work in two modes producing intersecting or disjoint covers. Rule induction in the intersecting mode produces covers that can logically intersect with those of other classes over "don't care" areas of the attribute space. On the other hand, rule induction in the disjoint mode produces covers that do not intersect at all with covers of other classes. As a consequence, rules produced in the intersecting mode are more general than rules produced in the disjoint mode.

In our experiments, we learned rule descriptions of texture classes both in intersecting and disjoint modes, mainly to compare recognition results. The output of the AQ14 algorithm consisted of the discrimination rules which were transferred to the texture recognition phase. The program is controlled by a set of parameters and they proper use has the great influence on system recognition effectiveness. Any change of these parameters must be done very carefully.

4. Texture Recognition through Flexible Matching

Acquired rule descriptions of texture classes were then applied to recognize test texture datasets. This test data was prepared in the same way as learning data, but test events were chosen from other parts of images that were not presented to the system during the learning phase. The same methods as in the learning phase were applied to the computation of texture numeric attributes. One hundred testing samples were obtained for each class of texture, and for each separate method used in extracting texture attributes. Extracted test events were converted from their numeric representation to the symbolic form applying scaling process with parameters calculated during the learning phase. The recognition phase was then activated to check the effectiveness of this machine learning approach to texture recognition.

Generally, there are two methods for classifying the concept membership of an instance; i.e., the strict match and the flexible match. In the strict match, one tests whether an instance strictly satisfies the condition part of a rule (i.e., one of rule complex). Such matching gives the response of two possible logical values; i.e., true or false. In the flexible match, one determines the degree of closeness between the instance and the condition part. Such closeness is represented by a coefficient that can vary in range from 0 (does not match) to 1.0 (matches). In strict matching, one determines the most closely related concept description. Flexible matching must be applied when the concept does not have clear boundaries, the learning data was noisy, or the test data was obtained from other parts of the image. If image sections applied to the testing phase are different from sections applied to the learning phase then test data can include variations of concept occurrence caused by local differences in projection, illumination, and noise.

In the recognition phase, we incorporated a special tool, the ATEST program,[30] that had been developed to test the performance of a rule base. ATEST provides the domain expert with three very important capabilities. First, it allows the expert to rapidly test a rule base on numerous test examples under a variety of evaluation schemes. These evaluation facilities provide information about the overall performance of specific rules, and this feature is used in the texture recognition phase. Second, ATEST provides routines that check a rule base for consistence and completeness. Third, it applies flexible matching to determine classification decision.

ATEST views rules as expressions when applying them on a vector of attribute values. The result of this evaluation is a real number which is a degree of consonance between the conditional part of the rule and the event. In our case, the program worked on the separated sets of events, where each set was obtained for a single class of texture as described above. Each event was classified into one of the given texture concepts. There were three possible classification decisions for a single event, i.e., an event belongs only to the correct class (unique-classification), an event does not belong to the correct class (misclassification), and an event belongs to several classes where one of them is the correct class (multiple-classification). The final recognition decision is made based on counting the unique- and multiple-classified events for a single set of test data. Such classification is called *first rank,* and is our main measure of the effectiveness of recognition.

The performance of the ATEST program is controlled by a set of parameters. The user can adapt the program to a specific domain. The control parameters determine program performance in evaluating the rules on the testing events. The evaluating parameters provide definitions for logical operators (e.g., the AND operator may be evaluated as "minimum" or as "average"). They control the degree of conconance and the definition of selectors (e.g., a selector can be treated as boolean *true/false*, or as a function with a normalized value from 0 to 1). Any change of these parameters must be done very carefully because the effectiveness of system recognition can be decreased.

5. Experimental Comparison of Recognition Effectiveness

In this section, we present the performance of applied symbolic machine learning methodology to the problem of texture recognition. Obtained results are compared with the k-NN non-parametric pattern recognition method.

5.1. System Architecture

The traditional open-loop architecture, presented in Figure 5, was used incorporating learning and recognition systems and two recognition approaches (i.e., machine learning and pattern recognition). The system included the extraction of texture numeric attributes, conversion of these attributes to symbolic representation, a learning-from-examples module (AQ14), an inductive assertion module (ATEST program), and a pattern recognition classifier. While most of these modules/tools have been explained in preceding sections, the application of the pattern recognition classifier is discussed below.

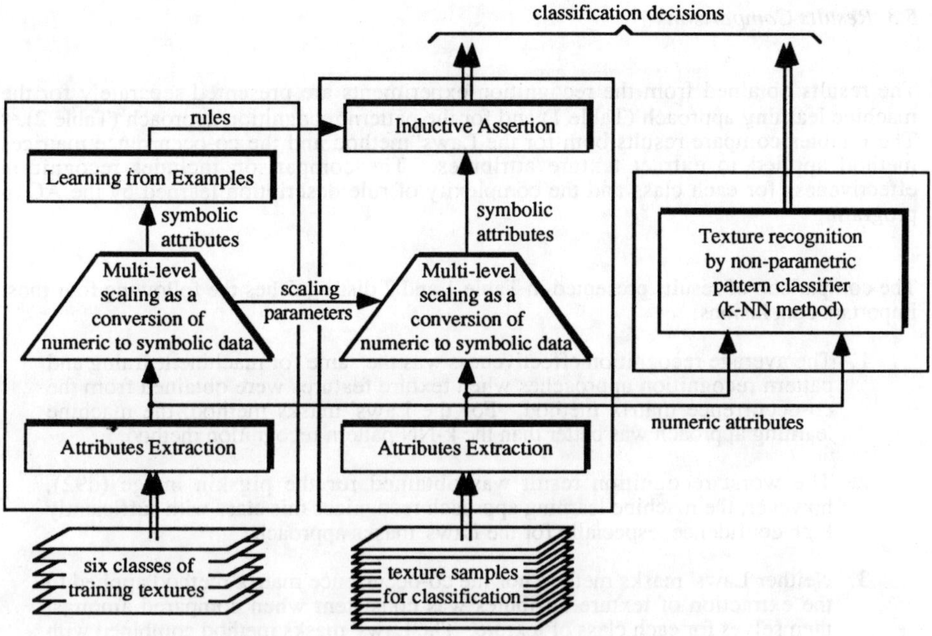

Fig.5 Open-loop architecture for texture concept acquisition and recognition incorporating symbolic machine learning and pattern recognition approaches

5.2. Recognition by Pattern Classifier

Several pattern recognition techniques have been considered for the texture recognition problem. The parametric methods, e.g., risk minimization using Bayes decision method, were excluded after testing the attribute space. The creation of parametric models of feature distribution was not satisfactory because the distribution of training data was irregular and difficult to estimate from parametric curves. Therefore, we chose the well-known k-NN non-parametric statistical pattern recognition method.[10]

In this method, learning examples are cumulated into sets corresponding to texture classes. During the recognition phase, a subset of k-nearest learning examples is selected from the set of all learning examples presented to the system. Thus, the classification decision is created indicating this class, for which most of the k-nearest learning examples were selected. The main advantage of this method is its handling of irregular and complex training data. However, the requirement of storing all learning data (or selected data from the most representative samples of the attribute domain distribution) for their use during the recognition phase is time and memory consuming. This disadvantage limits the use of such methods. The second problem is related to the choice of the most appropriate k value. In our comparison, we found k=10 and k=30 as most representative.

5.3. Results Comparison

The results obtained from the recognition experiments are presented separately for the machine learning approach (Table 1) and for the pattern recognition approach (Table 2).[26] These tables compare results both for the Laws' method and the co-occurrence matrices method applied to extract texture attributes. The comparison includes recognition effectiveness for each class and the complexity of rule description learned by the AQ14 program.

The comparison of results presented in Table 1 and 2 distinguishes the following four most important conclusions:

1. The average recognition effectiveness was the same for machine learning and pattern recognition approaches when texture features were obtained from the co-occurrence matrix method. For the Laws' masks method, the machine learning approach was better than the k-NN pattern recognition method.

2. The worst recognition result was obtained for the pigskin image (d92), however, the machine learning approach recognizes this class with sufficiently high confidence, especially for the Laws' masks approach.

3. Neither Laws' masks method nor the co-occurrence matrix method applied to the extraction of texture attributes was consistent when compared amongst themselves for each class of texture. The Laws' masks method combined with the machine learning approach to the acquisition of texture description was generally worse considering both the number of generated complexes of the rules and the recognition results. But in the case of the fifth class, the number of complexes in the rule was significantly lower and the recognition rate was higher.

4. The average deviation of the recognition, defined as $d = 1/N \sum |\underline{x} - x_i|$, was lower for the machine learning approach than for the pattern recognition approach. Such lower deviation with high average recognition rate makes the system more stable, i.e., each texture class can be recognized with the same level of confidence despite the similarity to another class.

The recognition results shown in Table 1 were obtained for the intersection cover mode of the inductive learning algorithm. The generation of rules for the intersecting mode was much faster than for the disjoint cover mode. The average recognition effectiveness was also better. The results show that for approximately the same number of complexes generated in the intersecting and disjoint modes, the recognition results were better for the disjoint mode. This tendency is presented in Table 3 for class d4 of texture. On the other hand, the recognition rate for class d37 of texture is also included, to show the tremendous decrease in recognition rate for the disjoint mode, where a large number of complexes was generated.

	Texture description method					
	Laws' masks			Cooccurrence matrices		
	Number of generated complexes	Recognition		Number of generated complexes	Recognition	
		First rank	Unique		First rank	Unique
d4	38	**72 %**	37 %	26	**88 %**	78 %
d9	35	**74 %**	45 %	31	**78 %**	57 %
d19	23	**81 %**	57 %	24	**90 %**	78 %
d37	8	**93 %**	91 %	6	**99 %**	96 %
d92	31	**73 %**	45 %	46	**57 %**	35 %
d93	23	**87 %**	64 %	13	**84 %**	78 %
Average recognition		**80 %**	56 %		**83 %**	70 %
Averaged deviation		**7 .**	14.2		**10.**	16.2

Table 1 Recognition results for machine learning approach (AQ14) - intersection mode

	Texture description method			
	Laws' masks		Cooccurrence matrices	
	Recognition for k = 10	Recognition for k = 30	Recognition for k = 10	Recognition for k = 30
d4	53 %	49 %	88 %	90 %
d9	73 %	68 %	75 %	77 %
d19	88 %	91 %	95 %	97 %
d37	99 %	97 %	98 %	96 %
d92	40 %	38 %	49 %	46 %
d93	69 %	57 %	97 %	93 %
Average recognition	**70 %**	**66 %**	**83 %**	**83 %**
Averaged deviation	**16.3**	**18.7**	**14.4**	**14.4**

Table 2 Recognition results for k-NN pattern recognition approach

	Texture description method							
	Laws' masks method				Co-occurrence matrices method			
	DC mode		IC mode		DC mode		IC mode	
	Number of complexes	Recog. result	Number of complexes	Recog. result	Number of complexes	Recog. result	Number of complexes	Recog. result
d4	38	85 %	38	72 %	26	89 %	26	88 %
d92	118	54 %	31	73 %	111	34 %	46	57 %

Table 3 Results comparison for two modes of rule generation - disjoint cover mode (DC), intersection cover mode (IC)

6. Increasing Recognition Effectiveness through the Optimization of Class Descriptions

6.1. Motivation and Methodology

In the experiment described above, the texture description was directly applied in the recognition phase to generate classification hypothesis for the test data. However, when one considers that the learning texture data can be noisy and imperfect, one has to agree that the derived descriptions of texture classes consequently will contain noisy components. The application of such descriptions in the recognition phase causes the missclassification of some testing data. If these texture descriptions are filtered (e.g., optimized in order to remove some noisy components) before the recognition phase is executed then applied matching of testing data with reduced concept prototypes should give better recognition results.

The above conclusion motivated us to perform the second experiment. This experiment investigated rule optimization in order to compress the description of a concept and to reduce some of its noisy/imperfect components. Obviously, this optimization must be executed after the learning and before the recognition phases. We used a method of rule optimization that is based on the two-tiered description of imprecise concepts introduced by Michalski et al.[22] A simple two-tiered concept description generates both the Base Concept Representation (BCR) of typical properties of a concept, as well as the Inferential Concept Interpretation (ICI) of allowed concept modifications. The SG-TRUNC method was applied to obtain a BCR through a sequence of generalization and specialization operations.[35] Initially, the SG-TRUNC method performs generalization to remove selectors from the complexes. After such removal, a complex is more general, i.e., it covers more examples. Then, a specialization operation removes the number of complexes. In this way, the description covers less examples.

The rule optimization processes performed by the SG-TRUNC method are based on rule characteristics described in Section 3. This rule characteristics is composed of two coefficients, the t-weight and the u-weight. The t-weight is the total number of examples covered by a complex, while the u-weight is the number of examples covered by the same complex and no other. Applyed optimization method preserves those complexes that have high t- and high u-weights and modifies these complexes with low t- and u-weights. The degree of rule optimization is controlled by two parameters. The first parameter controls the removal of selectors and the second one controls the reduction of complexes.[35,36] The increase in parameter values causes greater rule modification.

6.2. Recognition results

We applied the SG-TRUNC method as contained in the AQ16 algorithm. Relatively low parameter values were applied, both equal to 5 (where a single parameter has a value from 0 to 100), to control the removal of selectors and complexes. This means that the optimization of rules was low. Obtained recognition rates are presented in Table 4 and can be compared with the results presented in Table 1, where both tables illustrate results for the machine learning approach to texture recognition.[26] The analysis of these results gave us the following conclusions:

1. The description of texture classes has been compressed significantly without decreasing recognition effectiveness, and a secondary effect of the optimization was the increase of recognition speed.

2. The recognition effectiveness, both for the Laws' masks method and the co-occurrence matrix method, has been increased significantly.

3. The average deviation of the recognition rates has been reduced while the recognition effectiveness was increased.

	Texture description method					
	Laws' masks			Cooccurrence matrices		
	Number of generated complexes	Recognition First rank	Recognition Unique	Number of generated complexes	Recognition First rank	Recognition Unique
d4	6	**95 %**	1 %	5	**90 %**	42 %
d9	6	**88 %**	6 %	3	**77 %**	34 %
d19	3	**90 %**	25 %	3	**86 %**	61 %
d37	1	**92 %**	80 %	1	**94 %**	92 %
d92	7	**83 %**	2 %	12	**75 %**	32 %
d93	5	**98 %**	39 %	4	**96 %**	61 %
Average recognition		**91 %**	25 %		**86 %**	54 %
Averaged deviation		**4.1**	22.7		**7.**	17.7

Table 4 Recognition results for combined learning-from-examples and rule truncation method (AQ16 and ATEST programs) - intersection cover mode

The final summary of recognition effectiveness for three methods (machine learning with optimized rules, machine learning without applied optimization of rules, and k-NN pattern recognition method) is presented in Table 5.

Through the application of rules optimization, we were able to improve overall recognition results. The average recognition rate increased to 91% (from 80%) in the case of the Laws' masks method and to 86% (from 83%) for the co-occurrence matrix method applied to extract texture attributes. The recognition rate was significantly increased (up to 83% and 75% from 73% and 57%, respectively) for the recognition of the pigskin texture. In this way, the minimum recognition rate for both methods for extraction of texture attributes was improved. Moreover, the variation of recognition rates over texture classes has been reduced; i.e, the recognition rates have been smoothed. The smoothing effect has been observed for both methods of extraction of texture attributes.

	Texture description method					
	Laws' masks			Cooccurrence matrices		
	ML approach optimal rules	ML approach rules	PR approach k-NN method	ML approach optimal rules	ML approach rules	PR approach k-NN method
Average recognition rate	**91%**	**80%**	**70%**	**86%**	**83%**	**83%**
Highest recognition rate	98%	93%	99%	96%	99%	78%
Lowest recognition rate	83%	72%	40%	75%	57%	49%
Averaged deviation	**4.**	**7.**	**16.3**	**7.**	**10.**	**14.4**

Table 5 Summary of results for machine learning (ML) and pattern recognition (PR) approaches

6.3. Characteristics of Recognition Effectiveness

The optimization processes applied to modify rule descriptions are controlled by parameters. These parameters influence on the degree of performed removal of selectors and reduction of complexes. If they are increased then the description of texture classes are more compressed. However, it is difficult to set up these parameters a-priori. If the compressed descriptions perform better during the recognition phase then the question arises of what is the maximum value of each parameter that allows such optimized rules to perform better. To answer to this question, we decided to obtain characteristics of the recognition effectiveness versus the values of control parameters.

These characteristics were obtained for the average recognition rate calculated over the set of six classes of texture presented in Figure 1a and the Laws' method of texture attributes extraction. The individual characteristics of selected classes of texture were obtained, as well. We applied the AQ16 program[36] to perform the acquisition, optimization and recognition processes within its integrated software environment. Unfortunately, this integrated software package does not give us necessary flexibility in tuning the recognition part of the software system. The recognition processes of the AQ16 program cannot be controlled by externally defined parameters. In this way, the flexible matching implemented within the AQ16 program does not perform as well as implemented within the ATEST program that can be tuned to the specific domain of texture attributes (as applied in the above experiments.)

Bacause of mentioned limitations of the AQ16 program, results presented below cannot be compared to the first experiment with machine learning and pattern recognition approaches to the texture recognition problem. They are presented to illustrate the behavior of the recognition effectiveness versus increasing optimization parameters. Both optimization parameters responsible for removal of selectors and reduction of complexes were increased gradually by 1 from 0 to 16. Received recognition rates for six classes of images (Figure 1a) were then averaged and are presented graphically in Figure 6.

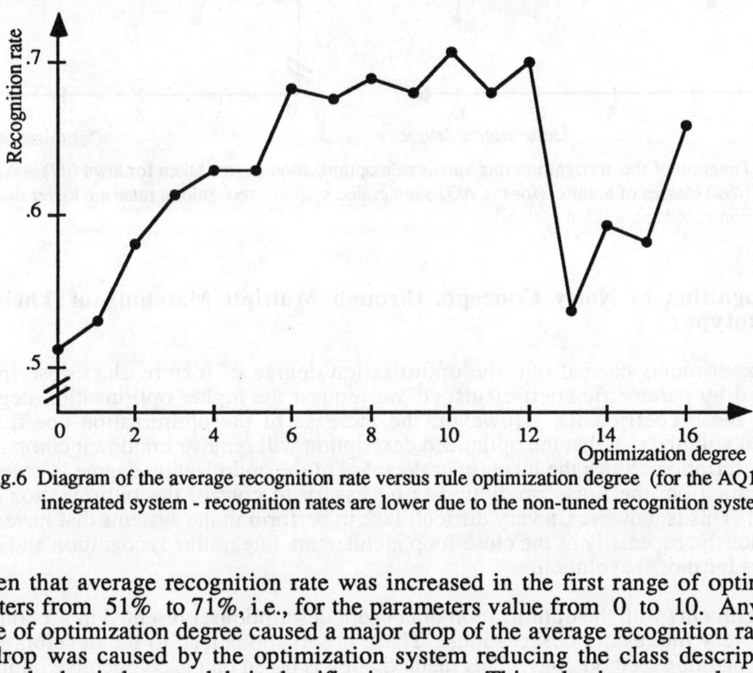

Fig.6 Diagram of the average recognition rate versus rule optimization degree (for the AQ16 integrated system - recognition rates are lower due to the non-tuned recognition system)

It is seen that average recognition rate was increased in the first range of optimization parameters from 51% to 71%, i.e., for the parameters value from 0 to 10. Any further increase of optimization degree caused a major drop of the average recognition rate. This sharp drop was caused by the optimization system reducing the class descriptions so significantly that it decreased their classification power. This reduction removed correct and significant components of class descriptions. This tendency is presented in Figure 7 for selected characteristics of two classes: lawn (d9) and the worst performer pigskin class (d92). These characteristics confirm our observation that the optimization of noisy concepts can increase recognition effectiveness, but the applicable degree of such optimization is limited to the lower values and it is followed by a sharper decrease of the recognition power of applied methodology. Therefore, the increase in optimization parameters must be watched carefully.

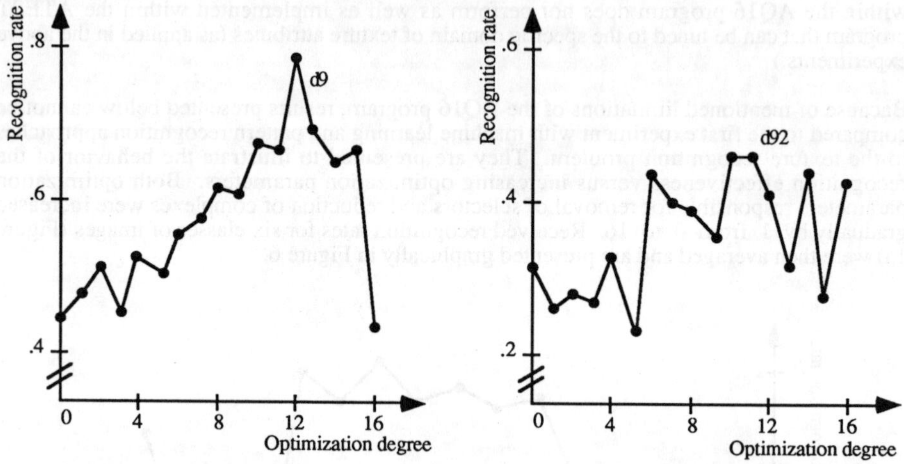

Fig. 7 Diagram of the recognition rate varsus rule optimization degree taken for lawn (d9) and pigskin (d92) classes of texture (for the AQ16 integrated system - recognition rates are lower due to the non-tuned recognition system)

7. Recognition of Noisy Concepts through Multiple-Matching of Their Prototypes

In the experiments carried out, the optimization degree of texture class descriptions is controlled by parametric coefficients. If we request the higher optimization degree, we increase these coefficients. However, the increase of the optimization coefficients is combined with the risk that the optimized description will remove important components of a class description. Since the increase or decrease of the optimization degree is based on the noise evaluation, the noise prediction is necessary to control the optimization process properly. This is, however, a very difficult task to perform in the systems that have to work autonomously, especially of the close-loop architecture integrating recognition and learning processes for model evolution.

The experiments with the optimization of concept descriptions, presented in section 6, gave us some very interesting observations. These observations lead to the development of a novel recognition methodology. This methodology is based on the consecutive optimization of concept prototypes and matching these prototypes with test data. The novelty of this method is that the recognition decision is based on the observation of the behavior of recognition curves over a sequence of optimization degree rather than on a single matching of concept prototypes.

The idea of optimization of a concept prototype in order to remove noisy/imperfect components has great advantage in such situations when the quality of learning examples presented to the system is low. This low quality can be caused not only by the low quality of image data or inefficient methods for the extraction of attributes; but it can be mostly caused by the system inefficiency in providing learning examples. It means that the system is working without teacher assistance, and it has to activate learning processes when the concept prototype must be updated according to a new information. Thus, the quality of updated concept descriptions depends on the system capability of selecting learning examples corectly.

7.1. Multiple-Matching System Architecture

A series of new experiments[1] has been executed incorporating modified architecture presented in Figure 8. The modifications (when one compares with the former architecture in Figure 5) include the extraction of texture attributes, re-designing of the recognition system, implementation of a new decision making methodology, and the creation of a closed control loop of the iterative optimization of rules and the recognition of test data. The experiments were carried out with the second set of texture images (see Figure 1b) and with the simplest method of attributes extraction.

The learning system is composed of the AQ15 inductive learning program. Acquired rules were applied to the recognition phase, where testing texture samples have been shown to the system. The recognition process has been redesigned and arranged into the closed-loop. This loop consists of three modules: rules optimization, inductive assertion (ATEST program), and a module of control and decision making. The loop is controlled by an optimization parameter. For a given parameter, the system optimizes rule descriptions and forwards them to the inductive assertion module. Inductive assertion processes are performed each time for newly optimized descriptions and on the same set of test data.

The system increases the optimization degree for each iteration loop. The decision making module completes partial classification rates computed for each optimization loop. In this way, a recognition curve is created for each set of partial descriptions of learned classes versus the optimization degree. Finally, the classification decision is made based on the characteristics of the recognition curves.

Since our goal is focused on the development of a novel approach to the recognition problem that deals with concept descriptions acquired from the noisy training data, we were not looking for a more powerful method applied for the extraction of texture attributes. Contrary to the development of more powerful, i.e., noise reducing methods, we have applied the simplest extraction of texture attributes that preserves the negative influence of noise and imperfectness of attribute acquisition on the learning phase of texture class descriptions from provided examples. This extraction of texture attributes was executed in the following way. A 5x5 window was applied to compute eight local attributes. Each attribute was calculated as a difference between the central pixel p(i,j) and one of the pixels belonging to the window's edge (i.e., p(i-2,j), p(i-2,j+2), p(i,j+2), p(i+2,j+2), p(i+2,j), p(i+2,j-2), p(i,j-2) and p(i-2,j-2).) Two hundred randomly selected training events were presented during the learning phase, while the recognition phase was performed with one hundred testing events.

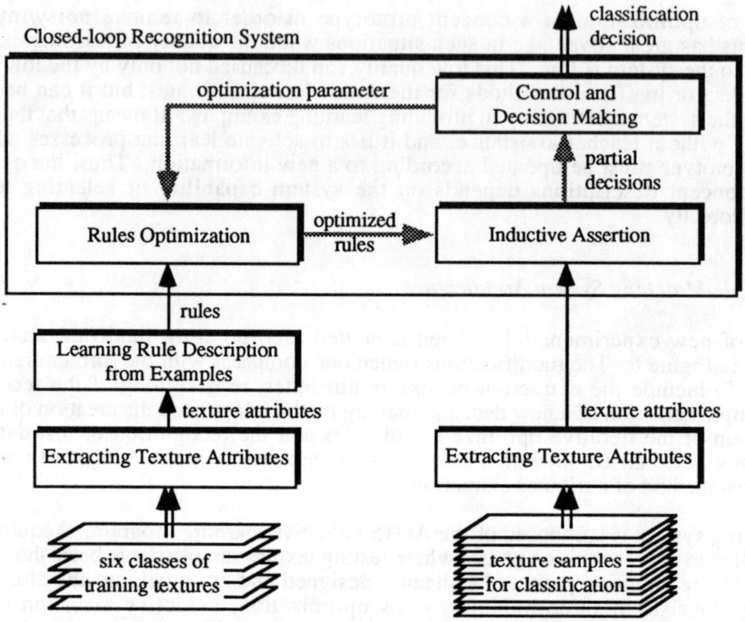

Fig.8 Closed-loop architecture for texture recognition incorporating iterative optimization of concepts prototypes and their matching with testing data

7.2. Recognition Methodology

Traditional approach to the concept recognition assumes that to recognize a concept one needs to match the observed data with a stored concept prototype. We have already discused the issue of matching a concept prototype with testing data. Flexible matching was suggested as a preferred method rather then the strict matching. In flexible matching, an approximate match is computed and used as a measure of confidence in recognizing particular concept among others. A concept prototype can have different less or more optimal forms depending on the applied optimization degree. We have already shown that the recognition result depends on the applied optimization degree and that the a-priori set up of the optimization parameters is very difficult to achieve. The solution of this problem can be seen in the application of a multiple-matching when testing data is matched several times with the gradually reduced (optimized) concept prototypes. This idea, when applied on experimental data, gave us very interesting observations.

First of all, we simplified the optimization method of rule descriptions. A new method takes into consideration the t-weight characteristics of complexes within a rule. This t-weight characteristics of a complex may be interpreted as a measure of typicality or the representativeness of this complex within a concept description. A complex with high t-weight is viewed as describing the most typical concept properties, while a complex with very low t-weight is viewed as describing concept exceptions from the typical properties. Such an exception from the typical properties can be viewed as a noisy component. So, the removal of complexes characterized by very low t-weight can be interpreted as the removal of noisy components from a concept description.

The rule complexes are ordered according to the decreasing values of the t-weight. Our approach to the optimization processes applies then the removal of a given number of the less significant (the lightest) complexes from the descriptions of all candidate concepts. The optimization of rule description by removing its lightest components is called complex truncation. Thus, the simplest definition of a truncation degree (as an optimization parameter) points out the number of the lightest complexes to be removed from the original description. How such complex truncation can have the influence on the improvement of recognition effectiveness? The answer is simple - because we remove noisy components from the concept description, i.e., we increase the probability that the testing data will be classified through flexible matching to a more typical complex.

Let us proceed further and ask whether it is possible to reverse the classification hypothesis. It means that the system classifies a set of testing data to one class for the non-optimized concept descriptions, and after applying optimization processes, it classifies the same testing data to the other class. It means that, the improvement of the class description that the test data was classified incorectly removes noisy components from the class description and in that way decreases the recognition rate for this class. In the same time, the optimization processes remove noisy components from the class description that should be recognized correctly. If these optimization processes are balanced properly, they can reverse the classification hypothesis.[1]

Such a possibility of reversing classification decision is difficult to achieve and it highly depends on the distribution of learning examples in the attribute space, the form and characteristics of generated complexes, and noise overlaped with learning and testing events. Decision making based on the recognition rates obtained for a single matching approach is very difficult, however, the observation of recognition curves obtained through the increasing truncation degree can help to support final classification decision. Based on the behavior of the recognition curve, we develop a novel recognition methodology as a complement to the traditional single match approach.

A new recognition algorithm incorporates the characteristics of the recognition curves obtained for matching test data with the gradually optimized descriptions of different classes. The recognition curve has "uptrend pattern" if the recognition rate increases with the increase of the truncation parameter, or it has "downtrend pattern" if the recognition rate decreases with the increase of the truncation parameter. Such characteristics is being built only for the low truncation degree because the higher optimization can cause the removal of significant complexes as well. Incorporating such characteristics of recognition curves, we define the following three step recognition algorithm:

Step 1. Label the first section of each recognition curve as the "uptrend" section or "downtrend" section,

Step 2. Select these recognition curves that have the "uptrend" pattern only, and

Step 3. Make the final decision indicating this class for which the "uptrend" pattern runs through the highest recognition rates. In case of no "uptrend" recognition curves, return "NO-DECISION".

7.3. Experimental Results

The recognition curves obtained incorporating introduced system architecture and recognition methodology are illustrated in Figure 9 (for the image data from Figure 1b).

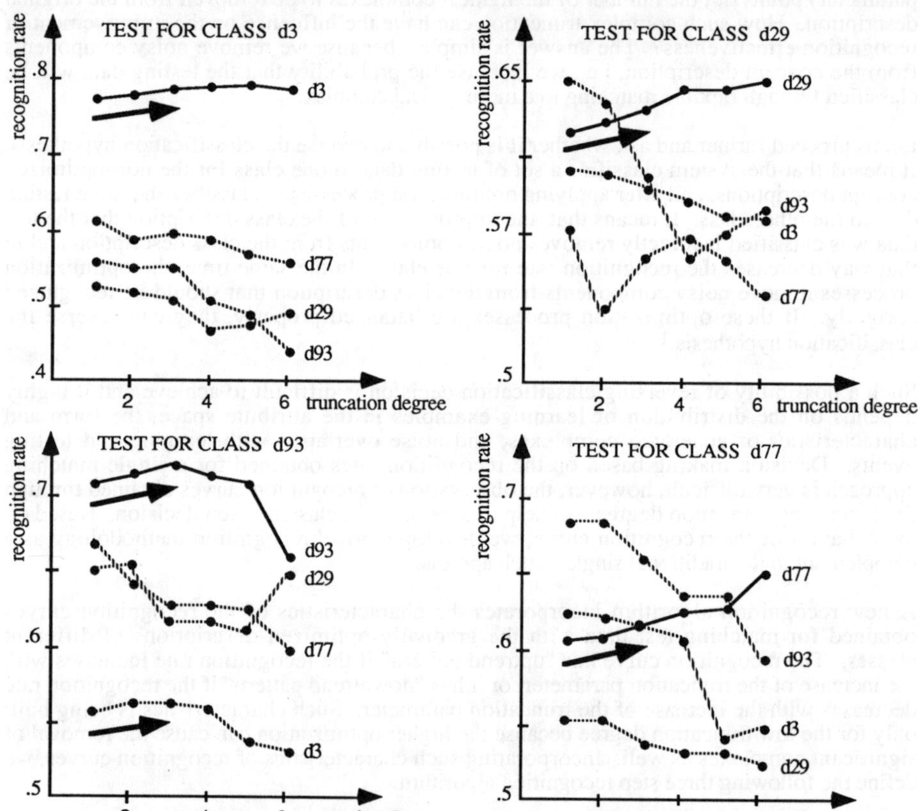

Fig.9 Recognition curves for the experiment with the closed-loop recognition incorpotating an idea of multiple-matching (an arrow indicates recognition curves with uptrend pattern, solid line - a class that should be recognized and was recognized correctly by the system, dotted line - other classes)

During the learning phase, the system completed four sets of training examples (corresponding to texture classes d3, d29, d77 and d93), each set consisting of 200 examples selected randomly from training sections of texture images. Acquired rule descriptions of these four classes were applied to recognize test sets of data, each set consisted of 100 testing events but selected from different sections of the texture image.

In Figure 9, obtained recognition curves (for each set of testing events) are marked by an arrow if they have "uptrend" pattern. The solid curve represents the class that should be recognized correctly and it was recognized by our method. Other classes are characterized by dotted lines. The truncation degree corresponds to the number of lightest complexes of a rule that were removed before the matching step. The recognition rate was computed as the number of test samples recognized to a given class divided by the total number of test samples of a given dataset. It should be pointed out that a single testing event can be recognized to more than one class when it is covered by complexes of different classes or the distance to such complexes is the same.

Applied recognition methodology recognized correctly all four classes of texture, while the traditional single-match approach failed to recognize two of them; i.e., cotton canvas (d77) and beach sand (d29). Both of these classes have a typical uptrend pattern of the recognition curve, while other curves have the "downtrend" pattern. The decision making process, however, performed for the fur class (d93) considers two "uptrend" recognition curves; i.e., classes d3 and d93. The final decision points out the class d93 because its recognition rates for the uptrend section of the recognition curve are higher than for the class d93.

Textures of cotton canvas (d77) and beach sand (d29) are not recognized if one applies traditional single match method. However, they are successfully recognized applying presented in this paper methodology based on the analysis of dynamic behavior of the recognition curve. It is seen that introduced new methodology can give us more effective recognition results in the case of noisy and imperfect data applied to learn class descriptions, what is in fact typical for engineering domain and for autonomous systems where a system (rather than a teacher) has to choose training data.

8. Conclusions and Future Work

This paper has presented the application of symbolic machine learning approach to the texture recognition problem. Learning from examples methodology has been incorporated to acquire texture concept descriptions. Then, we applied an idea of optimization of concept prototypes in order to remove noisy components from acquired class descriptions. The recognition of texture testing data has been performed through flexible matching rather than through strict matching. We have presented recently developed multiple matching methodology[1] that allows to recognize concepts applying noisy/imperfect class descriptions.

Three separate experiments have been presented to illustrate principles of introduced methods. The first experiment has compared the effectiveness of applied machine learning methodology with the k-NN pattern recognition method. The second experiment has illustrated characteristics of recognition effectiveness depending on the applied optimization degree. Finally, the third experiment has demonstrated how the characteristics of recognition rate was incorporated to develop a new recognition methodology that is based on the dynamic characteristics rather than static characteristics of the recognition rates computed for each texture class.

The results presented in the paper cannot be compared effectively to other works. They were received for different experiments. In some cases, they even differ between themselves because of applied different learning and/or matching tools, and different parameters that control these tools. However, these results confirm introduced approaches, what was the main goal to achieve.

The paper has presented our introductory experiments with the application of symbolic machine learning to typical numeric domain such as texture recognition problem. These experiments and developed methods gave us the following conclusions as future research topics:

- the development of a new machine learning tool (i.e., learning from examples) to support the input of numerical attributes that is typical for engineering domain --- or to develop a model and a method for learning hierarchical concept description incorporating multi-level scaling method (i.e., quantization of numeric data into symbolic intervals on different levels of resolution),
- further development of the iterative matching recognition methodology including the development of optimization methods of concept descriptions and a method for characterizing dynamics of recognition curves,
- the integration of traditional single match and proposed multiple match methods for generation classification hypotheses and local unification of these hypotheses in order to segment texture image,
- the application of the incremental learning mode for the evolution of texture concept over time that is caused by variable texture occurrences perceived for different external conditions (e.g., noise, illumination) and projection parameters (e.g., resolution).

Presented in this paper methods and experiments demonstrates our introductory effort in the area of the integration of computer vision and machine learning. We believe that the future progress in both areas will be increased through their theoretical and experimental integration. We expect to contribute in the computer vision area through the implementation of more flexible and powerful tools on higher levels of vision architecture to build smart vision systems capable to adapt to a new task and variable environment. On the other hand, acquired experience in such integration will be used to develop new machine learning methodologies and to test them on specific applications.

Acknowledgement

The author wishes to thank Ryszard Michalski, Azriel Rosenfeld and Yiannis Aloimonos for constructive discussion under the issues of the integration of learning methodologies with machine vision area, Janet Holms and Michael Hieb for their help in the preparation of this article. Special thanks are to Jerzy Bala for his cooperation, especially for caring out experiments presented in Section 7, and for technical help.

This research was done in the Artificial Intelligence Center of George Mason University. Research activities of the Center are sponsored in part by the Defense Advanced Research Projects Agency under grant, administrated by the Office of Naval Research, No. N00014-87-K-0874, in part by the Office of Naval Research under grant No.N00014-88-K0226, and in part by the Office of Naval Research under grant No. N00014-88-K-0397.

References

1. J.W. Bala. and P.W. Pachowicz, "Recognition of Noisy and Imperfect Texture Concepts via Iterative Optimization of Their Rule Descriptions", Report No. 90-25, MLI-90-12, Center for Artificial Intelligence, George Mason University, 1990.
2. B. Bhanu and J.C. Ming, "TRIPLE: A Multi-Strategy Machine Learning Approach to Target Recognition", Proc. DARPA Image Understanding Workshop, Cambridge, pp.537-547, 1988.
3. B. Bhanu, "Automatic Target Recognition: State of the Art Survey", *IEEE Transaction on Aerospace & Electronic Systems,* AES-22 (4), pp.364-379, 1986.
4. R.C. Bolles and R.A. Cain, "Recognizing and Locating Partially Visible Objects: the Local-Feature-Focus Method", in Robot Vision, A. Pough (Ed.), Springer-Verlag, pp.44-81, 1983.
5. P. Brodatz, "A Photographic Album for Arts and Design", Toronto, Dover Publishing Co., 1966.
6. G.R. Cross and A.K. Jain, "Markov Random Field Texture Models", *IEEE Trans. on Pattern Analysis and Machine Intelligence*, Vol.PAMI-5, No.1, pp.149-163, 1983.
7. L Davis, S. Johns and J. Aggarwal,"Texture analysis using generalized co-occurrence matrices", *IEEE Trans. Pattern Anal. Mach. Intell.*, Vol. PAMI-1, pp. 251-259, 1979.
8. H. Derin and H. Elliot, "Modeling and Segmentation of Noisy and Textured Images Using Gibbs Random Fields", *IEEE Trans. on Pattern Analysis and Machine Intelligence*, Vol.PAMI-9, No.1, pp.39-55, 1987.
9. J.M.H. DuBuf, M. Kardan and M. Spann, "Texture Feature Performance for Image Segmentation", *Pattern Recognition*, Vol.23, No.3-4, pp.291-309, 1990.
10. O.R. Duda and P.E. Hart, Pattern Classification and Scene Analysis, John Wiley&Sons, 1973.
11. J. Hong, I. Mozetic and R.S. Michalski, "AQ15: Incremental Learning of Attribute-Based Descriptions from Examples, the Method and User's Guide", ISG 86-5, UIUCDCS-F-86-949, Department of Computer Science, University of Illinois, Urbana, 1986.
12. J.Y. Hsiao and A.A. Sawchuk, "Unsupervised Textured Image Segmentation Using Feature Smoothing and Probabilistic Relaxation Techniques", *Computer Vision, Graphics and Image Processing*, Vol.48, pp.1-21, 1989.
13. K. Ikeuchi and T. Kanade, "Modeling Sensors and Applying Sensor Model to Automatic Generation of Object Recognition Program", Proc. DARPA Image Understanding Workshop, Cambridge, pp.697-710, 1988.
14. K.I. Laws, "Textured Image Segmentation", PhD Thesis, Dept. of Electrical Engineering, University of Southern California, Los Angeles, 1980.

15. C.-E. Liedtke and M. Ender, "A Knowledge Based Vision System for the Automated Adaptation to New Scene Contents", Proc. 8th Int. Conf. on Pattern Recognition, Paris, pp.795-797, 1986.
16. S.-S. Liu and M.E. Jernigan, "Texture Analysis and Discrimination in Additive Noise", *Computer Vision, Graphics and Image Processing*, Vol.49, pp.52-67, 1990.
17. D. Marr, "Vision", Freeman, San Francisco, 1982.
18. T. Matsuyama, "Expert Systems for Image Processing: Knowledge-Based Composition of Image Processes", *Computer Vision, Graphics and Image Processing*, Vol.48, pp.22-49, 1989.
19. R.S. Michalski, "AQVAL/1 - Computer Implementation of a Variable-Valued Logic System VL1 and Examples of its Application to Pattern Recognition", Proc. of the First Int. Joint Conf. on Pattern Recognition, pp.3-17, Washington, DC, 1973.
20. R.S. Michalski, "A Theory and Methodology of Inductive Learning", in Machine Learning: An Artificial Intelligence Approach, TIOGA Publishing, Palo Alto, CA, pp 83-134, 1983.
21. R.S. Michalski, I. Mozetic, J. Hong and N. Lavrac, "The AQ15 Inductive Learning System: An Overview and Experiments", ISG 86-23, UIUCDCS-R-86-1260, Department of Computer Science, University of Illinois, Urbana, 1986.
22. R.S. Michalski, "Two-Tiered Concept Meaning, Inferential Matching and Conceptual Cohesiveness", in "Similarity and Analogy", S. Vosniadou and A. Orton (Eds.), Cambridge University Press, 1987.
23. K.S. Narendra, (Ed.), "Adaptive and Learning Systems", Plenum Press, 1986.
24. B. Neumann, "Towards Computer Aided Vision System Configuration", in Artificial Intelligence II: Methodology, Systems, Applications, Ph. Jorrand and V. Sgurev (eds), pp.385-393, Elsevier Pub., 1987.
25. P.W. Pachowicz, "Low-level Numerical Characteristics and Inductive Learning Methodology in Texture Recognition", Proc. of the IEEE International Workshop on Tools for AI, Washington, D.C., pp.91-98, October 1989.
26. P.W. Pachowicz, "Integrating Low-Level Features Computation with Inductive Learning Techniques for Texture Recognition", *International Journal of Pattern Rec. and Artificial Intell.*, Vol.4, No.2, pp.147-165, 1990a.
27. P.W. Pachowicz, "Learning-Based Architecture for Robust Recognition of Variable Texture to Navigate in Natural Terrain", Proc. IEEE International Workshop on Intelligent Robots and Systems '90, Tsuchiura, Ibaraki, Japan, pp.135-142, 1990b.
28. S. Pinker, "Visual Cognition", MIT Press, 1985.
29. T. Poggio, J. Little, E. Gamble, W. Gillett, D. Geiger, D. Weinshall, M. Villalba, N. Larson, T. Cass, H. Buelhoff, M. Drumheller, P. Oppenheimer, W. Yang, A. Hurlbert, "The MIT Vision Machine", Proc. DARPA Image Understanding Workshop, Cambridge, pp.177-198, 1988.
30. R.E. Reinke, "Knowledge Acquisition and Refinement Tools for the ADVICE META-EXPERT System", ISG 84-4, UIUCDCS-F-84-921, Department of Computer Science, University of Illinois, Urbana, 1984.
31. S.J. Roan, J.K. Aggarwal and W.N. Martin, "Multiple resolution imagery and texture analysis", *Pattern Recognition*, Vol.20, No.1, pp.17-31, 1987.
32. L. Van Gool, P. Dewaele and A. Oosterlinck, "Texture Analysis Anno 1983", *Computer Vision, Graphics and Image Processing*, Vol.29, pp.336-357, 1985.
33. H. Wechsler and L. Zimmerman, "2-D Invariant Object Recognition Using Distributed Associative Memory", IEEE Transaction on Pattern Analysis and Machine Intelligence, Vol.10, No.6, pp.811-821, 1988.

34. J.S. Weszka, C.R. Dyer and A. Rosenfeld, "A Comparative Study of Texture Measures for Terrain Classification", *IEEE Trans. Systems Man Cybern.*, Vol.SMC-6, No.4, pp.269-285, 1976.
35. J. Zhang and R.S. Michalski, "Rule Optimization via SG-TRUNC Method", Proc. of the Forth European Working Session on Learning, Pitman, Morgan Kaufamnn, pp.251-262, Montpellier, Dec. 1989.
36. J. Zhang, "Learning Flexible Concepts from Examples: employing the idea of Two-Tiered Concept Representation", Ph.D. Thesis, Computer Science Department, University of Illinois, 1990.

Applications of Learning and Planning Methods
edited by N. G. Bourbakis

CHAPTER 6

AUTOMATING TECHNOLOGY ADAPTATION IN DESIGN SYNTHESIS

JAMES R. KIPPS and **DANIEL D. GAJSKI**
Department of Information and Computer Science
The University of California, Irvine
Irvine, CA 92717, USA

ABSTRACT

One of the key issues yet to be fully addressed in knowledge-based design is that of *technology adaptation*, which refers to the task of maintaining the integrity of a design system against changes in the material parts from which it builds designs. For domains such as VLSI design, where fabrication technologies change rapidly, design systems that cannot adapt to material upgrades face quick obsolescence. In this paper, we describe a knowledge-acquisition approach to automating technology adaptation with respect to the **Design and Technology Adapatation System (DTAS)**. DTAS synthesizes generic logic components from a library of technology-specific cells. Using knowledge of fundmental principles for logic design, the **Logic Learning Assistant (LOLA)** generates design rules that map generic components into library cells. LOLA enables DTAS to take advantage of new library cells as they become available, keeping DTAS robust to technology changes.

1. Introduction

1.1. IC Design Synthesis

Synthesis means the combining of parts to form a whole. *Design synthesis* refers to the process of combining material parts to form the design of an artifact. The artifact is describe in the abstract by a behavioral or functional specification and by constraints on such aspects as its cost, performance, and manufacturability. The material parts represent the basic building blocks from which designs are constructed.

The task of automating the design synthesis process for digital logic and integrated circuits (ICs) has been a primary objective of the computer-aided design (CAD) community for several decades. In this domain, the artifact to be synthesized is the IC system; the material parts from which it is synthesized are (ultimately) the transistors printed on a chip. Current trends are to factor IC design into layered subtasks. For instance, behavioral synthesis maps system behavior to structure; logic synthesis decomposes and optimizeds structure; and layout synthesis maps structure to silicon. The material parts of one level become the abstract specification for the next.

1.2. Technology Independence

One of the major problems facing IC design synthesis is that of maintaining technology independence. For instance, if a design system can synthesis an IC system using CMOS technology can it also synthesis this system in Gallium-Arsenide? The technology used in fabricating an IC system, as well as the foundary that manufactures the chips, dictates the physical details, such as fan-in, load, placement, and routing, that must be considered in order to generate high-quality designs. There are many fabrication technologies and foundries available today for manufacturing ICs. The characteristics of these technologies change often; they can be expected to continue changing well into the next century.

Early IC synthesis systems[1-4] demonstrated the potential for automation but used a "one-size-fits-all" fixed-architecture approach to technology independence. The weakness of this approach is its inability to take advantage of the technology-specific knowledge that human designers use to generate high-quality designs. As a results, the synthesized designs were too large and slow to be commercially viable.

The current approach to technology independence is to localize dependence within a library of simple and commonly used IC cells, such as one- and two-level Boolean gates, and then synthesize designs from these cells. Maintaining such libraries against technologies changes is a relatively simple task. The weakness of this approach is that it requires designs to be decomposed to a very fine grain, i.e., the level of Boolean gates. The Boolean description of even a modest-sized IC is enough to overwhelm design systems that take this approach. While there are heuristic methods for reducing this complexity, they compromise design quality.

1.3. Knowledge-Based Design

The artificial intelligence (AI) community is studying the process of human design for the purpose of moving design into the machine.[5] During the 1980's, AI researchers demonstrated the utility of knowledge-based models of design for the synthesis task.[6,7] Knowledge-based design systems (KBDS) have been applied in domains such as computer systems,[8] circuit boards,[9] integrated circuits,[10] paper handling systems,[11] and others.

Because these systems are knowledge-intensive, researchers in AI and Machine Learning have been pursuing techniques for automating the acquisition of design knowledge. Techniques that improve system performance by shortening design and avoid design flaws have received the greatest attention. This is referred to as *learning control knowledge*. Both inductive[12] and analytic[13,14] methods have been applied to this problem. There is also research that shows how acquiring too much control knowledge can hinder performance.[15] Analytic methods have also been applied to the problem of learning to design novel artifacts;[13] sometimes referred to as *learning implementation knowledge*.

One issue that this research has failed to clearly address is technology independence. Mittal and Frayman[16] cite that many knowledge-based design systems assume a "*fixed, pre-defined set of components* [i.e., material parts]." Such a "fixed parts" assumption may explain why knowledge aquisition techniques have not been applied to technology independence, but this assumption is unrealistic and will severly limit the useful lifetime of a design tool. In domains such as VLSI, where fabrication technologies change rapidly, any design synthesis system that cannot adapt to material upgrades faces quick obsolescence. With the exception of Birmingham and Sieworek,[17] little or no research exists on the problem of acquiring knowledge that allows a design system to adapt to changes in its material parts.

1.4. Technology Adaptation

We refer to the task of maintaining technology independence as *technology adaptation* and to techniques that automate technology adaptation in knowledge-based systems as *learning construction knowledge*. We view the task of learning construction knowledge as distinct from both learning control knowledge and learning implementation knowledge. In our research, we have developed an approach to learning construction knowledge for an IC design synthesis system.

The **Design and Technology Adaptation System (DTAS)** is a rule-based system for functional synthesis of regular-structured data path components, such as adders and comparitors, ALUs and multipliers, and shift registers and counters. Our objective is to generate high-quality designs using complex functional library cells, from multiplexers, adders, and comparators up to n-bit ALUs, multipliers, and counters. For instance, synthesizing a generic 12-bit ALU using three 4-bit library adders plus external function-select logic. Because library cells at the MSI and LSI level subsume large configurations of simple cells, DTAS avoids the complexity of designing entirely with Boolean gates and, thus, can scale up to large components, such as 64-bit mulitpliers. Because complex library cells are optimized to provide improved layout over functionally equivalent configurations of simple cells, DTAS can generate designs of higher-quality than synthesis systems relying on simple cells alone.

Like other synthesis systems, we localize technology dependence in an library of physical cells. Unlike other systems, we do not limit ourselves to simple cells but use complex cells; such cells are not uniformly supported across libraries. Our approach to learning construction knowledge focuses on acquiring design rules that allow DTAS to take advantage of the complex cells in a given ASIC library or to new cells as they become available. We use knowledge of fundamental principles of logic design to automatically generate library-specific design rules. As new rules are generated, the rule base is evaluated; design rules that do not produce high-quality designs under certain conditions are appropriately constrained. This approach has been implemented in the **Logic Learning Assistant (LOLA)**. LOLA is an adjunct system that runs as a DTAS preprocessor. When changes occur in a given cell library, LOLA automates the acquisition of new design rules that reflect those changes and that enable DTAS to take advantage of new cells.

1.5. Summary

In this chapter, we review the current approaches and limitations in logic synthesis and describe a new model of IC design synthesis called derivational processing, discussing its strengths and weaknesses. We also overview the DTAS synthesis system that implements this model. Next we describe our approach to technology adaptation and discuss our implementation of this approach in the LOLA knowledge acquisition subsystem of DTAS. We demonstrate how LOLA is used to upgrade DTAS's design knowledge as its material parts evolve from simple Boolean cells to more complex functions.

2. Design Synthesis in VLSI

2.1. Silicon Compilation and Logic Synthesis

IC design methods have not kept pace with advances in IC fabrication technology. Manufacturers are capable of integrating highly complex circuits of 100,000 to one million transistors on a single chip. Relying on human design engineers to deal with this level of integration has created a bottleneck in the IC development cycle. Silicon compilation is an emerging technology intended to resolve this bottleneck.[18] A silicon compiler is a software system supporting chip layout synthesis starting from a behavioral system description.[19]

In a design methodology based on silicon compilation,[20] the design engineer begins by describing the behavior of a system in a hardware description language. *Behavioral synthesis* refers to the task of reducing the high-level description to a structural design, described at the register-transfer level (RTL) and consisting of control logic, registers and memory, and data path components, such as decoders and multiplexers, adders and comparitors, and ALUs and counters. *Logic synthesis* refers to the task of mapping the RTL components into an optimized network of library or layout cells that can then be passed to a phase of *layout synthesis* for placement and routing.

2.2. Approaches

The primarily focus of current approaches to logic synthesis is on *technology mapping* and *optimization.* RTL components are reduced to Boolean equations, either manually or with module generators, and mapped into a library of Boolean cells. Graph-matching[21–23] and rule-based search[24–26] techniques are used to transform the initial configuration into an optimal (or at least high-performance) configuration.

By mapping designs to simple cells, such as one- and two-level Boolean gates or cells that are widely available in application-specific IC (ASIC) libraries, existing logic synthesis tools can be ported to new libraries with relatively little effort. Further, Boolean gates that are unavailable in one library can be expressed in terms of supported gates, e.g., implementing an AND gate with a NAND and an INV. In terms of design quality, the literature reports performance measures equal to and sometimes better than those of human designers. Several logic synthesis tools have been so successful that they are now commercial products.

Current approaches to logic synthesis work well when applied to control logic and gate arrays, both of which are naturally described with Boolean equations. However, such components make up only about twenty percent of the RTL design.[27] By enforcing a limitation to simple cells, these approaches are actually trading off range and quality for technology independence.

2.3. Scaling Up to Large Components

Although all logic components can be described with Boolean equations, it is not always desirable to do so. As the number of inputs and outputs grows, Boolean descriptions become increasingly complex. The upper bound on the number of minterms in a Boolean equation grows exponentially with each new input, until minimization of the equation becomes computationally intractable. While each output requires only one Boolean equation, the complexity of factoring a set of equations is also bounded by the number of minterms that must be considered.

To emphasize the problems this creates, consider synthesizing the arithmetic component shown in Fig. 1. This component can perform four arithmetic operations on two n-bit inputs A and B: addition, subtraction, increment, and decrement. The n-bit output of the operations is generated at F. S is a 2-bit function select line; C_{in} is an input carry; and C_{out} is an output carry.

In the case of $n = 2$, there are 7 inputs and 3 outputs, requiring three Boolean equations with a possible 127 minterms each. When $n = 4$, there are 11 inputs and 5 outputs, requiring five equations with a possible 2048 minterms each.

S_0	S_1	F
0	0	A ADD B
0	1	A SUB B
1	0	INCR A
1	1	DECR A

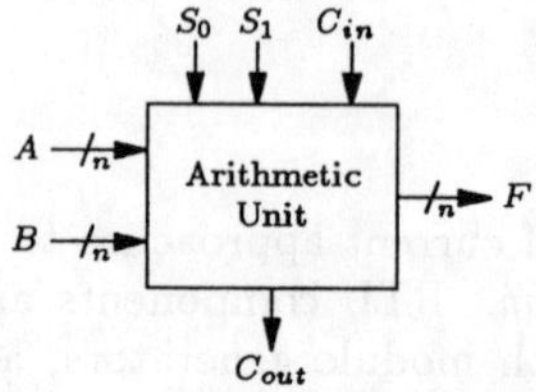

Fig. 1. Arithmetic component

It is not too unrealistic to assume that existing logic synthesis tools can handle either of these cases. On the other hand, when $n = 64$, which is a realistic figure for today's processors, there are 131 inputs and 65 outputs, requiring 65 equations with a possible 2^{131} minterms each. While a Boolean description of this size can be computed, it cannot be minimized and factored by existing synthesis tools without the availability of considerable computing resources and without incurring considerable delay.

Many of the components in a high-level design are special-purpose combinational and sequential circuits such as the one shown above. Logic synthesis tools based on Boolean minimization and "peep hole" optimizations do not scale up to the complexity of these regular-structured components. Human designers can implement components much larger than this with little difficulty. While existing synthesis tools can be used for portions of the design, they require a wider range of design knowledge and capabilities to synthesize the design in its entirety.

2.4. Simple vs. Complex Library Cells

ASIC library cells provide optimized layouts for commonly occurring logic circuits. We view library cells as falling into two classes: *simple* and *complex*. Simple cells are at the SSI level; they include one-and two-level Boolean gates, 2-bit multiplexers and and 1- and 2-bit full adders. Complex cells are at the MSI and LSI level; they include > 2-bit decoders and multiplexers, > 2-bit adders and comparators, as well as multipliers and ALUs, shift registers and counters. Complex cells are intended to provide better performance (e.g., smaller, faster, more powerful), than functionally equivalent configurations of simple cells. The use of complex library cells during synthesis can ultimately improve design quality. Complex cells are not used by existing logic synthesis tools because of their size and nonuniform library support.

Synthesis tools that take the graph-matching approach represent library cells in a canonical Boolean form that is matched against a directed-acyclic graph representing the Boolean description of the cells being synthesized. For simple library cells, the complexity of graph-matching is relatively low, so the technique works well. As the number of inputs and outputs increases, the complexity grows by orders of magnitude, making the task computationally intractable.

Synthesis tools that take the rule-based approach can avoid the complexity of graph matching by specifying the component configuration being matched in terms of other complex components—as opposed to a configuration of Boolean gates. For example, there might be a rule for replacing a configuration of four 1-bit adders with a 4-bit adder from an ASIC library. The problem with this approach is that complex cells have more features than simple cells. More features means wide variations in library support, which ultimately increases maintenance costs when the library a tool uses is upgraded or when the tool is moved to a different library.

3. The Derivational-Process Model

3.1. Designing by Function

Human designers overcome the problems that limit current approaches to logic synthesis by viewing components as functions, rather than as sets of Boolean equations. By applying a top-down design strategy, they use fundamental principles of logic design to iteratively decompose high-level components into a hierarchy of increasingly smaller subcomponents. When there are alternative ways to decompose a component, knowledge of design styles is used to select the decomposition that best fits the design constraints. Decomposition stops when the design reaches a level of granularity that can be implemented with cells from a given ASIC library. When library cells do not precisely fit, the same fundamental design principles used in decomposing the design can be applied from the bottom-up to augment or modify the library cells until their physical characteristics match the requirements for implementation of the generic design.

For example, consider the arithmetic component discussed in earlier. Figure 2 illustrates a top-down design method that decomposes a 16-bit arithmetic component based on its function. In Fig. 2 (a), the arithmetic unit is decomposed into a 16-bit adder (FA) with an external combinational circuit (CC1) on attached to its inputs. In Fig. 2 (b), the combination circuit is further decomposed into 16 identical circuits (CC2), one for each data input, and another combination circuit (CC3) controlling the carry input. CC2 and CC3 can be described by three simple Boolean equations and further synthesized from these. In Fig. 2 (c), the 16-bit adder is decomposed into four 4-bit adders. In this case, a serial design style is selected to optimize for area, but, if time were the critical constraint, a parallel design style could be used. The 4-bit adders can be further decomposed into four 1-bit adders, each of which can be implemented with a simple configuration of Boolean gates.

If the target ASIC library contained adder cells, then it is not necessary to completely decompose the generic adders to Boolean gates. For instance, if the library contained 4-bit adders, then decomposition could stop at this level, implementing the four 4-bit generic adders with four of the library adders. If,

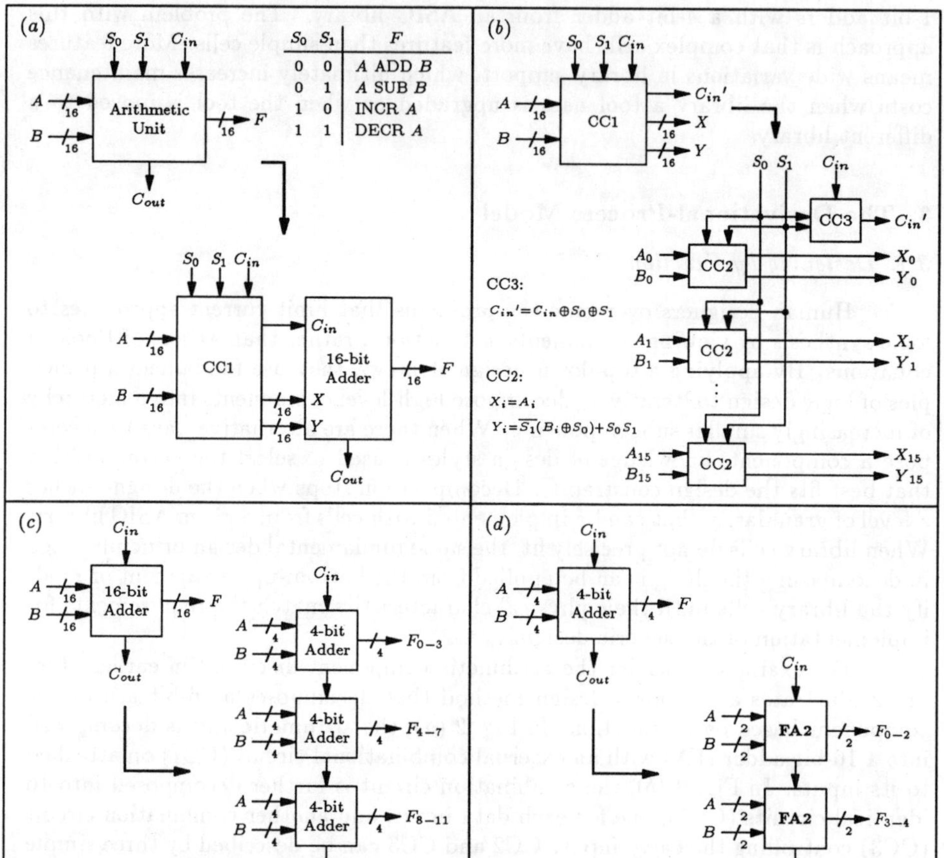

Fig. 2. Functional design

on the other hand, the library only contains 2-bit adders (FA2), then the human designer can use his knowledge of adders to construct the generic 4-bit adder from two 2-bit library adders, as shown in Fig. 2 (*d*).

We have developed the derivational-process model as an approach to synthesis that reflects the human design process. This model extends the traditional view of logic synthesis in several unique ways, the four most significant of which include: the addition of a top-down design phase, *functional decomposition*; the use of *design styles* and *derivation-driven search* to generate a candidate set of designs that closely satisfy design constraints; a *model of technology adaptation* to ensure robustness to library changes; and the instantiation of *fundamental principles of design* to aid in acquiring both design and mapping knowledge.

3.2. Synthesis as Search

In the derivational-process model, synthesis is viewed as search through a two-dimensional space of functionally-equivalent designs.[28] Along one dimension, designs vary by their degree of functional abstraction; search moves from the most abstract design, i.e., the initial functional specification, to the most specific design, i.e., a netlist of library cells. Along the other dimension, designs vary in their structural configuration; search moves between designs at corresponding levels of abstraction, e.g., the level of library cells, looking for designs that conform to given performance constraints.

In traversing the design space, Gajski and Brewer[29] identify three issues that must be addressed by a synthesis tool: *style selection*, *technology mapping*, and *optimization*. Style selection is the problem of choosing appropriate design styles for meeting design constraints (e.g., serial vs. parallel implementation). Technology mapping is the problem of selecting a physical instantiation that provides the required functionality and meets performance constraints. Optimization is the problem of applying functionally equivalent structural transformations to the design in order to meet violated constraints. Style selection and technology mapping correspond to search along the dimension of abstraction; optimization corresponds to search along the dimension of configuration.

Existing logic synthesis tools are primarily concerned with optimization, focusing search on the dimension of configuration; little search effort is spent on technology mapping. In the derivational-process model, we are also concerned with style selection, focusing search along the dimensions of abstraction as well.

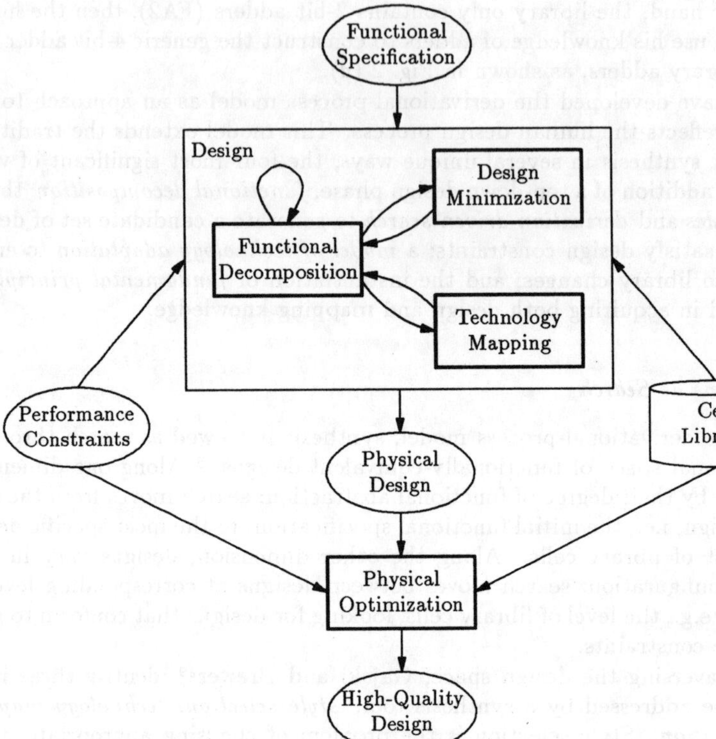

Fig. 3. Derivational-process model

3.3. The Derivational-Process Model

As depicted in Fig. 3, synthesis in the derivational-process model is factored into a *design phase* and an *optimization phase*. The design phase is further factored into interacting processes of *functional decomposition*, *design minimization*, and *technology mapping*. The purpose of these processes is summarized below.

- **Functional decomposition** is a process of top-down hierarchical design. Selecting a design style appropriate to constraints, the functional decomposition process outputs a netlist of connected subcomponents. Some subcomponents can be described by Boolean equations; these are passed to the design minimization process. Others are functionally "close" to library cells, in which case they are treated as "leaf" nodes and passed to the technology mapping process.
- **Technology mapping** is a process that instantiates the leaf nodes of the generic design with configurations of library cells. In simple cases, this

merely involves a one-to-one mapping. In other cases, the technology mapping process implements the leaf node with library cells augmented by generic subcomponents. The generic subcomponents can be passed to the functional decomposition process for further refinement and design.

- **Design minimization** is a process that subsumes a variety of graph reduction processes, such Boolean minimization, state reduction, and protocol analysis. Eventually the design phase outputs a physical design, implemented with ASIC library cells, which is then passed to the physical optimization phase for fine-tuning.
- **Physical optimization** is a process of refining the physical design along critical paths in order to meet performance constraints.

The input to the model is a set of functional specifications for generic hardware components. Additional inputs include performance constraints on the design and a library of physical cells. The output is an optimized netlist of library cells that constitutes a physically-realizable design.

Our primary focus in implementing the derivational-process model of design synthesis is on functional decomposition and technology mapping. The extended capabilities resulting from these processes is one of the aspects of this research that sets it apart from other efforts in logic synthesis.

3.4. Derivation-Driven Search

The processes of functional decomposition and technology mapping are controlled by *derivation-driven search*, the goal of which is to generate a set of candidate library-specific designs whose physical characteristics either meet or approximate the given design constraints. Derivation-driven search uses design rules to explore the space of generic and physical designs along the dimension of abstraction. Each design rule has a head that matches generic component specifications, a set of constraining conditions that must by satisfied by a specification for the rule to be applicable, and a body of actions that implements the specified component by generating a netlist of connected subcomponents.

There are two types of design rules, distinguished by their role in the design process, i.e., whether they are used during functional decomposition or technology mapping. Design rules for functional decomposition are called *decomposition rules*. Decomposition rules define how to achieve the functionality of a generic component in terms of connected subcomponents. Several decomposition rules that are applicable to the same generic component reflect alternative design styles. Design rules for technology mapping are called *construction rules*. Construction rules define how to implement a generic component by augmenting the functionality of a physical library cell. Decomposition rules can be viewed as operating from the top-down, designing increasingly specific components, while construction rules can be viewed as operating from the bottom-up, designing increasingly abstract components. Eventually, the most specific decomposition rules merge with the most abstract construction rules, denoting the transition from functional decomposition to technology mapping.

4. DTAS: An IC Design Synthesis System

4.1. System Architecture

We have implemented the derivational-process model in the **Design and Technology Adaptation System (DTAS)**. DTAS is a rule-based system for functional synthesis of regular-structured data path logic, including combinational (e.g., decoders, multiplexers, and function generators), arithmetic (e.g., adders, comparators, multipliers, and ALUs), and sequential (e.g., shift registers and counters). Designs are hierarchically decomposed into netlists of cells from a given ASIC vendor's library. These can be simple Boolean cells or complex functional cells, from multiplexers, adders, and comparators up to n-bit ALUs, multipliers, and counters. DTAS compares alternative design styles to find candidate designs that best fit performance constraints. Designs can be output in a structural VHDL[30] and input to logic synthesis and layout tools.

The connection between DTAS and behavioral synthesis is through a library of generic microarchitecture components, called GENUS.[31] High-level synthesis systems translate behavioral system descriptions into structural netlists of instantiated GENUS components. GENUS components are highly parameterized templates, where parameters include *type* (e.g., multiplexer, comparator, ALU), *functionality* (e.g., an ALU that performs addition, increment, and less than), *input/output characteristics* (i.e., the number, type, and size of ports), and other attributes. A synthesis system instantiates these generic templates by supplying parameter values. The advantage of using GENUS components is that it supports a truly "generic" view of structural elements.

DTAS was developed to support technology mapping of regular-structured components. Its top-level structure is outlined in Fig. 4. The input to DTAS is a netlist of instantiated GENUS components specified using structural VHDL. This netlist is translated into an internal representation that is then passed through a phase of functional decomposition and technology mapping. The result is a set of hierarchical netlists that trace the top-down design of the input netlist into subcomponents. Leaves of the hierarchy implement the design with cells drawn from a given ASIC vendor's library. Netlists vary by the design styles and library cells used in their construction. The hierarchical netlists can be output in structural VHDL and passed to tools for performance analysis and simulation, logic-level optimizations, floor-planning and layout, and so on.

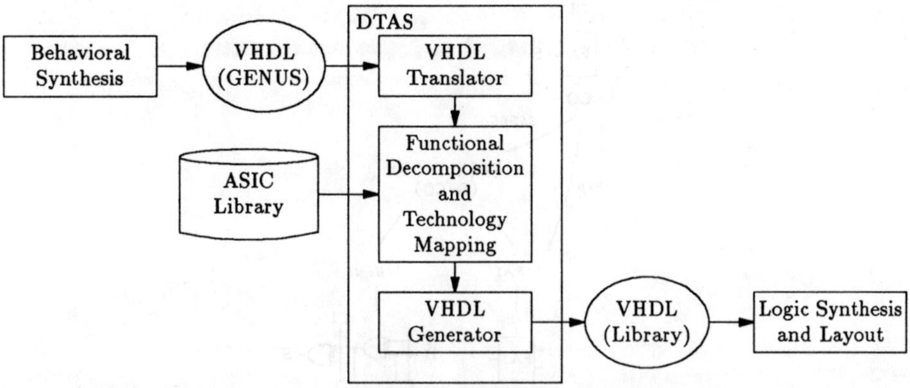

Fig. 4. DTAS system structure

4.2. The Design Space

In DTAS, functional decomposition and technology mapping are implemented by a rule-based system that searches a space of design decompositions. The design space is represented as an acyclic graph; its structure is outlined in Fig. 5. The nodes of the graph alternate between functional component specifications (*cspecs*) and competing component designs (*comps*), each of which implement the cspec. Each comp corresponds to a library cell or a netlist. The netlist decomposes the comp into connected subcomponents (*modules*). A module represents an instance of a comp; its functionality is described by a cspec, but it is mapped to only one of the cspec's comps.

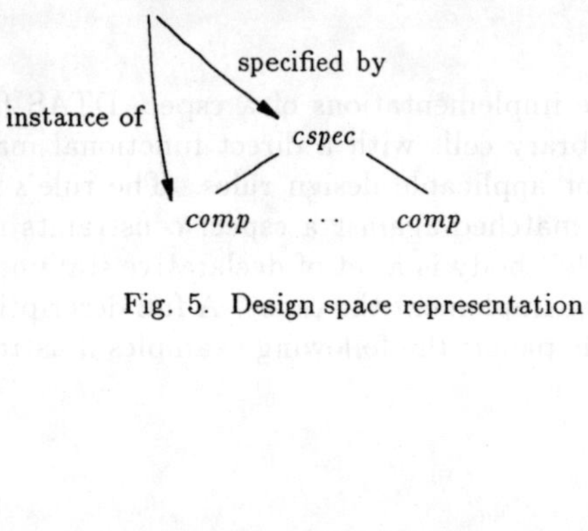

Fig. 5. Design space representation

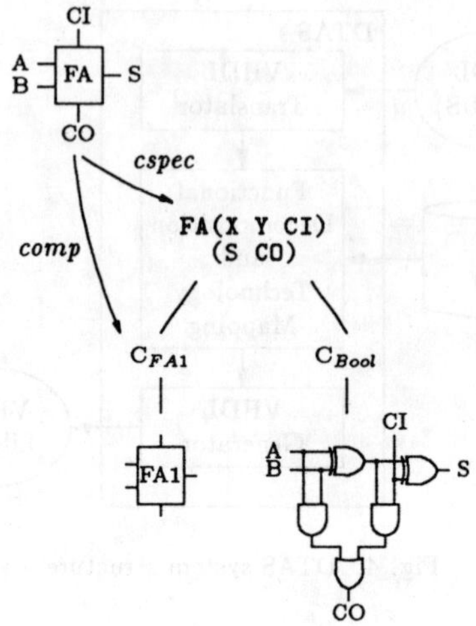

Fig. 6. Design space for 1-bit FA

A partial design space shown in Fig. 6. Its root is a module whose functionality is specified by a cspec for a 1-bit full adder (FA), indicated by the link labeled *cspec*. The cspec has two alternative comps (implementations). One corresponds to library adder (FA1); the other to a netlist of five connected modules that implement the FA with Boolean gates. The link labeled *comp* indicates that the root module is an instance of the library component.

This example is extended one level in Fig. 7. We are designing a 4-bit adder module. The cspec has three alternative implementations, indicated by comps C_{ripple_1}, C_{ripple_2}, and C_{carry}. The first two comps use ripple carry; the third uses carry look-ahead. The FA modules in the netlists of C_{ripple_1} and C_{ripple_2} are copies of the root module in Fig. 6. C_{ripple_1} uses all library FAs and C_{ripple_2} is uses Boolean gates. C_{carry} is implemented by a netlist of one carry look-ahead generator (CLA) and four 1-bit FAs with carry propagate and generate outputs. The cspec for the four FAs is decomposed into Boolean gates; there are no appropriate library FAs. The cspec for the CLA will be decomposed next.

4.3. Design Rules

To find alternative implementations of a cspec, DTAS first examines the given ASIC library for library cells with a direct functional match. DTAS then examines the rule base for applicable design rules. The rule's head is a parameterized template that is matched against a cspec; constraints may be placed on parameter values. The rule's body is a set of declarative statements for specifying and configuring modules to implement the cspec. A full description of design rules is beyond the scope of this paper; the following examples illustrate some features.

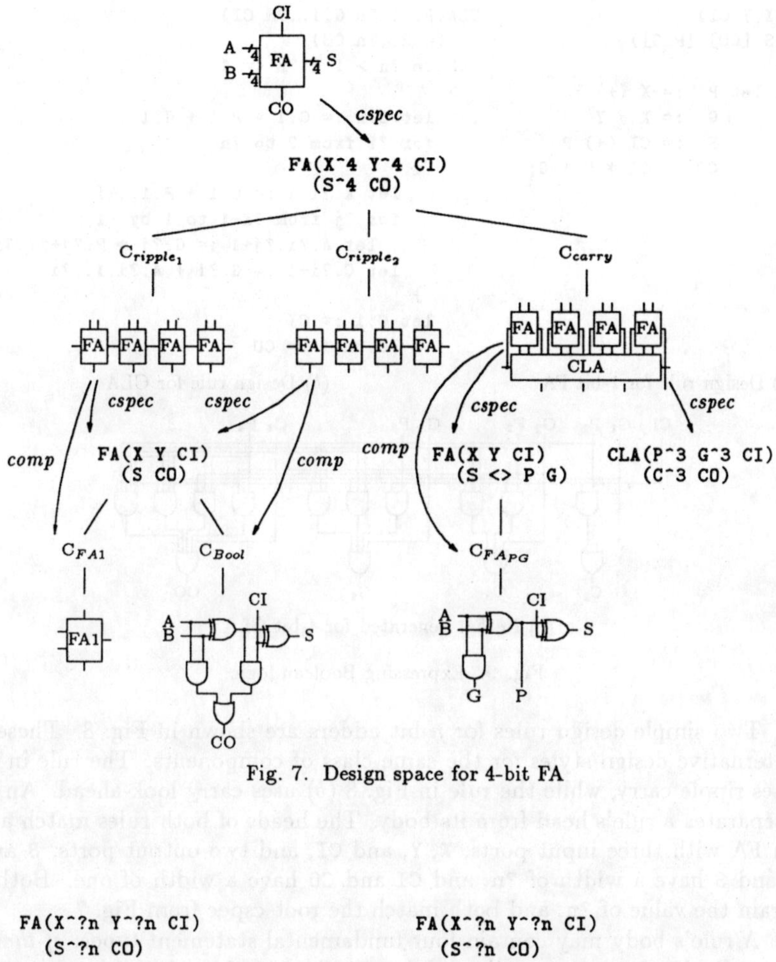

Fig. 7. Design space for 4-bit FA

```
FA(X^?n Y^?n CI)
  (S^?n CO)
where ?n > 1
->
  for ?i from 1 to ?n
    FA(X[?i] Y[?i] C.?i)
      (S[?i] C.?i+1)
  let C.1 := CI
      C.?n+1 := CO;
```

(*a*) Serial style: ripple carry

```
FA(X^?n Y^?n CI)
  (S^?n CO)
where ?n > 1 & ?n < 4
->
  for ?i from 1 to ?n
    FA(X[?i] Y[?i] C.?i)
      (S[?i] <> P.?i G.?i)
  CLA(P.1..?n G.1..?n C.1)
     (C.2..?n CO)
  let C.1 := CI
```

(*b*) Parallel style: carry look-ahead

Fig. 8. Alternative design rules for *n*-bit FA

```
FA(X Y CI)
  (S [CO] [P G])
  ->
     let P  := X (+) Y
         G  := X * Y
         S  := CI (+) P
         CO := CI * P + G;
```

(*a*) Design rule for 1-bit FA

```
CLA(P.1..?n G.1..?n CI)
    (C.2..?n CO)
where ?n > 1 & ?n <= 4
->
   let C.2 := C.1 * P.1 + G.1
   for ?i from 2 to ?n
   {
      let A.?i.1 := C.1 * P.1..?i
      for ?j from ?i-1 to 1 by -1
        let A.?i.?j+1 := G.?j * P.?j+1..?i
      let C.?i+1 := G.?i + A.?i.1..?i
   }
   let C.1 := CI
       C.?n+1 := CO
```

(*b*) Design rule for CLA

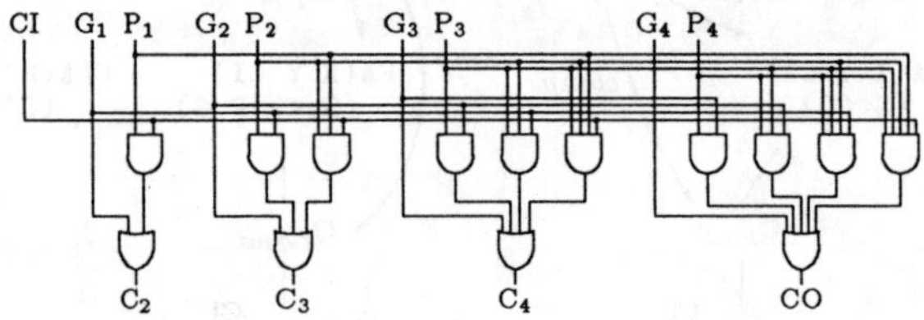

(*c*) Netlist generated for 4-bit CLA

Fig. 9. Expressing Boolean logic

Two simple design rules for n-bit adders are shown in Fig. 8. These rules use alternative design styles for the same class of components. The rule in Fig. 8 (*a*) uses ripple carry, while the rule in Fig. 8 (*b*) uses carry look-ahead. An arrow (`->`) separates a rule's head from its body. The heads of both rules match a cspec for an FA with three input ports, `X`, `Y`, and `CI`, and two output ports, `S` and `CO`. `X`, `Y`, and `S` have a width of `?n`, and `CI` and `CO` have a width of one. Both rules constrain the value of `?n`, and both match the root cspec from Fig. 7.

A rule's body may contain four fundamental statement types: *if-then-else*, *for-loop*, *Boolean equation*, and *module specifier*; nesting of *if-then-else* and *for-loop* statements is permitted. Module specifiers and Boolean equations designate modules to be generated when the rule is fired; *for-loops* cascade modules. Both rules in Fig. 8 generate `?n` 1-bit FA modules; that in Fig. 8 (*b*) also generates a CLA module. In this example, Boolean equations (introduced by `let`) simply rename pin connections.

The two design rules shown in Fig. 9 illustrate further use of Boolean equations for an adder and CLA. Equations are minimized and mapped to generic cspecs for the appropriate Boolean gates; these gates are later mapped to corresponding library cells. The netlist generated by the CLA rule is shown in Fig. 9 (*c*).

```
To design netlist:
  for module of netlist,
    unless module is already
      an instance of a comp
    design module

To design module:
  if cspec (of module) has comps,
    comp = select one
  else
  if cspec has matching
      library cell
    comp = library cell
    add comp to comps of cspec
  else
    select design rule that matches cspec
    netlist = fire design rule
    create comp for netlist
    add comp to comps of cspec
    design netlist
  make module instance of comp
```

(*a*) Initiating design

```
To redesign netlist:
  for module of netlist,
    if newmod = redesign module
      return copy of netlist using newmod

To redesign module:
  if cspec (of module) has comps of which
      module has not been an instance,
    comp = select one
  else
  if cspec has matching library cell
      not already comp of cspec
    comp = library cell
    add comp to comps of cspec
  else
    select unfired rule matching cspec
    netlist = fire design rule
    create comp for netlist
    add comp to comps of cspec
    design netlist
  if comp selected,
    newmod = copy of module
    make newmod instance of comp
    return newmod
```

(*b*) Finding alternatives

Fig. 10. Procedures for DTAS design process

4.4. The Design Process

DTAS takes as input a netlist of instantiated GENUS components and generates a set of candidate designs mapped into cells from a given ASIC library. Candidate designs can be distinguished by the design styles used in their decomposition and the type and complexity of library components appearing at their leaves. After fine tuning the candidates with an optimization process, a design can be selected that best fits performance constraints.

In generating the candidate designs, DTAS follows a process of design and redesign, as outlined below.

1. *Initialize design space.* Makes input netlist root of design space and calls procedures **design netlist** and **design module** outlined in Fig. 10 (*a*). These procedures make each netlist module an instance of a comp that satisfies their cspec. This is repeated recursively for the netlist of each comp, initializing the space in a depth-first manner.
2. *Expand design space.* This step successively generates variations of the initial design, thus expanding the design space. A variation is generated with the procedures **redesign netlist** and **redesign module** outlined in Fig. 10 (*b*). This step is repeated until there are no new alternative designs.

3. *Select candidate set.* This step examines the design space and recovers the best designs based upon a comparison of the designs to performance constraints (e.g., area and delay). These are returned as the candidate designs.

The candidate designs are represented as paths through the design space, i.e., as a hierarchy of netlists, modules, and comps. The leaves of the designs correspond to library cells. DTAS provides support facilities for extracting the Boolean description from a design, for measuring area and delay, for flattening a design into a netlist of library cells, and for outputting a design in structural VHDL.

5. TECHNOLOGY ADAPTATION

5.1. The Robustness Problem

The use of complex library cells and a rule-based design strategy can improve the state of the art in logic synthesis, but only at a cost to robustness against technology changes. The effectiveness of the derivational-process model of design synthesis can be measured against existing logic synthesis tools in terms of competence, quality, and robustness. Because our model scales up to regular-structured microarchitecture components, we will have improved competence. Because our model considers alternative design styles and complex library cells, we will have improved design quality. However, because our model utilizes library-specific knowledge, we fall short in regards to robustness to library changes or to changes in the fabrication technology. We refer to this shortcoming as the *robustness problem.* Addressing the robustness problem in IC design synthesis is a major goal of our research.

Construction rules that operate on simple cells, such as Boolean gates, will typically transfer easily between libraries, while rules that operate on complex library cells, such as n-bit adders, ALUs, and counters, will have little or no transfer. Because decomposition rules work on generic components they are likely to have better transfer. However, they may still embody design styles that are appropriate to one technology or application area and ignore alternative styles that are appropriate for another.

This leads to a significant drawback in using the DTAS synthesis system. Generating new design rules by hand is time consuming and error prone. It also requires a certain amount of expertise with the DTAS rule language and experience with rule interactions. As a result, the level of effort required to maintain and upgrade the DTAS knowledge base could be prohibitively expensive. We are attempting to overcoming this bottleneck by automating technology adaptation with a model of learning and acquisition for construction knowledge.

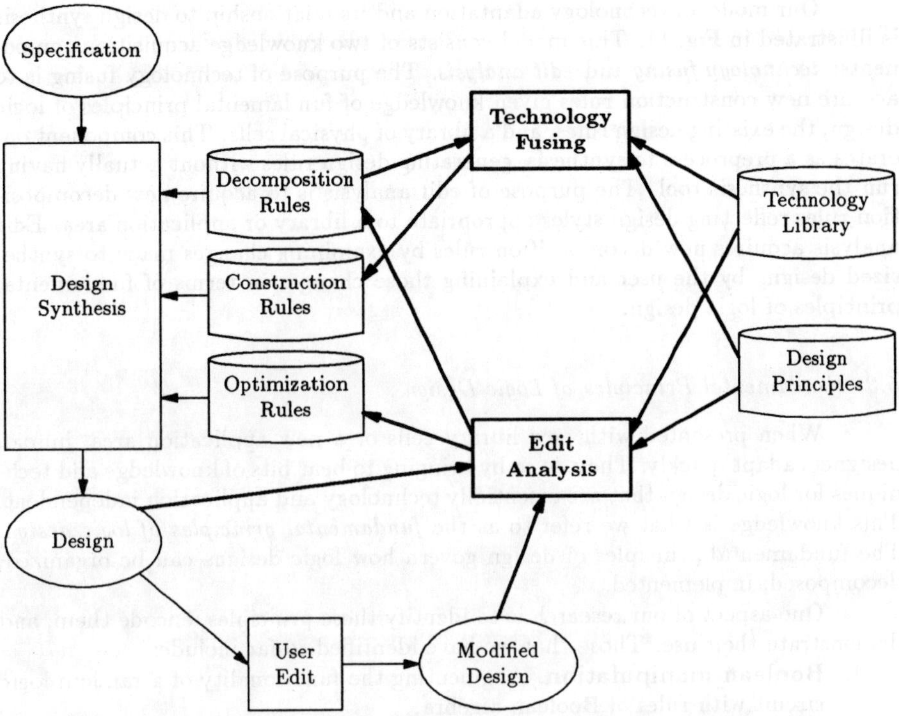

Fig. 11. Model of technology adaptation

5.2. Technology Adaptation as Learning

The robustness problem is special case of the knowledge acquisition bottleneck encountered in developing expert systems. The AI literature reports many efforts to reduce cost and increase performance of knowledge-based systems with semi-automated tools for aiding the knowledge acquisition process. For instance TEIRESIAS[32] interactively repairs and extends the knowledge base of the medical diagnostic system MYCIN;[33] LEX[34] learns search control heuristics for problem solving in integral calculus; LEAP[13] learns new design rules for the VEXED[10] IC design system; and CGEN[17] interactively captures design expertise for the MICON[35] computer design system.

As with these systems, our approach to alleviating the robustness problem is to develop an adjunct model of knowledge aquisition. However, we actually view the task of automating technology adaptation in a knowledge-based design system as a distinct machine learning task, which we have already refered to as learning construction knowledge. We take an analytic approach to this task, using knowledge of fundamental principles of logic design as the domain theory.

Our model of technology adaptation and its relationship to design synthesis is illustrated in Fig. 11. This model consists of two knowledge acquisition components: *technology fusing* and *edit analysis.* The purpose of technology fusing is to acquire new construction rules given knowledge of fundamental principles of logic design, the existing design rules, and a library of physical cells. This component operates as a preprocess to synthesis, generating design rules without actually having run the synthesis tool. The purpose of edit analysis is to acquire new decomposition rules reflecting design styles appropriate to a library or application area. Edit analysis acquires new decomposition rules by examining changes made to synthesized designs by the user and explaining those changes in terms of fundamental principles of logic design.

5.3. Fundamental Principles of Logic Design

When presented with new library cells or a new application area, human designers adapt quickly. They do so by bringing to bear bits of knowledge and techniques for logic design that are essentially technology and application independent. This knowledge is what we refer to as the *fundamental principles of logic design.* The fundamental principles of design govern how logic designs can be organized, decomposed, implemented.

One aspect of our research is to identify these principles, encode them, and demonstrate their use. Those that we have identified so far include:

1. **Boolean manipulation**. Augmenting the functionality of a random logic circuit with rules of Boolean algebra.
2. **Factoring**. Combining like components in a tree, which can be balanced, left-branching or right-branching.
3. **Cascading**. Combining (or splicing) components of the same type in sequence.
4. **Multiplexing.** Combining dissimilar components through a multiplexer or bus.
5. **Exclusion**. Ignoring inputs and outputs;
6. **Externalization**. Augmenting the inputs and outputs of a component with combinational circuits.

Use of exclusion, cascading, externalization, and multiplexing is demonstrated by example in Fig. 12. For this example, assume that the physical cell is a simple 4-bit ALU (ALU4) with four operations: addition, subtraction, nand, and nor.

To generalize the data width of the ALU4 so as to implement a generic n-bit ALU with the same four operations, we can apply the first and third principles. When $n < 4$, as in Fig. 12 (a), the ALU4 can be used to implement the n-bit ALU by setting the $4 - n$ least significant input pins to low and grounding the $4 - n$ least significant output pins. When $n > 4$, as in Fig. 12 (b), the ALU4 can be used to implement the n-bit ALU by cascading $\lfloor \frac{n}{4} \rfloor$ ALU4's, rippling carries and

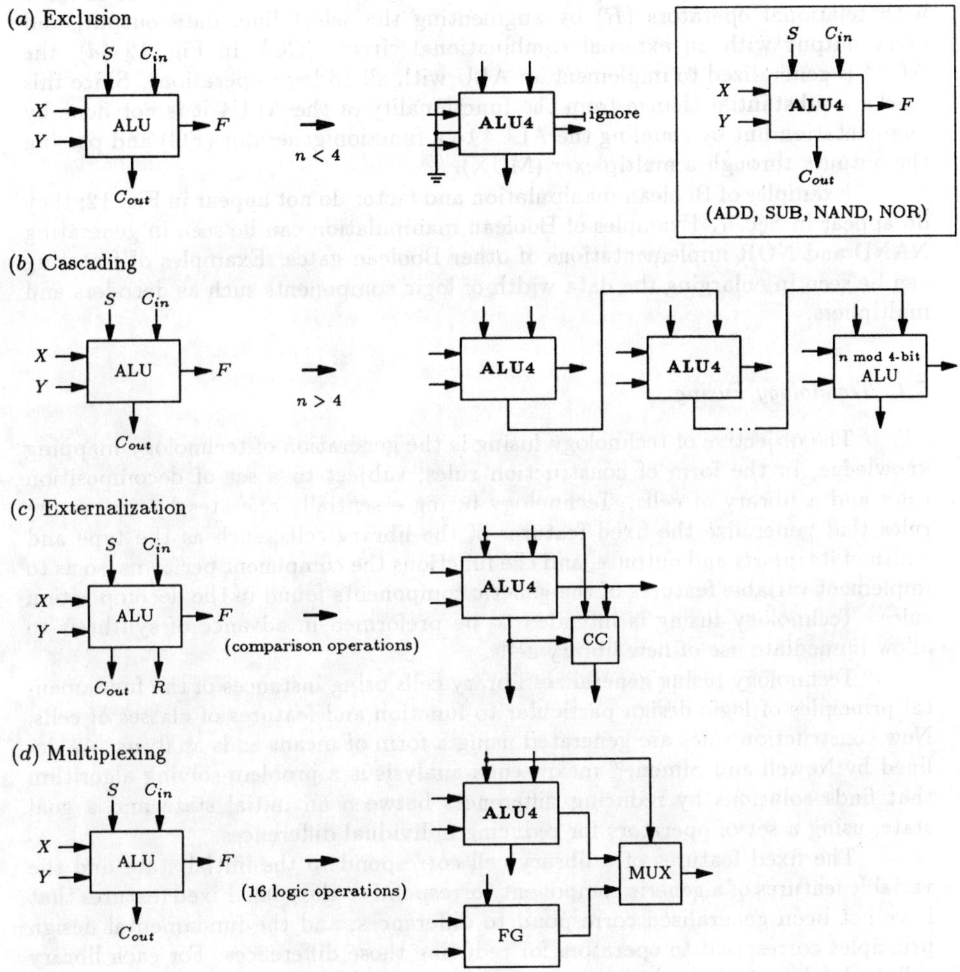

Fig. 12. Fundamental principles of logic design

attaching a generic mod$(n,4)$-bit ALU at the tail. To generalize the functionality of the ALU4, we can apply the second and fifth principles, augmenting I/O when the change in functionality is slight and multiplexing when a more complicated change is required. In Fig. 12 (c), the ALU4 is generalized to implement an ALU with relational operators (R) by augmenting the select line, data output, and carry output with an external combinational circuit (CC). In Fig. 12 (d), the ALU4 is generalized to implement an ALU with all 16 logic operations. Since this marks a substantial change from the functionality of the ALU4 it is not done by augmentation but by coupling the ALU4 to a function generator (FG) and passing the outputs through a multiplexer (MUX).

Examples of Boolean manipulation and factor do not appear in Fig. 12; they do appear in Sec. 7. Examples of Boolean manipulation can be seen in generating NAND and NOR implementations of other Boolean gates. Examples of factoring can be seen in enlarging the data width of logic components such as decoders and multipliers.

5.4. Technology Fusing

The objective of technology fusing is the generation of technology mapping knowledge, in the form of construction rules, subject to a set of decomposition rules and a library of cells. Technology fusing essentially operates by generating rules that generalize the fixed features of the library cells, such as the type and width of its inputs and outputs, and the functions the component performs, so as to implement variable features of the generic components found in the decomposition rules. Technology fusing is intended to be preformed in advance of synthesis to allow immediate use of new library cells.

Technology fusing generalizes library cells using instances of the fundamental principles of logic design particular to function and features of classes of cells. New construction rules are generated using a form of means-ends analysis. As defined by Newell and Simon,[36] means-ends analysis is a problem-solving algorithm that finds solutions by reducing differences between an initial state and a goal state, using a set of operators for reducing individual differences.

The fixed features of a library cell correspond to the initial state and the variable features of a generic component correspond to the goal. Fixed features that have not been generalized correspond to differences, and the fundamental design principles correspond to operators for reducing those differences. For each library cell, technology fusing selects a generic component that is "close" in functionality to the cell. Technology fusing then iteratively generates a sequence of construction rules that implements the generic component using the physical cell. On each iteration, technology fusing selects a difference between the two components and a design principle particular the component class of the library cell that can reduce that difference and generates a construction rule based on that principle.

Each generated construction rule implements a component that is a little closer in functionality to the generic component. Eventually, a construction rule is generated that will implement the generic component and iteration stops. The technology fusing process then continues to the next library cell.

Technology fusing can be viewed as goal-directed search and is similar to learning macro-operators.[37] As search, we use the encoded principles of logic design as operators. The objective of the search is to find a path from available material parts to possible artifacts. Generalization along the solution path generates design knowledge that is both principled and appropriately biased. The design rules generated in this manner can be viewed as macro-operators because they correspond to a planned sequence of steps for instantiating the generic component.

5.5. Edit Analysis

The objective of edit analysis is the acquisition of design knowledge, in the form of decomposition rules, subject to user modifications to synthesized designs. Input to the edit analysis process includes the top-level component specification and a list of changes made to the synthesized design. The output of is a set of decomposition rules that generalize an explanation of how the modified design implements the original component specification. Edit analysis is intended to be preformed as a postprocess to synthesis to allow users to fill in gaps in the knowledge base as they become apparent.

Edit analysis operates by capturing new design knowledge when deficiencies are discovered in the knowledge base of the synthesis tool. There are various reasons for these appearing in the knowledge base at all. The knowledge base of the synthesis tool may have been tailored for one application area and inappropriate for others. Another possibility is the need for new design styles after the encoding of the knowledge base. Likewise, the synthesis tool could be introduced to a ASIC library containing physical cells for which no generic component is a "close" fit, i.e., a situation requiring new decomposition rules for technology fusing to have been useful.

Deficiencies are said to be "detected" when the user finds it necessary to modify synthesized designs. Such modifications would trigger edit analysis in the hopes of filling in the missing design knowledge. Working interactively with the user, edit analysis would reconstruct the steps in the decomposition of the design that explains how the user reached the new design in terms of the fundamental principles of design. The decomposition would form a proof that can be generalized to provide new decomposition rules. These new decomposition rules would be able to reproduce the steps taken by the user when designing similar components.

Edit analysis is similar to the learning apprentice approach taken in LEAP. In our model of synthesis, however, the user is decoupled from the design process and cannot lead the synthesis tool through the correct sequence of steps in decomposing the design. Analytic techniques can still be applied during edit analysis to acquire new decomposition rules. By treating the fundamental principles of design as a theory of the domain, edit analysis has a set of guide lines for reconstructing the steps taken by the user in developing the modified design. This reconstruction also acts as a proof of the new design's soundness. Each step in the reconstruction of the design can be related to a decomposition rule. If a rule for the step is not already present in the knowledge base, then the step can be generalized to give a new decomposition rule that can be used for similar situations occurring in the future.

5.6. Comparison to LEAP

Our research into IC design synthesis and technology adaptation shares certain surface similarities to Mitchell and Steinberg's work with LEAP and VEXED. One way to characterize the differences is to say that our work developed from a CAD perspective rather than an AI perspective. In other words, Mitchell and Steinberg looked to VLSI as an example domain in which to develop general approaches knowledge-based design and knowledge acquisition, while we looked to AI as a source of technology with which to solve particular problems in IC design synthesis.

LEAP was conceived as a "learning apprentice system" for VEXED, a knowledge-based consultant for designing simple circuits. LEAP would observe the actions of an expert designer, verify the functionality of the design using using Boolean logic, and then generalize the proof into design rules. What is missing is a "sense of purpose" in regards to what is being learned and why. Perhaps this can be explain to some degree by the model of design appearing in VEXED, which simplified IC design to decomposition and constraint propagation and largely ignored the hard issues of technology mapping and optimization.

Our model of learning and acquisition was developed in order to alleviate the robustness problem in the derivational-process model of IC design, which *does* include processes for technology mapping and optimization. Constraint propagation is not carried out during the design phase, since performance estimates are tentative at best before technology mapping. Instead, the optimization phase is used to fine-tune the design so as to meet performance constraints. We factor learning and acquisition into two components, technology fusing and edit analysis, and each component focuses on acquiring a particular type of knowledge by a distinct mechanism. Our model uses a knowledge of fundamental principles of logic design, which subsumes Boolean logic and provides a richer domain theory.

As a difference, LEAP is intended to acquire technology-independent design knowledge, while our model is indented to acquire design knowledge that is technology specific. This is an important distinction. Once captured, either by knowledge engineering or acquisition techniques, technology-independent design knowledge changes slowly, if at all. Technology-specific design knowledge, on the other hand, changes with the fabrication technology (which is quite frequent for the IC domain). Thus, the expected lifetime and utility of an acquisition system for technology-specific design knowledge should be greater than that of an acquisition system for technology-independent design knowledge.

6. LOLA: A SYSTEM FOR TECHNOLOGY FUSING

6.1. System Architecture

We have implemented our model of technology fusing in the **Logic Learning Assistant (LOLA)**; we plan to pursue an implementation of edit analysis at a later time. LOLA operates as a preprocess to synthesis. Its inputs include the generic components that DTAS can design, which are drawn from the initial set of decomposition rules, the physical cells available in the ASIC library, and knowledge of the fundamental principles of logic design. The output consists of library-specific design rules for implementing generic components with library cells. This process is performed in advance to synthesis to allow DTAS immediate use of the library.

The relationship of LOLA to DTAS is shown in Fig. 13. The input to DTAS is a high-level component specification and the output is a physical design using cells from a given ASIC library. The design is generated as a hierarchical netlist using decomposition rules and mapped into library cells using construction rules. The purpose of LOLA is to ensure that DTAS's construction rules are appropriately biased towards the available library cells. This interaction allows DTAS to take full advantage of the library and generate high-quality designs.

LOLA generates construction rules by generalizing the fixed physical features of the library cells to match to the variable features of one or more generic components. The physical features of a cell correspond to the type and width of its inputs and outputs and the functions it performs. For example, a library might contain a 4-bit adder, which is represented as a cell with two 4-bit data inputs, one 4-bit data output, a carry input, and a carry output, and performs the function ADD. This cell can be used to implement a generic adder with two n-bit data inputs, one n-bit data output, and optional carry input and output.

After generating new construction rules, LOLA evaluates the entire rule base. Evaluation is used to identify the conditions under which existing or newly generated rules fail to generate high-performance designs, i.e., in terms of area or delay. These conditions become constrains on the unproductive rules, limiting their applicability. This process eliminates large portions of the design space that DTAS would otherwise have to explore.

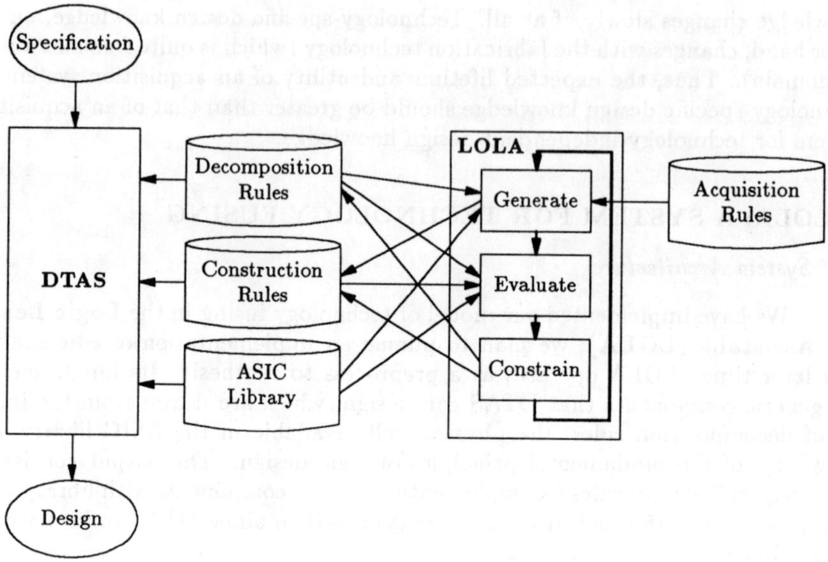

Fig. 13. Relation of LOLA to DTAS

6.2. Acquisition Rules

In LOLA, the knowledge of fundamental principles of logic design is encoded as rules, called *acquisition rules.* Unlike decomposition and construction rules, which are used for component decompositions, LOLA's rules generate construction rules, which can be used by DTAS for component decomposition.

As with design rules, an aquisition rule has a head and a body. The head of an acquisition rule is a parameterized template and parameters can be constrained. The head is matched against the component specifications (cspecs) describing cells available in the ASIC library and the component generators (cgenrs) describing components that appear in the existing design rules. The body of an acquisition rule is a set of declarative statements for specifying the construction rules that can be generated when the rule is fired.

The head of an acquisition rule has three parts:

- a list of cspec templates for cells that must be included in the library;
- a list of cspec templates for cells that must not be in the library;
- a list of cgenr templates for components that must be used in a design rule.

The latter two parts are optional; when the templates for matching design rule components are excluded, the writer of the rule is making an assumption about the structure of DTAS's design rules.

Two simple acquisition rules for a NAND-implementation of Boolean components are shown in Fig. 14. The first rule, Fig. 14 (a), generates a construction

```
Library:  NAND(I^2)
               (O)
Excludes: INV(I)
              (O)
  INV(I)
    (O)
  ->
     NAND(<I PWR>)
        (O)
      :source LIBRARY;
```

(*a*) NAND-to-INV rule

```
Library:  NAND(I^2)
                (O)
Excludes: AND(I^$)
               (O)
  AND(I^?n)
     (O)
  ->
     NAND(I^?n)
          (Z)
     INV(Z)
        (O);
```

(*b*) NAND-to-AND rule

Fig. 14. Acquisition rules for NAND implementations

rule for implementing an inverter (INV) with a 2-bit NAND cell when the library does not already include an INV cell. Similarly, the second rule, Fig. 14 (*b*), generates a construction rule for implementing an n-bit AND from an n-bit NAND when the library contains a 2-bit NAND cell and no AND cells of any size. Variables in an acquisition rule are introduced by dollar sign ($). When the acquisition rule generates construction rules, the dollar sign variables are replaced by the value to which they are bound. Design rule variables, introduced by a question mark (?), are treated as literals and are not bound nor evaluated by LOLA.

The rules in Fig. 14 do not examine generic components appearing in the design rules; instead, the assumption is being made that DTAS will invariably try to use INV and n-bit AND gates. The second rule also makes the assumption that if the library contains an 2-bit NAND gate that it will also contain an INV or that a INV construction rule will be generated. The second rule makes the further assumption that there will be construction rules for generating an n-bit NAND.

These rules also illustrate the simple instantiation two of the fundamental design principles. By setting one input of the library NAND gate to high (`PWR`), the rule in Fig. 14 (*a*) uses the principle of exclusion to implement INV. By augmenting the output of the n-bit NAND, the rule in Fig. 14 (*b*) uses the principle of externalization.

A more complex acquisition rule is shown in Fig. 15. The head of this rule matches any library cell (`$FN`) of `$N` inputs and 1 output, where `boolean_p($FN)` return true if `$FN` is a Boolean function. The body generates two construction rules. The first is generated using the principle of exclusion for constructing an `?n`-bit `$FN` from the library cell when `?n` is less than `$N`. The second is generated using the principle of factoring for the cases where `?n` is greater than `$N`; in this case, factoring between `$N` smaller functions. This rule can be used to generate the construction rule for decomposing the n-bit NAND required in Fig. 14 (*b*) into 2-bit library NANDs.

```
Library: $FN(I^$N)
              (O)
              where boolean_p($FN) & $N > 2
  {
    $FN(I^?n)
        (O)
      where ?n < $N
      -> $FN(I.1..$N)
           (O)
           :source LIBRARY
           for ?i from 1 to ?n
             let I.?i := I[?i]
           for ?i from ?n+1 to $N
             let I.?i := boolean_identity($FN);
    let $BF := boolean_factor($FN)
    $FN(I^?n)
       (O)
      where ?n > $N
      -> let ?d := floor(?n/$N)
             ?r := mod(?n,$N)
           for ?i from 1 by ?d as ?j from 1 to $N-?r
             $BF(I[?i..?i+?d-1])
                 (I.?j)
           for ?i from 1+?d*($N-?r) by ?d+1
             as ?j from $N-?r+1 to $N
             $BF(I[?i..?i+?d])
                 (I.?j)
           $FN(I.1..$N)
               (O)
             :source LIBRARY;
  }
```

Fig. 15. Acquisition Rule for n-bit Boolean Function

6.3. The Acquisition Process

When DTAS explores the design space, it expands all applicable design rules. Search is not controlled by a global cost function because such a function should be defined in terms of design performance, which can not be determined until the design has been mapped into library cells. Instead, search is controlled by constraints on the applicability of design rules.

LOLA acquires new construction rules through a process of generate, evaluate, and constrain. As new construction rules are generated, LOLA runs a series of tests to evaluate the quality of designs generated with these rules as opposed to designs generated by existing design rules. For each case in which a new or existing rule does not generate a design that outperformed another rule (e.g., in terms for area or time), the case is generalized and appended as a constraint to the rule, which ensures that the rule will not be fired under conditions when it is known to generate inferior designs.

For example, consider the example seen above where an n-bit NAND gate can be constructed from a 2-bit library NAND. If the library also contained a 4-bit NAND that is smaller and faster than the configuration of 2-bit NAND gates generated by the acquired construction rule, then LOLA will add the constraint that ?n not be bound to 4.

The generate-evaluate-constrain cycle eliminates many overly general design rules that would otherwise force DTAS to do unnecessary search. Construction rules that generate designs using complex library cells will typically subsume design rules that decompose into functionally equivalent configurations of simple cells. In most cases, this means that the design space will be successively shallower with the introduction of complex library cells. As DTAS is supplied libraries of increasingly complex library cells, it should actually generate fewer candidate designs; it should generate designs of higher quality; and it should generate them faster.

7. DEMONSTRATION

7.1. Initial Design Rules

In this section, we step through an application of LOLA to an ASIC library as it evolves from an initial state through two library upgrades. With each upgrade, LOLA is called to generate design rules that allow DTAS to decompose designs to increasingly more complex library cells. The library of cells is drawn from the macro-cell library of LSI Logic Inc.[38]

The initial set of design rules provided to DTAS for this example allow it to decompose components to the level of generic inverters (INV) and 2-bit AND, OR, and XOR Boolean gates. Among the components that DTAS can design are n-bit Boolean gates, full adders (FA), n-bit multiplexers (MUX), and n-by-m bitwise MUX. Generic design rules for these components are depicted in Fig. 16.

7.2. Stage 1: NAND Implementation

We begin the example with a simple cell library that only contains a 2-bit NAND gate. The generic components in DTAS which are closest in functionality to the library cell are the INV gate and the 2-bit AND, OR, and XOR gates. The difference between the NAND gate and each of these gates is the Boolean function computed. This difference can be reduced by the principle of Boolean manipulation using the two aquisition rules seen earlier in Fig. 14, as well as similar rules for OR and XOR. The construction rules generated from the application of these rules are depicted in Fig. 17. From this set of additional design rules, DTAS can now decompose any design to library cells.

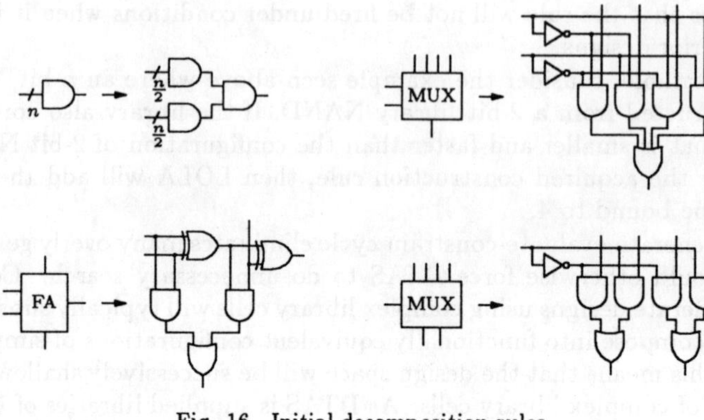

Fig. 16. Initial decomposition rules

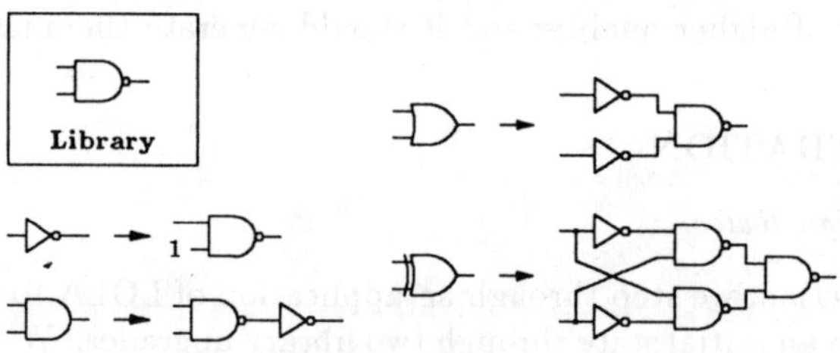

Fig. 17. NAND-implementation construction rules

7.3. Stage 2: Adding Boolean Gates

The second stage in the evolution of the library is the addition of an INV gate and 2-bit, 3-bit, and 4-bit AND, OR, and XOR gates. These physical cells are close in functionality to the n-bit Boolean gates found in DTAS. The difference in each case is the number of data inputs, which can be reduced by the principle of factoring into balanced trees using acquisition rules similar to that appearing in Fig. 15. The design rules generated by the application of this principle to the 3-bit and 4-bit AND gate are depicted in Fig. 18. Similar design rules are generated for the OR and XOR gates.

The new library cells are smaller and faster than the designs generated using the construction rules acquired in Stage 1. This is detected by LOLA in the generate-evaluate-constrain cycle. LOLA will constrain the older rules so that they are no longer applicable and will never be applied.

7.4. Stage 3: Adding Half Adders and Multiplexers

The third and final stage in the library's evolution is the addition of two functional cells, a half adder (HA) and a 2-bit MUX. The HA is close in function-

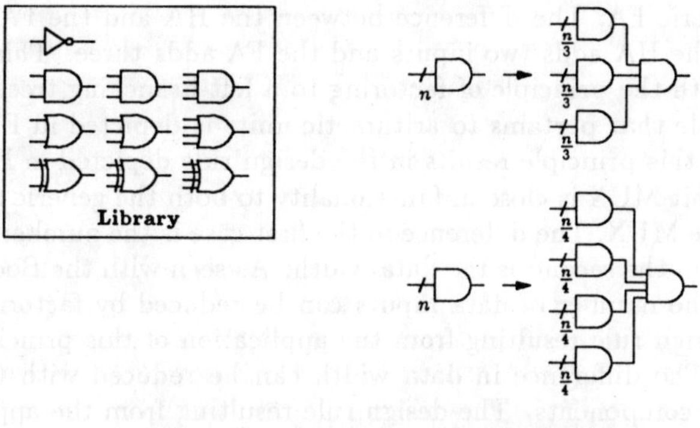

Fig. 18. Boolean construction rules

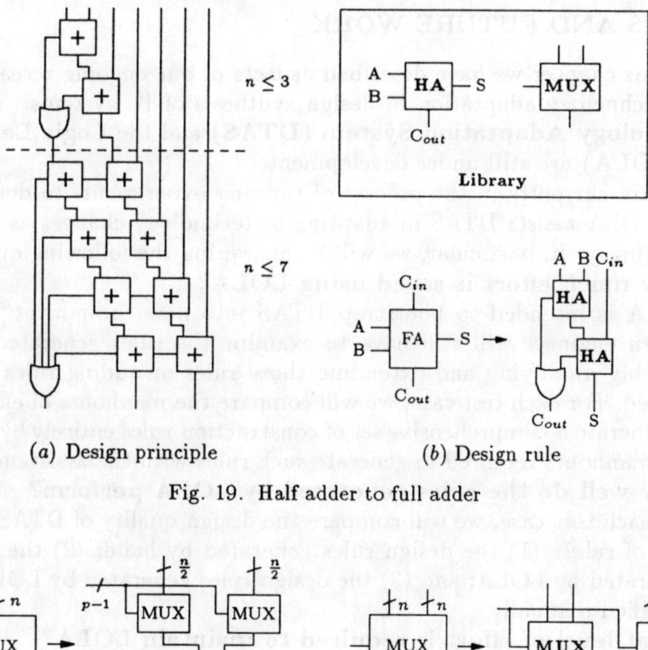

(*a*) Design principle (*b*) Design rule

Fig. 19. Half adder to full adder

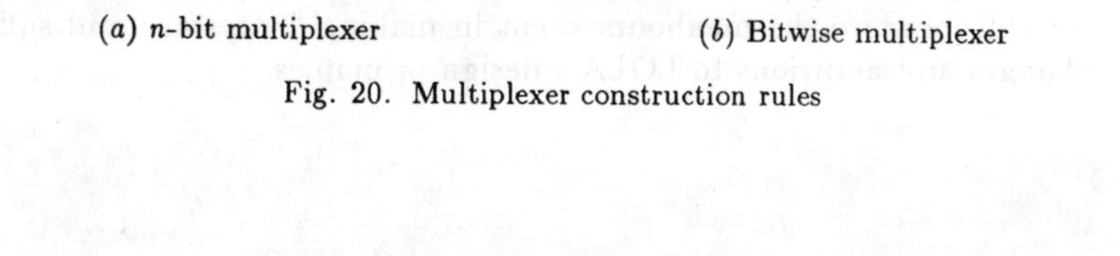

(*a*) n-bit multiplexer (*b*) Bitwise multiplexer

Fig. 20. Multiplexer construction rules

ality to a generic FA. The difference between the HA and the FA is the number of operands; the HA adds two inputs and the FA adds three. This difference can be reduced with the principle of factoring to a left-branching tree. The instances of this principle that pertains to arithmetic units is depicted in Fig. 19 (*a*). The application of this principle results in the design rule depicted in Fig. 19 (*b*).

The 2-bit MUX is close in functionality to both the generic n-bit MUX and n-by-m bitwise MUX. The difference in the first case is the number of data inputs; the difference in the second is the data width. As seen with the Boolean gates, the difference in the number of data inputs can be reduced by factoring to balanced trees. The design rule resulting from the application of this principle is shown in Fig. 20 (*a*). The difference in data width can be reduced with the principle of cascading like components. The design rule resulting from the application of this principle is shown in Fig. 20 (*b*).

8. STATUS AND FUTURE WORK

In this chapter we have described aspects of our ongoing research into automating technology adaptation in design synthesis of IC systems. The **Design and Technology Adaptation System (DTAS)** and the **Logic Learning Assistant (LOLA)** are still under development.

We are currently in the process of running experiments to determine how effectively LOLA assists DTAS in adapting to technology changes as they appear in a given library. In particular, we will be answering the following questions:

- **How much effort is saved using LOLA?**
  LOLA is intended to bootstrap DTAS into new component libraries. A design engineer will still have to examine the rules generated by LOLA, possibly modifying and extending those rules or adding rules that LOLA missed. For each test case, we will compare the manhours of effort required to generate a comprehensive set of construction rules entirely by hand verses the manhours required to generate such rules with the assistance of LOLA.
- **How well do the rules generated by LOLA perform?**
  For each test case, we will compare the design quality of DTAS using three sets of rules: (1) the design rules generated by hand; (2) the design rules generated by LOLA; and (3) the design rules generated by LOLA and then modified manually.
- **What level of effort is required to maintain LOLA?**
  This is a measure of the generality of LOLA's design principles. It can be expected that LOLA's design principles will require periodic upgrades; this should be especially true while running the experiments. For each test case, we will measure the manhours spent in making "necessary and sufficient" changes and additions to LOLA's design principles.

The initial base of generic design rules provided to DTAS will allow it to decompose GENUS components to four simple generic Boolean gates: inverters and 2-input AND, OR, and XOR gates. Among the components that DTAS will be able to design are n-bit and bitwise Boolean gates, multiplexers, ripple and carry look-ahead adders, comparators, ALUs, tree and matrix multipliers, shift registers, and parallel up-down counters.

For these experiments, we will provide DTAS with a component library consisting of a simple subset of components from a commercially available ASIC library (i.e., LSI Logic's macro-cell databook) and then run LOLA to generate the initial set of construction rules. We will then advance the complexity of the library through several stages of evolution, where each stage constitutes a test case for LOLA. After completing the integration of the LSI Logic library into DTAS, we will transfer DTAS to a new library (i.e., Texas Instrument's TTL 74-Series databook). This library will be initialized at a higher level of complexity than with the LSI Logic library and will also be evolved to a higher level.

These experiments and the results we gather are intended to demonstrate the utility of our approach to acquiring construction knowledge in a realistic situation in which technology adaptation is required to maintain the performance of a design synthesis tool.

References

1. T. D. Friedman and S. C. Yang, "Methods Used in an Automatic Logic Design Generator (ALERT)," *IEEE Transactions on Computers*, **C-18**, (1969), pp. 593–614.
2. D. Johannsen, "Bristle Blocks: A Silicon Compiler," *Proceedings of the 17th Design Automation Conference*, (1979), pp. 310–313.
3. S. W. Director, A. C. Parker, D. P. Siewiorek, and D. E. Thomas, "A Design Methodology and Computer Aids for Digital VLSI Systems," *IEEE Transactions on Circuits and Systems*, **CAS-28(7)**, (1981), pp. 634–645.
4. J. R. Southard, "MacPitts: An Approach to Silicon Compilation," *COMPUTER*, **16(12)**, (1983), pp. 74–82.
5. J. Mostow, "Toward Better Models of the Design Process," *The AI Magazine*, **6(1)**, (1985), pp. 44–57.
6. M. Stefik, "Planning with Constraints (MOLGEN: Part 1)," *Artificial Intelligence*, **16(2)**, (1981).
7. D. C. Brown and B. Chandrasekaran, "Knowledge and Control for a Mechanical Design Expert System," *IEEE Computer*, **19(7)**, (1986), pp. 92–100.
8. J. McDermott, "R1: A Rule-Based Configurer for Computer Systems," *Artificial Intelligence*, **19(1)**, (1982).

9. W. P. Birmingham and D. P. Siewiorek, "Automated Knowledge Acquisition for a Computer Hardware Synthesis System," EDRC 18-06-88, Department of Electrical and Computer Engineering, Carnegie Mellon University, (1988).
10. L. Steinberg, "Design as Refinement Plus Constraint Propagation: The VEXED Experience," *Proceedings of the Sixth National Conference on AI (AAAI-87)*, (1987), pp. 830–835.
11. S. Mittal, C. L. Dym, and M. Morjaria, "PRIDE: An Expert System for the Design of Paper Handling Systems," *IEEE Computer*, **19(7)**, (1986), pp. 102–114.
12. R. H. Lathrop and R. S. Kirk, "A System Which Uses Examples to Learn VLSI Structure Manipulation," *Proceedings of the Fifth National Conference on AI (AAAI-86)*, (1986).
13. T. M. Mitchell, S. Mahadevan, and L. I. Steinberg, "LEAP: A Learning Apprentice for VLSI Design," *Proceedings of the Ningh International Joint Conference on Artificial Intelligence (IJCAI-85)*, (1985), pp. 573–580.
14. C. Tong and P. Franklin, "Tuning A Knowledge Base of Refinement Rules to Create Good Circuit Design," *Proceedings of the Eleventh International Joint Conference on AI (IJCAI-89)*, (1989).
15. S. Minton, *Learning Effective Search Control Knowledge: An Explanation-Based Approach.* PhD thesis, Computer Science Department, Carnegie Mellon University, Pittsburg, PA, (1988).
16. S. Mittal and F. Frayman, "Towards a Generic Model of Configuration Tasks," *Proceedings of the Eleventh International Joint Conference on AI (IJCAI-89)*, (1989), pp. 1395–1401.
17. W. P. Birmingham and D. P. Siewiorek, "Capturing Designer Expertise: The CGEN System," *Proceedings of the 26th Design Automation Conference*, (1989), pp. 610–613.
18. D. D. Gajski and D. E. Thomas, "Introduction to Silicon Compilation," *Silicon Compilers*, ed. D. D. Gajski, (Addison-Wesley, Reading, MA, 1988), pp. 1–48.
19. J. Rabaey, H. de Man, J. Vanhoof, G. Goossens, and F. Catthoor, "CATHEDRAL-II: A Synthesis System for Multiprocessor DSP Systems," *Silicon Compilers*, ed. D. D. Gajski, (Addison-Wesley, Reading, MA, 1988), pp. 311–360.
20. D. Smith, "What is Logic Synthesis?," *VLSI Design*, (1988), pp. 18–26.
21. R. K. Brayton and T. M. E. Detjens, S. Krishna, "Multiple-Level Logic Optimization System," *Proceedings of the International Conference on Computer-Aided Design (ICCAD-86)*, (1986), pp. 356–359.
22. K. Keutzer, "DAGON: Technology Binding and Local Optimization by DAG Matching," *Proceedings of the 24th Design Automation Conference*, (1987), pp. 341–347.

23. R. A. Bergamaschi, "Automatic Synthesis and Technology Mapping of Combinational Logic," *Proceedings of the IEEE International Conference on Computer-Aided Design (ICCAD-88)*, (1988), pp. 466–469.

24. A. J. de Geus and D. J. Gregory, "The Socrates Logic Synthesis and Optimization System," *Design Systems for VLSI Circuits: Logic Synthesis and Silicon Compilation*, eds. G. De Micheli, A. Sangiovanni-Vincentelli, and P. Antognetti, (Martinus Nijhoff Publishers, Boston, MA, 1986), pp. 473–498.

25. W. H. Joyner, Jr., L. H. Trevillyan, D. Brand, T. A. Nix, and S. C. Gundersen, "Technology Adaptation in Logic Synthesis," *Proceedings of the 23rd Design Automation Conference*, (1986), pp. 94–100.

26. N. Vander Zanden and D. D. Gajski, "MILO: Microarchitecture and Logic Optimizer," *Proceedings of the 25th Design Automation Conference*, (1988), pp. 403–408.

27. A. J. de Geus, "Logic Synthesis Speeds ASIC Design," *IEEE Spectrum*, **26(8)**, (1989), pp. 27–31.

28. J. R. Kipps and D. D. Gajski, "The Role of Learning in Logic Synthesis," *International Journal of Pattern Recognition and Artificial Intelligence*, **4(2)**, (1990), pp. 167–180.

29. D. D. Gajski and F. D. Brewer, "Towards Intelligent Silicon Compilation," *Design Systems for VLSI Circuits: Logic Synthesis and Silicon Compilation*, eds. G. De Micheli, A. Sangiovanni-Vincentelli, and P. Antognetti, (Martinus Nijhoff Publishers, Boston, MA, 1986), pp. 365–383.

30. *IEEE Standard VHDL Language Reference Manual.* New York, NY, (1988).

31. N. Dutt, "GENUS: A Generic Component Library for High-Level Synthesis," Tech. Rep. 88-22, Department of Information and Computer Science, University of California, Irvine, (1988).

32. R. Davis, "Applications of Meta Level Knowledge to the Construction, Maintenance and Use of Large Knowledge Bases," *Knowledge-Based Systems in Artificial Intelligence*, eds. R. Davis and D. B. Lenat, (McGraw-Hill, New York, NY, 1982).

33. E. H. Shortliffe, *Computer-Based Medical Consultations: MYCIN*, (Elsevier Publishers. New York, NY, 1976).

34. T. M. Mitchell, P. E. Utgoff, and R. B. Banerji, "Learning by Experimentation: Acquiring and Refining Problem-Solving Hueristics," *Machine Learning*, eds. R. S. Michalski, J. G. Carbonell, and T. M. Mitchell, (Tioga Publishing Company, Palo Alto, CA, 1983).

35. W. P. Birmingham, A. P. Gupta, and D. P. Siewiorek, "The MICON System for Computer Design," *Proceedings of the 26th Design Automation Conference*, (1989), pp. 135–140.

36. A. Newell and H. A. Simon, "GPS: A Program that Simulates Human Thought," *Computers and Thought*, eds. E. A. Feigenbaum and J. Feldman, (McGraw-Hill, New York, NY, 1963).
37. R. E. Korf, *Learning to Solve Problems by Searching for Macro-Operators*, (Pitman Advanced Publishing Program. Boston, MA, 1985).
38. LSI Logic, Inc., Milipitas, CA, *CMOS Macrocell Manual*, (1987).

Applications of Learning and Planning Methods
edited by N. G. Bourbakis

CHAPTER 7

Connectionist Production Systems in Local and Hierarchical Representation

ANDREW SOHN
and
JEAN-LUC GAUDIOT
Department of Electrical Engineering - Systems
University of Southern California
Los Angeles, California 90089-0781, U.S.A.

Abstract

Much effort has been expended on developing special architectures dedicated to the efficient execution of problems in production systems. While artificial neural networks (ANNs) offer the promise of solving various optimization problems, we demonstrate here that the connectionist approach can be applied to artificial intelligence production systems. Characteristics of the production system paradigm are identified along neural networks, based on which two types of representation techniques are studied: local and hierarchical representations. To demonstrate the use of a neural network approach, the mapping of a generic production system is explained in detail. The results of a deterministic simulation demonstrate that the three layers of network architecture can be an efficient processing mechanism for the AI production system paradigm.

1. Introduction

A major obstacle in the processing of artificial intelligence (AI) applications lies in the large search/match time. In rule-based production systems, for example, it is often the case that the rules and the database needed to represent a particular production system in a certain problem domain would be on the order of hundreds to thousands of rules and assertions. It is thus known that simply applying software techniques to the matching process would yield intolerable delays. Indeed, as Forgy[3] pointed out, the time taken to match patterns over a set of rules can reach 90% of the total computation time spent in production systems. The need for faster execution of production systems has spurred research in both the software[3,6,12] and hardware domains[5,9,16,17].

In an effort to find an efficient mechanism for the production system paradigm, several attempts have been made along the line of neural networks[4,14,18,19]. Multiple layers of networks have been proposed to suit the production system paradigm[4,14]. In order to represent the state of a search space, local representation is used along with multilayer feedforward network[4]. Drawbacks in local representation are identified in and a distributed representation is suggested[7,18]. Among the suggested architectures, an issue regarding variable binding is of important concern. Some architectures[4,14] have limitations on variable binding while others[15,18] allow it.

This paper explores the applicability of a neural network approach to the production system paradigm. Our study employs multilayer feedback network with two representation techniques: local and hierarchical representations. We shall start our discussion in section 2 by finding some characteristics as well as problems in the production system paradigm. Several modifications to the paradigm are suggested. An architecture is proposed in section 3, which consists of three layers of neurons. Section 4 discusses the mapping of production systems in local representation method. In section 5, an attempt on a hierarchical representation is made to map the production system to a feature space which is built from a set of features derived from the production system. Section 6 gives simulation results based on the proposed architecture and representation method. The target problem for our simulation is in STRIPS[2]. The last section summarizes this work and outlines directions for future research.

2. Production Systems

Problems in production systems are discussed from the parallel and distributed processing perspectives. Several possible solutions to the problems are suggested in the neural network environment.

2.1. Production systems

A production system paradigm depicted in Fig.1 consists of *production memory* (PM), *working memory* (WM), and *inference engine* (IE). PM (or rulebase) is composed entirely of conditional statements called productions (or rules). These productions perform some predefined actions when all the necessary conditions are satisfied. The left-hand side (LHS) is the condition part of a production rule, while the right-hand side (RHS) is the action part. Both LHS and RHS consist of one to many elements, called *patterns* (p). The condition and action patterns will be referred by u and v, respectively.

The productions operate on WM which is a database of assertions called *working memory elements* (WMEs or w's). Both patterns and WMEs have a list of elements, called *attribute-value pairs* (AVPs). The value to an attribute can be either *constant* or *variable*: the former in lower case and the latter in upper case (which are similar to the *Prolog* representation). A number of AVPs in a pattern p_i (or a WME w_i) is denoted as $\eta(*)$.

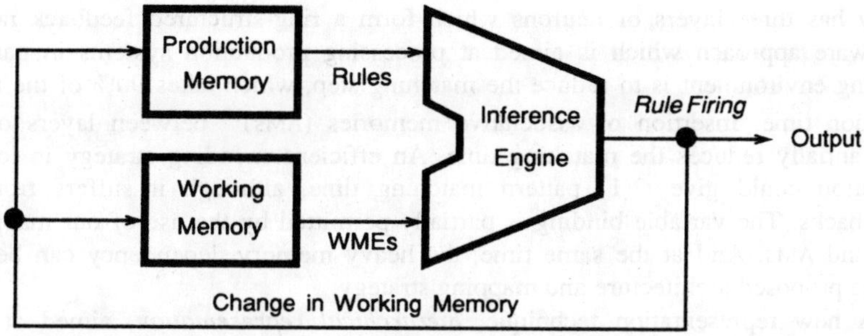

Fig.1: An architecture of the production system paradigm, consisting of three modules: a production memory, a working memory, and an inference engine.

The inference engine repeatedly executes an inference cycle which typically consists of three steps: pattern matching, conflict resolution, followed by rule firing:

- *Pattern Matching*: The condition patterns of all the production rules are matched against the current WMEs to determine the set of satisfied productions.
- *Conflict Resolution*: If the set of satisfied productions is non-empty, one rule is selected for execution in the next step.
- *Rule Firing*: The action patterns specified in the RHS of the selected production are performed on WM.

The inference engine will halt the production system either when there are no satisfied productions or when the desired solution is found.

2.2. Problems and approaches

The production system paradigm described above presents several inefficiencies. The processing mechanism itself inherently renders a sequential processing. It requires that the three steps be performed in sequence by an inference engine. The definition embedded in the production system prevents its evaluation on a parallel machine.[5,16,17]

Another inefficiency is in heavy memory dependency. All the patterns and WMEs to be matched must be repeatedly stored and recalled whenever a new inference cycle is started. The memory dependency existing in WM between inference cycles in turn gives less-adaptability to complex problems and new environments where robustness and self-learning are essential.

It is our goal in this study to develop a certain execution model so that the production system paradigm can be fully utilized and therefore executed efficiently in a parallel processing environment. The architecture of a neural network proposed in this study has three layers of neurons which form a ring-structured feedback network. A hardware approach which is aimed at processing production systems in parallel processing environment is to reduce the matching step, which takes 90% of the total computation time. Insertion of associative memories (AMs)[10] between layers of neurons substantially reduces the matching time. An efficient encoding strategy in local representation could give $O(1)$ pattern matching time, although it suffers from several drawbacks. The variable binding is partially permitted by the use of our mapping strategy and AMs. And at the same time, the heavy memory dependency can be flattened by the proposed architecture and mapping strategy.

A new representation technique, *hierarchical representation*, aimed at changing the underlying concepts of production systems is developed to overcome the barrier in local representation. Information hiding in different levels of the problem domain is derived, based on which we build a 3-dimensional space called *feature space*. A pattern is uniquely defined in the feature space as a pattern vector $p(x,y,z)$. In this paper we discuss a way to represent production systems in feature space for neural network processing.

3. The Ring-Structured Architecture

The architecture proposed for processing production systems consists of three layers, (1) *condition-*, (2) *rule-*, and (3) *action-layers*, as well as three associative memories, AM1, AM2, and AM3. Fig.2(a) shows the logical organization of the architecture. Neurons in the three layers connected through a ring structure feed information forward to naturally follow the logical organization of production system paradigm. Upon completion of an inference cycle, the newly created state of WM is fed backward for another inference cycle.

A detailed architecture is depicted in Fig.2(b). The bottom condition-layer (*X-field*) which consists of n neurons $x_1,...,x_n$, $x_i \in \{0,1\}$, holds a set of WMEs, reflecting the current state of WM. When a local representation is used, x_i represents the state of a particular WME. If $x_i=1$, the corresponding WME is present at WM. Otherwise, the WME is absent. An input of the initial state of a production system is presented to this layer in the form of a vector X with n elements. The middle rule-layer (*Y-field*) holds a set of rules. There are p neurons $y_1,...,y_p$, $y_i \in \{0,1\}$ in the layer, each of which represents one of several instantiations of the rules. All the possible instantiations of the rules are represented in this layer. The activation of a neuron indicates that the corresponding rule is to instantiate. The top action-layer (*Z-field*) with $z_1,...,z_m$, $z_i \in \{0,1\}$, holds a set of WMEs indicating how and what operations it will enforce on WM. The representation of neurons in the action-layer is similar to that in the condition-layer but their roles are completely different.

The first associative memory AM1 holds dependence relations between a set of WMEs and a set of conditions of a particular rule. Given a set of WMEs, a set of rules will be instantiated, thereby turning on/off the corresponding neurons in the rule layer. The second associative memory AM2 holds dependence relations between a set of actions and a set of WMEs of a particular rule. Upon selection and firing of a rule, a set of WMEs in the action layer will be changed, thereby turning on/off the corresponding neurons in the action layer. The third associative memory AM3 inserted between the action and the condition layers holds dependence relations between the previous and next states of WM. When WMEs are changed due to the rule firing, the neurons in the action layer must correctly reflect the current state. To ensure this correctness, AM3 keeps the information regarding the previous state such that the new state will be correctly reflected to the neurons in the action layer.

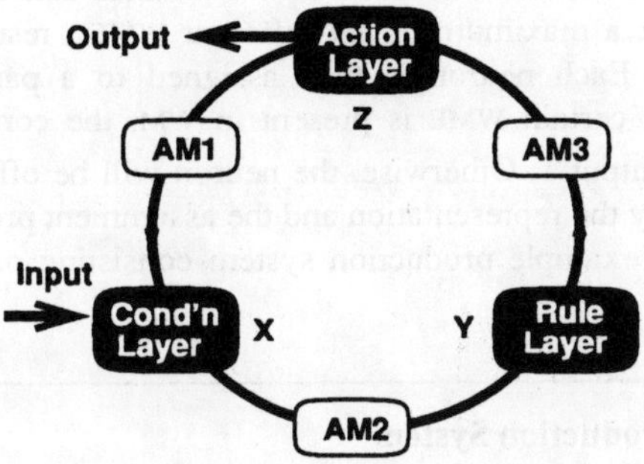

(a)

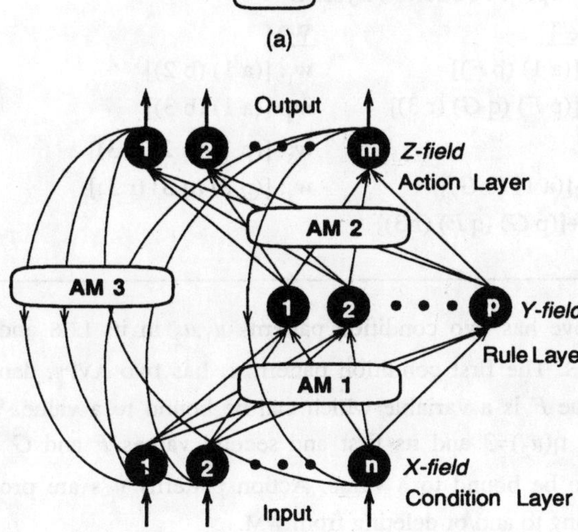

(b)

Fig.2: The ring-structured architecture. (a) Logical organization, (b) The detailed architecture consisting of three layers.

4. Local Representation

Among the two types of representation techniques, the local representation is described in this section to model production systems along with an example.

4.1. Representing working memory

WM at any point in time holds a subset of all the possible instantiation of condition patterns. A straightforward way of representing the state of WM is a local representation, where each pattern or WME is assigned to a neuron. Let p_i=[(a_1 v_1),...,(a_n v_n)] be a pattern consisting of n AVPs. Suppose that p_i has m variables and $n-m$ constants among n value-parts, and that each of m variables can be bound to d different values. There would be a maximum of md different WMEs, resulting in md neurons in each X- and Z-layers. Each neuron is then assigned to a particular instantiation of the pattern p_i. When a certain WME is present in WM, the corresponding neuron will be activated and will output 1. Otherwise, the neuron will be off and will output 0. An example will best clarify the representation and the assignment process.

Consider below an example production system consisting of a rule *Rule1* and four WMEs w_1,...,w_4:

Example Production System

Rule 1	WM
u_1: [(a 1) (b *F*)]	w_1: [(a 1) (b 2)]
u_2: [(p *F*) (q *G*) (r 3)]	w_2: [(a 1) (b 3)]
→	w_3: [(p 1) (q 2) (r 3)]
v_1: -[(a 1) (b *G*)]	w_4: [(p 2) (q 3) (r 3)]
v_2: +[(p *G*) (q *F*) (r 3)]	

RULE1 above has two condition patterns u_1,u_2 in its LHS and two action patterns, v_1,v_2 in its RHS. The first condition pattern u_1 has two AVPs, denoted as $\eta(u_1)$=2, and its second value F is a variable which can be bound to a value. The second condition pattern u_2 has $\eta(u_2)$=3 and its first and second values F and G are variables each of which can also be bound to a value. Action patterns v_i's are proceeded by '+' or '–' indicating adding to and/or deleting from WM.

Suppose that a set of values to which the variables F and G can be bound is {1,2,3}. To represent u_1, three neurons are required since the variable F can be bound to 1, 2, or 3. The WME w_1 is an instance of u_1 with F bound to 2 whereas w_2 is with F

bound to 3. For u_2, nine neurons are required to represent all the possible instances. In total, 12 neurons (3 for u_1 and 9 for u_2) are needed to represent any state of WM in the X-layer or in the Z-layer.

For the example production system, two vectors X and Z are used to model each of the two layers of the network. Suppose that all the 12 possible WMEs are assigned to neurons $x_1,...,x_{12}$ and $z_1,...,z_{12}$ as shown in Table 1. The initial state of WM shown above can then be represented in the X-layer as $X = (x_1,...,x_{12}) = (011010001000)$.

Neurons	WMEs assigned	Neurons	WMEs assigned
x_1,z_1:	[(a 1) (b 1)]	x_7,z_7:	[(p 2) (q 1) (r 3)]
x_2,z_2:	[(a 1) (b 2)]	x_8,z_8:	[(p 2) (q 2) (r 3)]
x_3,z_3:	[(a 1) (b 3)]	x_9,z_9:	[(p 2) (q 3) (r 3)]
x_4,z_4:	[(p 1) (q 1) (r 3)]	x_{10},z_{10}:	[(p 3) (q 1) (r 3)]
x_5,z_5:	[(p 1) (q 2) (r 3)]	x_{11},z_{11}:	[(p 3) (q 2) (r 3)]
x_6,z_6:	[(p 1) (q 3) (r 3)]	x_{12},z_{12}:	[(p 3) (q 3) (r 3)]

Table 1: Assignment of all the possible WMEs to neurons of X- and Z-layers. Neurons x_i and z_i are assigned to the same WME here and can be assigned to any WMEs.

Given the initial state $w_1,...,w_4$, RULE1 will be instantiated by w_1 and w_4 with F any G bound respectively to 2 and 3. Since there is no other satisfied rules, RULE1 will be selected and fired. Firing of RULE1 will affect two WMEs [(a 1) (b 3)] and [(p 3) (q 2) (r 3)] according to v_1 and v_2 of RULE1. The fact that two WMEs will be influenced by the rule firing is represented in the Z-layer in the form of turning on (addition of WME) or off (deletion of WME) the corresponding neurons. Since the WME, [(p 3) (q 2) (r 3)], is added to WM, z_{11} is turned on. The deletion of WME [(a 1) (b 3)] turns z_3 off, marked Ø (*null*) to indicate its deletion from WM. The Z-layer is then represented as $Z = (z_1,...,z_{12}) = (00\varnothing 000000010)$.

4.2. *Representing production memory*

Assignments of neurons for the X- and the Z-layers are based on the *state* of WM. Assignment of neurons for the Y-layer is made based on *how* a rule is instantiated. Information regarding as which variables bind what values will establish a basis for the encoding of the Y-layer.

Consider again the example production system. RULE1 has two condition patterns u_1 and u_2. To satisfy u_1 and u_2, the variable F of u_1 and the variable F of u_2 must be bound to the same value v. The variable G can be bound to any value v' such that

$v \neq v'$. For the given 4 WMEs, the variables F and G are bound respectively to 2 and 3 to satisfy RULE1. The fact that F and G are bound to 2 and 3 constitutes a particular state of the Y-layer. For RULE1 there can be a maximum of *nine* different combinations of variable bindings, requiring *nine* neurons in the Y-layer. A simple way of assigning variable bindings to the neurons $y_1,...,y_9$ can be made as listed in Table 2.

	F	G		F	G		F	G
y_1:	1	1	y_4:	2	1	y_7:	3	1
y_2:	1	2	y_5:	2	2	y_8:	3	2
y_3:	1	3	y_6:	2	3	y_9:	3	3

Table 2: Assignment of variable binding information to neurons of the *Y*-layer.

The neuron y_1 is assigned to the variable binding (F=1,G=1), the neuron y_2 is assigned to (F=1,G=2), y_3 to (F=1,G=3), and so on. When the rule is instantiated with the variable bindings (F=2, G=1), the Y-layer is represented by turning y_4 on, resulting in $Y = (y_1,...,y_9) = (000100000)$.

4.3. Encoding of the three associative memories

In what follows, we shall continue to use the example production system shown in the previous section. As explained earlier in section 3, AM1 holds pairs of information regarding how the rule is satisfied with respect to the current state of WM. An association for AM1 can be derived from the X- and Y-layers. The current state of WM for the given four WMEs of the example production system is represented in the X-layer as $X_1 = (x_1,...,x_{12})$ = (011010001000). At the same time, the corresponding instantiation of the rule for the given WMEs is represented in the Y-layer as $Y_1 = (y_1,...,y_9)$ =(000100000). We then find an association for AM1 as

$$A(X_1,Y_1)_{AM1} = \{(011010001000),(000100000)\}.$$

Associations for the second associative memory AM2 can be similarly obtained from the Y- and the Z-layers. The AM2 holds the dependence relations between a set of WMEs and a set of action patterns of a particular rule. Given RULE1 and $w_1,...,w_4$, the Y-layer will have the fourth neuron y_4 turned on, resulting in Y_1=(000100000). According to the action patterns of RULE1, the Z-layer will have the following state (as we discussed in section 4.1): Z_1= (00∅000000010). These two states of the Y- and Z-layers will form an association for AM2:

$A(Y_1,Z_1)_{AM2} = \{(000100000),(00\varnothing000000010)\}$.

Associations for AM3 hold information regarding as to how to merge the previous state of the X-layer and the current state of the Z-layer. After firing of RULE1, WM must contain the following WMEs:

WM = {[(a 1) (b 2)], [(p 1) (q 2) (r 3)], [(p 2) (q 3) (r 3)], [(p 3) (q 2) (r 3)]}

According to Table 1, an appropriate representation for this new state of WM is (010010001010), which must be subsequently represented in the X-layer for the next inference cycle. A simple way of doing this update is a combination of AND and OR operations on two states of the X- and Z-layers, X_{prev} and Z_{prev}. For the plain binary values, a straight OR operation is performed whereas for the null (∅) values an AND operation is carried out. The result of this AND-OR operation will be a new state of the X-layer, X_{new}, as listed below:

- Previous state, X_{prev} = (011010001000)
- Current state, Z_{prev} = (00∅000000010)
- New state, X_{new} = (010010001010)

An association of AM3 for the above two states and operation is, therefore, represented as

$A(Z_1,X_1)_{AM3} = \{(00\varnothing000000010),(010010001010)\}$.

Applying the three procedures described above, other remaining associations for the three AMs can be easily obtained.

5. Hierarchical Representation

A hierarchical representation is aimed to ease the mapping of the production system paradigm to neural networks. In this section we shall develop a way to represent production systems in feature space for neural network processing.

5.1. Feature hierarchy in production systems

In one end of the spectrum of representation techniques used in ANNs is the local representation[4,15], where a concept is assigned to a neuron. While it is quite straightforward to implement and easy to understand its mapping, the local representation reveals several major drawbacks such as difficulty in capturing information regarding *variable bindings*, weakness in *fault tolerant* computing, *varying* number of neurons for different problem size, and so on.

On the other end of the spectrum, the distributed representation[7] is used, which assigns a concept to a pattern of activity of many neurons. This method which is often said to resemble the way in which the brain works provides a graceful degradation for faulty neurons, thereby giving fault tolerant computing. However, it is very difficult to visualize the mapping and to understand how it works. Without a good learning algorithm, it is almost impossible for a large problem size to be appropriately trained for practical applications[1].

To overcome the difficulties in local and distributed representations, we introduce a *hierarchical* representation which attempts to combine the two techniques at different levels of the information hierarchy. In the high level of the information hierarchy that we obtain from the problem domain, the local representation is used to partition the problem while in the low level, the distributed representation is used to implement neurons' mapping. A set of features we can derive from the production system domain is listed below in order of decreasing importance:

- f_1: A number of AVPs in a pattern
- f_2: Similarity in *attribute* parts of patterns
- f_3: Similarity in *value* parts of patterns

The value parts in the third feature include not only the constants assigned to attributes but also variables. The three features f_1,f_2,f_3 above are used as criteria to partition the patterns into groups of patterns. A tree for each feature f_i is constructed from the groups of patterns. Merging three trees into one, we obtain a *partitioned pattern tree* (PTT). Upon constructing PPT, we build a 3-dimensional space $\Re$, called *feature space*, where a pattern p_i of the production system can then be uniquely defined as a pattern vector $p_i(f_1,f_2,f_3)$.

5.2. *Representing PM and WM in feature space*

Let $p_i=[(a_1\ v_1),\ldots,(a_n\ v_n)]$ be a pattern (condition pattern, action pattern, or WME) in a production system, where a_i is an attribute and v_i is a fixed value or a variable. Suppose that a production system has a set of m patterns, $P=\{p_1,\ldots,p_m\}$. We shall define the three features, f_1,f_2,f_3, as follows: (1)Given a pattern $p_i=[(a_1\ v_1),\ldots,(a_n\ v_n)]$, it has n number of AVPs, denoted by $\eta(p_i)=n$. (2)Two patterns $p_i=[(a_1\ v_1),\ldots,(a_n\ v_n)]$ and $p_j=[(a_1'\ v_1'),\ldots,(a_n'\ v_n')]$ are said to be *attribute similar*, denoted by $p_i\approx p_j$, if $\forall i,\ a_i=a_i'$. (3) p_i and p_j are said to be *value similar*, denoted by $p_i\cong p_j$, if $\forall i,\ v_i=v_i'$. From (2) and (3), we define that p_i is *equivalent* to p_j, denoted by $p_i\equiv p_j$ if $p_i\approx p_j$ and $p_i\cong p_j$.

The first feature f_1 partitions P into n subsets $P_1,\ldots,P_n$ such that for $p_i \in P_k$, $\eta(p_i)=k$ for all i. After partitioning, each pattern p_j in P_k will have k AVPs. The second feature f_2 partitions P into m subsets $P_1,\ldots,P_m$ such that for $p_i,p_j \in P_k$, $p_i \approx p_j$ for all i,j, $i \neq j$. After partitioning, patterns in P_k will be similar in attributes. The third feature f_3 which uses the *value* similarity measure partitions P into r subsets $P_1,\ldots,P_r$ such that for $p_i,p_j \in P_k$, $p_i \equiv p_j$ for all i,j, $i \neq j$. After partitioning, patterns in P_k will be similar in values.

Consider two rules with 6 condition patterns, $p_1,\ldots,p_6$, listed below:

Rule 1	**Rule 2**
p_1: [(a 1) (b 2)]	p_4: [(p 2) (q 3) (r 3)]
p_2: [(a 1) (b 3)]	p_5: [(a 3) (q 2)]
p_3: [(p 1) (q 2) (r 3)]	p_6: [(t 3)]
→	→
[Make (c 1) (d 2)]	[Remove 1st pattern]

Let $P=\{p_1,\ldots,p_6\}$ be the six condition patterns. The first feature f_1 partitions P into $\{p_6\}$, $\{p_1,p_2,p_5\}$, and $\{p_3,p_4\}$ since $\eta(p_6)=1$, $\eta(p_1)=\eta(p_2)=\eta(p_5)=2$, and $\eta(p_3)=\eta(p_4)=3$, as depicted in Fig.3(a). Using f_2, we obtain $\{p_6\}$, $\{p_1,p_2\}$, $\{p_5\}$, and $\{p_3,p_4\}$, as shown in Fig.3(b), since $p_6 \neq p_1 \approx p_2 \neq p_5 \neq p_3 \approx p_4$. The third feature f_3 partitions P into $\{p_1\}$, $\{p_2\}$, $\{p_3\}$, $\{p_4\}$, $\{p_5\}$, and $\{p_6\}$, as shown in Fig.3(c). Numbers on arcs indicate the group a pattern belongs to. Merging (a), (b), and (c) of Fig.3 yields a partitioned pattern tree, as depicted in Fig.3(d).

Once the partitioned pattern tree is built, the assignment process for each pattern to the feature space is automatic. The numbers next to arcs in Fig.3 are assigned to coordinate values (f_1,f_2,f_3) of the feature space. From Fig.3(d), we obtain the following coordinates: $p_1(2,2,2)$, $p_2(2,2,3)$, $p_3(3,4,5)$, $p_4(3,4,6)$, $p_5(2,3,4)$, and $p_6(1,1,1)$. Fig.4 shows the mapping of six patterns to corresponding points in the 2-dimensional feature space (the third feature is not shown for the simplicity of presentation).

It is interesting to note in Fig.4 that those patterns such as p_1,p_2 that have the same number of AVPs and are similar in attributes form a *cluster* in the 2-dimensional feature space. It is this cluster that we use to match patterns over WMEs. We are not going to implement this hierarchical representation in the proposed architecture. Instead, we shall briefly outline the way to implement the *pattern matching* procedure. Suppose that all the condition patterns are partitioned into c clusters, $C_1,\ldots,C_c$, in the 2-dimensional feature space $\Re^2$. A cluster C_i which represents a set of particular pattern vectors is assigned to a unit network N_i, consisting of m neurons $n_1,\ldots,n_m$. The

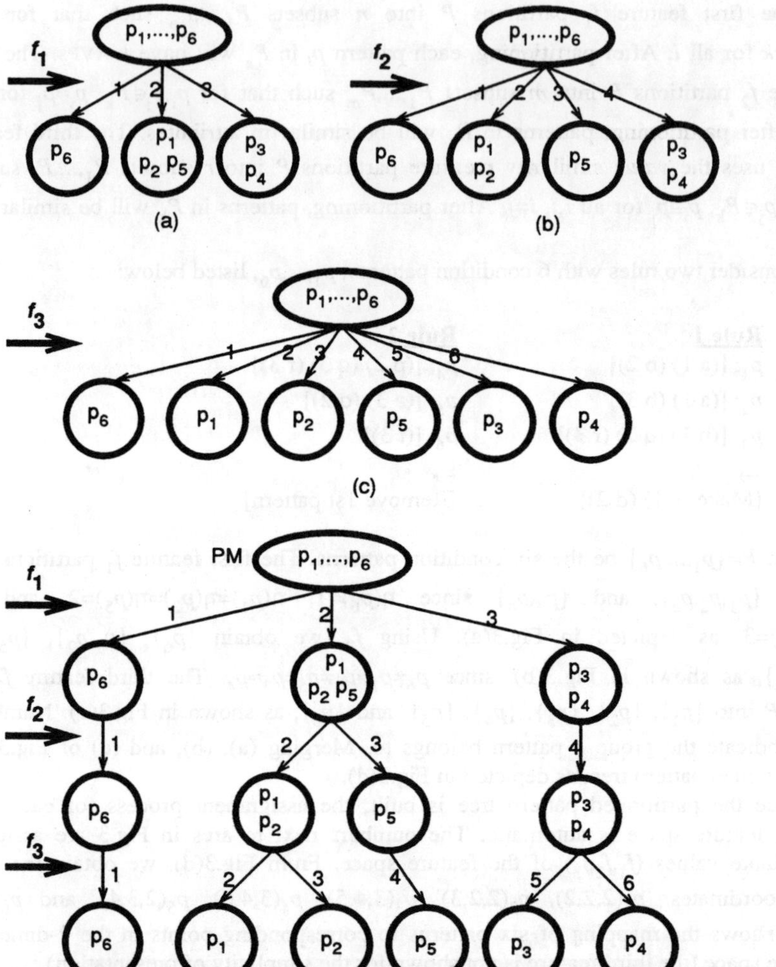

Fig.3: Partitioned Pattern Tree (PTT) for PM. (a) f_1 partitions P into three groups, (b) f_2 into 4 groups, (c) f_3 into 6 groups, (d) All three subtrees in (a),(b),(c) are merged to form a partitioned pattern tree for PM. Numbers next to arcs uniquely define each pattern in the feature space. One can construct (d) directly from patterns without going through (a),...,(c).

coordinate value (f_1,f_2) for C_i, denoted as $C_i(f_1,f_2)$, can then be trained in a unit network N_i such that when a WME w_j is presented to N_i, the corresponding cluster can be activated by N_i. By building a set of N_i, i=1,...,c and training them, we will be able to achieve $O(1)$ pattern matching time. Furthermore, by doing this the heuristics which enable the *good* or *optimal* selection of choice points in the search space can be efficiently represented in terms of energy function. Selection of a choice point in the context of search tree is basically the same as an optimization process in the context of artificial neural networks.

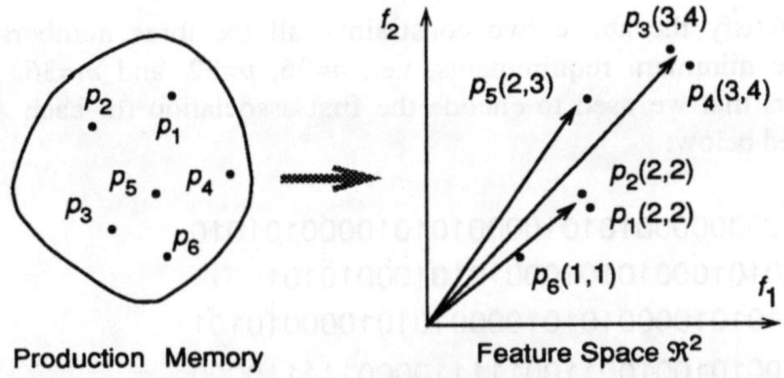

Fig.4: A production memory in 3-dimensional feature space. f_3 is not shown in this figure. Six patterns are completely mapped to the feature space, $\Re^2$, based on the features and the partitioned pattern tree.

6. Simulations

Our simulation using *local* representation is performed on STRIPS[2], a general problem solver. For our simulation purpose, we use *two* blocks (A,B) and adopt a forward reasoning strategy in this experiment since our interest is in executing a production system paradigm in parallel/distributed processing environment. Note also that in this simulation we view STRIPS as a production system *not* a planning system.

Assigning binary {0,1} or bipolar values {−1,+1} to associations seems easy and straightforward at first. However, in practice it is not the case as the number of associations becomes realistic. The actual procedure for the encoding of AMs is a rather difficult task for several reasons: First, the continuity assumption[10] must be satisfied to use AMs, i.e., $H(X_i,X_j)/n \approx H(Y_i,Y_j)/p \approx H(Z_i,Z_j)/m$, where $H(*,*)$ is a *Hamming* distance for two vectors and n,p,m is a number of neurons each in the X-, Y-, and Z-layers. Second, $H(*,*)$ for any two vectors of the *same* layer need be large enough to separate each other. It is especially the case for the Y-layer, where for our local representation $H(Y_i,Y_j)$=2 for i,j=1,...,p, $i \neq j$. Writing a program to train the network, we were able to enforce the following condition: $\forall i,j$, $H(X_i,X_j)>n/2$, $H(Y_i,Y_j)>p/2$,

$H(Z_i,Z_j)>m/2,\ i\neq j$.

For STRIPS with two blocks A and B, there can be a maximum of eight different instantiations for the four rules *Stack*, *Unstack*, *PickUp*, and *PutDown* (see Ref.2 for details). For example, an operator (Stack X Y) can have two different instantiations: (Stack A B) and (Stack B A). Other three rules have two different instantiations each, resulting in eight instantiations. There exists a maximum of 9 different instantiations of WMEs: (Hold A), (Hold B), (Clear A), (Clear B), (On A B), (On B A), (ArmEmpty), (OnTable A), and (OnTable B). The minimum numbers of neurons for the three layers are n=9, p=8, and m=9, respectively. A maximum number of associations, n_A, which can in principle be encoded in each AM is 8 since $n_A \leq \min(n,p,m)$. However, to satisfy the above two constraints, all the three numbers are augmented three times the minimum requirements, i.e., n=36, p=32, and m=36. As an example, the four vectors that we used to encode the first association for each AM in our simulations are listed below:

X_1= 100100000000101010000101010000101010
Y_1= 10110101000101010001010100010101
Z_1= 001010101000010101000010101000010101
X_1'= 000100101001001100111110000111110000

The first associations of AM1, AM2, and AM3 are derived each from the pair (X_1,Y_1), (Y_1,Z_1), and (Z_1,X_1'). The two vectors that we used for the initial and goal states of STRIPS with 2 blocks are listed below:

X_i= 101010101000010101000000101000010101
X_g= 000101110101110111011110111101111011

The three AMs are not listed in this paper due to the space constraint. Using our architecture and mapping strategy described thus far, matching condition patterns over WMEs reduces to $O(1)$. When the correct initial state is presented to the network, our architecture can find the goal state in four iterations for the problem with two blocks. When a disturbed initial state is presented to the network, irrelevant operators are selected and fired, resulting in an incorrect solution. This is due to the facts that the *heuristics* which prune the irrelevant search paths are not considered in this study, and that our study is mainly in reducing pattern matching time in production systems. The architecture and mapping strategy we developed in this work substantially reduces the pattern matching time and assimilate the memory dependency of production systems in parallel processing environment by flattening it into layers of neurons.

7. Conclusions

A neural network approach to production system paradigm is proposed in this paper. Problems in production system paradigm are identified. Possible solutions to the problems are suggested in three ways: First, the three layers of ring-structured architecture is proposed to map the production system paradigm to neural networks.

Second, the time spent to match patterns is reduced to $O(1)$ by using three associative memories between three layers of neurons. Third, the two representation techniques are studied: local and hierarchical representations. Local representation for our target production system is implemented and satisfactory to a limited extent. To overcome the problems in the local representation, a hierarchical representation is developed, which derives features from the production systems and transforms patterns to vectors of the feature space.

Our simulations on the target problem STRIPS with 2 blocks show that the proposed architecture, when a local representation is used, finds an optimal solution. It takes 4 iterations to find a solution when the correct input is given to the input layer of the network. For incorrect input sets, it does not generate a good solution and gets into a loop since *heuristics* to prune unpromising branches in the search tree and to avoid repeating the same action are not considered in this paper.

Instead of performing simulations using hierarchical representation, we demonstrated a way of implementing production systems in the hierarchical representation. Our current effort is in the direction in which we develop the hierarchical representation as a tool for the neural network processing.

References

1. S. Fahlman and G. E. Hinton, "Connectionist Architecture for Artificial Intelligence" in *IEEE Computer*, January 1987, pp.100-109.
2. R. Fikes and N. Nilsson, "STRIPS: a new approach to the application of theorem proving to problem solving," *Artificial Intelligence* **2** (1971) pp.189-208.
3. C. L. Forgy, "Rete: A Fast Algorithm for the Many Pattern/Many Object Pattern Match Problem," in *Artificial Intelligence* **19** (1982) pp.17-37.
4. S. Gallant, "Connectionist expert systems," in *Comm. of the ACM*, **31:2** (1988) pp.152-169.
5. J.-L. Gaudiot and A. Sohn, "Data-Driven Multiprocessor Implementation of the Rete Match Algorithm," in *Proc. International Conf. on Parallel Processing*, August 1988, pp.I256-260.
6. A. Gupta, C. L. Forgy, A. Newell, and R. Wedig, "Parallel Algorithms and Architectures for Rule-Based Systems," in *Proc. 13th International Symp. on Computer Architecture*, June 1986, pp.116-120.
7. G. E. Hinton, J. L. McClelland, and D. E. Rumelhart, "Distributed Representations," in J.L. McClelland and D.E. Rumelhart (Eds.), *Parallel Distriburted Processing: Explorations in the Microstructure of Cognition.* Vol 1: Foundations,

(MIT Press 1986).

8. G. E. Hinton and J. A. Anderson (Eds.), *Parallel Models of Associative Memory*, (Hillsdale, NJ, Lawrence Erlbaum Associates 1981).
9. M. A. Kelly, and R. E. Seviora, "A Multiprocessor Architecture for Production System Matching," in *Proc. AAAI National Conf. on AI*, August 1987, pp.36-41.
10. B. Kosko, "Bidirectional Associative Memories," *IEEE Trans. on Systems, Man, and Cybernetics* **18:1** (1988) pp.49-60.
11. B. Kosko, "Adaptive Bidirectional Associative Memories," *Applied Optics* **26:23** (1987) pp.4947-4960.
12. D. P. Miranker, "TREAT: A Better Match Algorithm for AI Production Systems," in *Proc. AAAI National Conf. on AI*, August 1987, pp.42-47.
13. J. Quinlan, "A comparative Analysis of Computer Architectures for Production System Machines," in *Proc. Hawaii International Conf. on System Sciences*, January 1986, pp.187-193.
14. T. Samad, "Towards Connectionist Rule-Based Systems," in *Proc. IEEE International Conf. on Neural Network*, June 1988, pp.II-525 - 532.
15. A. Sohn and J.-L. Gaudiot, "Connectionist Production Systems in Local Representation," in *Proc. International Joint Conf. on Neural Network*, January 1990.
16. S. J. Stolfo, "Initial Performance of the DADO2 Prototype" in *IEEE Computer* (January 1987) pp.75-83.
17. M. F. M. Tenorio and D. I. Moldovan, "Mapping Production Systems into Multiprocessors," in *Proc. IEEE International Conf. on Parallel Processing*, August 1985, pp.56-62.
18. D. S. Touretzky and G. E. Hinton, "Symbols Among the Neurons: Details of a Connectionist Inference Architecture," in *Proc. International Joint Conf. on AI*, August 1985, pp.238-243.
19. J. Voevodsky, "PLATO/ARISTOTLE: A Neural Net Knowledge Processor," in *Proc. IEEE International Conf. on Neural Network*, June 1987, pp.II-399 - 407.

PART II: PLANNING

Applications of Learning and Planning Methods
edited by N. G. Bourbakis

CHAPTER 8

A Parallel Architecture for AI Nonlinear Planning

Sukhan Lee
Department of Electrical Engineering - Systems
University of Southern California
Los Angeles, California 90089-0273

Kyusik Chung
Department of Electronics Engineering
Soong-Sil University
Seoul, Korea

Abstract

This chapter presents a resource-level conflict detection and conflict resolution scheme which is combined with a state-level backward planning algorithm and provides efficient conflict detection and global conflict resolution for nonlinear planning. The scheme keeps track of the usage of individual resources during planning, and constructs a Resource-Usage Flow (RUF) structure (based on which conflict detection and resolution is accomplished). The RUF structure allows the system to perform minimal and nonredundant operations for conflict detection and resolution. Furthermore, resource-level conflict detection and resolution facilitates problem decomposition in terms of resources, thereby providing easy implementation in a parallel and distributed processing environment.

Performance analysis indicates that the proposed architecture has a speed-up factor of the average depth of a plan network, $D(N_a)$, compared to the distributed NOAH, where N_a (the total number of action nodes at the completion of planning) and $D(N_a)$ are considerably larger than the number of resources involved in planning as well as the number of initial goal states.

1 Introduction

Planning is a process to develop a sequence of actions (operators) achieving a goal(s) from an initial world. AI planners can be classified into two according to the order in which conjunctive goals are to be achieved: *linear planner* and *nonlinear planner*. The former, such as STRIPS[3], HACKER[6], and INTERPLAN[7], makes goal ordering first and then achieves each goal one by one accordingly. This strategy is based on the linearity assumption that goals are independent, and thus can be sequentially achieved in an arbitrary order. The latter, such as NOAH[5], NONLIN[8], DEVISER[9], SIPE[11] and TWEAK[1], makes goal ordering only when it is absolutely necessary, based on a least commitment strategy. Assuming each goals are independent, subplans are generated to achieve each goal. As subplans are generated, interactions among subplans are detected and resolved.

Nonlinear planners are believed to be potentially more efficient than linear planners. This is because nonlinear planners can avoid searching of all possible plan orderings. Instead, nonlinear planners should pay for handling interaction problem during planning. This requires not only detecting of conflicts among actions in parallel branches but also assigning of the precedence relations between the conflicting actions. Nonlinear planners generate a partially ordered plan network while linear planners generate a totally ordered plan network. This makes nonlinear planning systems attractive to multi-agent environment where actions can be executed in parallel.

In dealing with nonlinear or parallel planning, the following issues should be taken into consideration: 1) the efficiency of a planning algorithm 2) the optimality of the resulting plan network in terms of the maximum parallelism or the minimum depth of a plan network, 3) the implementation of a planning algorithm in a parallel and distributed processing environment. The efficiency and optimality of parallel planning largely depend on how conflicts among parallel actions or goals are detected and resolved. Much effort has been given to the development of efficient conflict detection and resolution: One method developed for efficient conflict detection and resolution is based on the domain-independent rules for resolving contradictory states, *e.g.* the critics with the table of multiple effects (TOME) in NOAH[5] or the improved critics based on goal structure (GOST) in NONLIN[8]. Another method is based on the domain-specific heuristics for resolving conflicting resources shared by parallel branches, *e.g.* the assignment of priority to the predefined resource types in SIPE[11]. However, detecting and resolving conflicts has yet not been achieved with the desired efficiency[1], specially in complex system where there is a high frequency of conflicts. Furthermore, issues such as the optimality of the resulting plan network and the implementation in a parallel and distributed processing environment have not received proper attention yet. Although a distributed NOAH[2] was proposed for a distributed

[1]The conflict problem is handled in terms of *declobbering* in TWEAK[1]. Although it provides a theoretical principle regarding conflict detection and resolution, it does not address the implementational details for achieving efficiency in conflict detection and resolution.

processing environment, it lacks possible improvement in planning efficiency.

This chapter presents a resource-level conflict detection and resolution scheme which is combined with a state-level backward planning algorithm and provides efficient conflict detection and global conflict resolution for nonlinear planning. This scheme avoids the unnecessary comparisons in conflict detection as well as the redundant assignment of precedence relations in conflict resolution. Based on the conflict detection and resolution algorithm, a scheme for the decomposition of nonlinear planning is presented. Also presented is a parallel architecture that can be implemented in a parallel and distributed processing environment.

2 Problem Statements

2.1 Parallel Planning

Conventional parallel planners[8,9] generate a partially ordered plan network based on backward planning where a cycle of linking, expansion, and conflict detection/resolution[2] is repeated until all the goals generated are achieved.

Procedure *Parallel Planning*

```
1.    Begin
2.       Repeat  (*Backward Planning *)
3.          • Linking;
         (*If a goal (goal node) can be achieved by the effects
         of a node existing in the plan netwrok, a link is formed
         between the two nodes unless it results in a cycle
         of nodes in the plan network *)
4.          If Linking fails then
5.             Begin
6.                • Expansion;
             (*The goal is achieved by introducing a new action
             (action node), and the preconditions of the new
             action become new goals to be achieved. *)
7.                • Conflict detection-resolution;
             (*The new action generated from expansion is
             checked against all other parallel nodes
             currently existing in the plan network
             to detect conflicts, and any detected conflicts
             are subject to immediate resolution *)
8.             End
9.          Until no more goals to be achieved
10.   End  (*Parallel Planning *)
```

2.2 Approach

---

[2]These three operations correspond to *simple establishment*, *step addition*, and *declobbering*, respectively, in TWEAK[1].

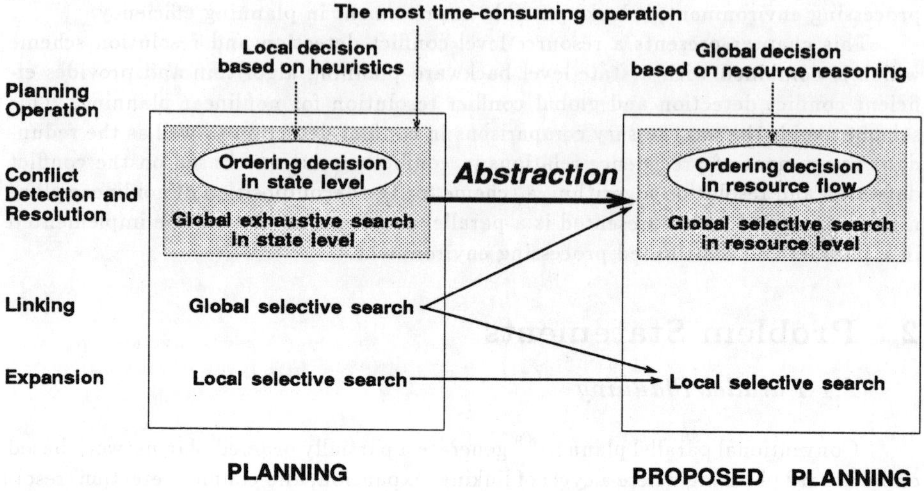

Figure 1. Approach of Abstraction

As shown above, planning involves an exponential search with possible backtracking during decision making in each planning operation. To reduce the time complexity involved in (optimal) planning requires not only the application of parallel and distributed processing to planning but also the exploration of intelligent heuristic planning strategies which constrains search space and avoids backtracking. Thus, we adopt the following approaches for the development of an efficient (optimal) planning: 1) development of an intelligent planning strategy by abstracting the conflict detection and resolution at the resource level to allow for efficient conflict detection and global conflict resolution, and 2) development of a parallel architecture by decomposing the planning into a set of subprocesses to allow even distribution of workload and less communication overhead among subprocesses.

3 Abstraction of Conflict Detection and Resolution

3.1 Why needs abstraction?

As shown in Fig.1, each planning operation has different search characteristics: the expansion operation requires a local selective search (select one after finding some or all candidates in the action table), the linking operation requires a global selective search (select one after finding some or all candidates in the plan network), and conflict detection-resolution requires a global exhaustive search (find all in the plan network) with decision making (select one of two alternative orderings between two actions).

For the selections in each planning operation, conventional parallel planner adopts local decisions based on heuristics. If the local decisions prove to be poor later, planning may backtrack. Particularly, poor decisions in linking or conflict detection-resolution are more harmful than those in expansion since they can directly lead to action deadlocks.

Linking and conflict detection-resolution operations are directly related as follows: (1) if a good linking is performed, it removes the necessity of detection and resolution of the interaction as a side-effect. (2) if a poor linking which leads to an action deadlock is performed, it makes conflict detection and resolution more complicated. Therefore, it is desirable for conflict detection and resolution to guide the linking operation. This necessitates an intelligent conflict detection and resolution scheme for parallel planning.

The analysis of time complexity in parallel planning (see the appendix) shows that conflict detection-resolution consumes the most of planning time, $O(N_a \times D(N_a))$, where N_a and $D(N_a)$ indicate, respectively, the total number of action nodes at the completion of planning and the average depth of the plan network.

The above qualitative and quantitative analyses of parallel planning illustrate that the development of an efficient conflict detection and intelligent (global) conflict resolution scheme is unavoidable in cases where the performance of parallel planning needs to be improved. One way to develop such a scheme is the abstraction of conflict detection and resolution at the resource level[3].

3.2 Conflict Detection and Resolution Based on Resource Reasoning

Consider a pair of two parallel actions a_i and a_j in different branches. A generic planner detects and resolves any possible conflict between a_i and a_j as follows: potential conflicts between a_i and a_j can be detected (1) *in the state-level* by crosschecking between the effects[4] of a_i and a_j, and/or (2) *in the resource-level* by reasoning the usage of resources for a_i and a_j. The conflicts detected from the former state-level case are resolved by applying domain independent rules such as the critics in NOAH[5], or by applying the procedures called promotion/separation/white knight in TWEAK[1]. The latter conflicts, the conflicts detected from the resource-level are resolved by applying domain-specific heuristics such as the assignment of priority to the predefined resource types in SIPE[11]. However, note that the above two schemes resolve conflicts immediately upon their detection based solely on a local decision. If the local

[3]This aims to reduce the exponent of N_a, not N_a itself, where N_a is an exponentially explosive number.

[4]Postconditions of an action and a part of preconditions denied at the end of the action

decisions prove to be poor later, then the schemes would result in the assignment of unnecessary precedences among actions, thereby missing the optimal plans (in terms of the maximum parallelism or the minimal depth of a plan network). In the worst case, deadlocks among actions would occur.

To improve the efficiency in conflict detection, and to avoid the undesirable effects due to the local decisions in conflict resolution, we propose the following schemes: 1) a resource[5] level conflict detection scheme and 2) conflict resolution based on a global decision.

Conflicts in parallel planning occur due to the competition or sharing of resources. Based on this fact, we can avoid unnecessary conflict detection as follows: unless any two parallel branches compete or share the same resource, conflict detection is not needed. When a new action is generated in a plan network, conflicts can be detected by checking whether any action in parallel branches use the same resource. To implement the proposed resource level conflict detection, a table called the Resource-Usage Flow Table (RUFT) is designed which records the usage of individual resource at run time. the RUFT, dynamically constructed during the planning process, enables avoidance of unnecessary checking of all parallel branches. When a conflict is detected between the newly generated action and any parallel branch, information regarding the identification of the conflicting branch for each resource is derived from the RUFT, called the Resource-Usage Flow Structure (RUF-structure). This structure provides a basis for conflict resolution based on a global decision.

There exists a distinct characteristic in the flow of the resources (objects) concerned in conflict detection and resolution: a flow of resource is formulated due to a causal[4] relationship between two succeeding actions associated with the resource. For example, see the flow of block "a" in Fig.2, Start - T9 - T6 - T2 -End. This is different from the flow of a tool such as a robot, which is formulated due to a temporal relationship between actions associated with the resource (scheduling). Since all objects (resources) involved in conventional parallel planning are types of single resources, there exists a global constraint for each resource: the resource flow from the initial state to the goal state in a plan should be serial in terms of a causal relationship. Note that this global constraint allows the process of conflict resolution to make a global decision.

Consider a final RUF-structure for a resource obtained from the successful planning. An intermediate RUF-structure for the resource, constructed dynamically during planning, is a partially ordered network of the actions requiring the resource. A branch of the network corresponds to a part of the final RUF-structure, under the assumption[6] that the plan network generated so far leads to a correct plan. As planning proceeds, the intermediate RUF-structure is expanded accordingly. Then, conflicts are detected and resolved by adding a precedence relation to the RUF-

[5]A resource is a reusable, physical object associated with an action or goal, like in SIPE[11]

[6]In case the assumption does not hold, *e. g.* a goal deadlock, plan revision (plan splicing[?]) can be also performed based on RUF-structure.

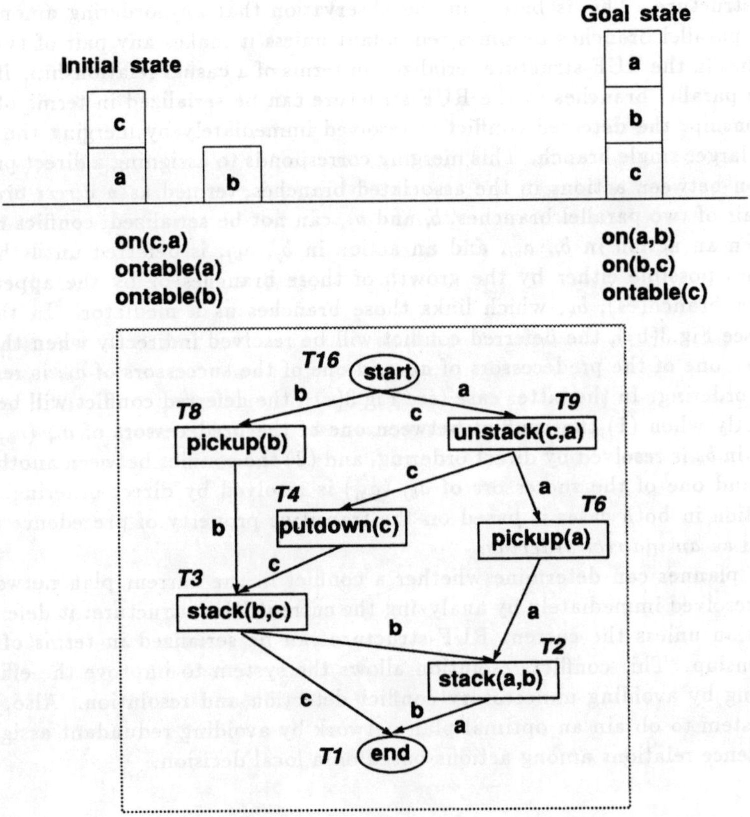

Figure 2. An example of a blocks world

structure. This is repeated until the intermediate RUF-structure becomes the final RUF-structure. Note that the transformation of the intermediate RUF-structure (a partially ordered network) into the final RUF-structure (a serialized network) is performed by conflict resolution.

The decision in conflict resolution can be made based on the following rule: the assignment of a precedence relation, *i.e.* ordering, is deferred unless it directly contributes to the transformation of the intermediate RUF-structure into the final RUF-structure. This is based on the observation that any ordering among actions in the parallel branches becomes redundant unless it makes any pair of two parallel branches in the RUF-structure serialized in terms of a casual relationship. If any pair of two parallel branches in the RUF-structure can be serialized in terms of a causal relationship, the detected conflict is resolved immediately by merging the branches into a larger single branch. This merging corresponds to assigning a direct precedence relation between actions in the associated branches, termed as *a direct ordering.* If any pair of two parallel branches, b_i and b_j, can not be serialized, conflict resolution between an action in b_i, a_{bi}, and an action in b_j, a_{bj}, is deferred until the merger becomes possible either by the growth of those branches or by the appearance of another branch(es), b_k, which links those branches as a mediator. In the former case (see Fig.3(b)), the deferred conflict will be resolved indirectly when the conflict between one of the predecessors of a_{bi} and one of the successors of a_{bj} is resolved by direct ordering. In the latter case (see Fig.3(c)), the deferred conflict will be resolved indirectly when (1) the conflict between one of the predecessors of a_{bi} (a_{bj}) and an action in b_k is resolved by direct ordering, and (2) the conflict between another action in b_k and one of the successors of a_{bj} (a_{bi}) is resolved by direct ordering. Conflict resolution in both cases is based on the transitive property of precedence relations, termed as *an indirect ordering.*

A planner can determine whether a conflict in the current plan network needs to be resolved immediately by analyzing the current RUF-structure: it defers conflict resolution unless the current RUF-structure can be serialized in terms of a casual relationship. This conflict resolution allows the system to improve the efficiency of planning by avoiding unnecessary conflict detection and resolution. Also, it allows the system to obtain an optimal plan network by avoiding redundant assignment of precedence relations among actions based on a local decision.

3.3 Procedure of Conflict Detection and Resolution

The following summarizes the procedure of the conflict detection/resolution:

Step 1. Resource level Conflict Detection : For each resource (object) allocated to a newly generated action, 1) update the corresponding RUF-structure by recording the usage of the resource in the corresponding entry of the RUFT. 2) For each updated RUF-structure, check whether it is a serial flow. If yes,

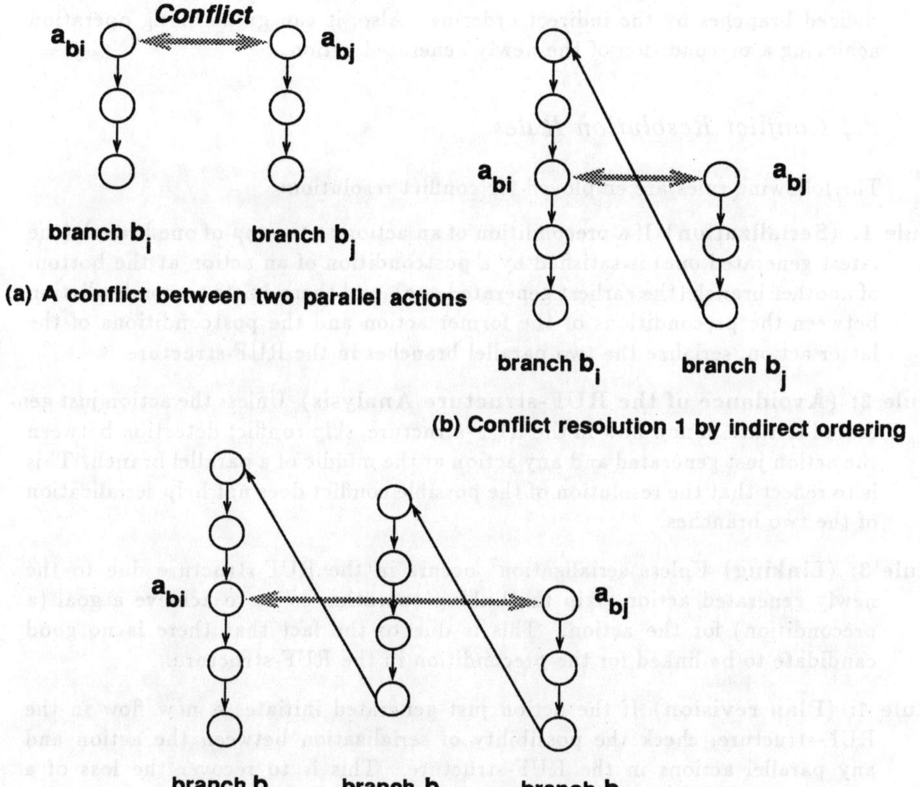

Figure 3. Conflict Resolution by Indirect Ordering

then skip Step 2 (no conflict exists).

Step 2. Global Conflict Resolution : For each updated RUF-structure, check whether the updated branch can be serialized with any of the parallel branches in terms of a causal relationship. If yes, resolve a conflict by merging the two branches into a larger one and update the RUF-structure accordingly. This conflict resolution may resolve some deferred conflicts associated with the serialized branches by the indirect ordering. Also, it can guide liking operation achieving a precondition of the newly generated action.

3.4 Conflict Resolution Rules

The following rules are employed for conflict resolution:

Rule 1: (Serialization) If a precondition of an action at the top of one branch (the latest generated one) is satisfied by a postcondition of an action at the bottom of another branch (the earliest generated one), and there exists no condradiction between the preconditions of the former action and the postconditions of the latter action, serialize the two parallel branches in the RUF-structure

Rule 2: (Avoidance of the RUF-structure Analysis) Unless the action just generated initiates a new flow in the RUF-structure, skip conflict detection between the action just generated and any action at the middle of a parallel branch. This is to reflect that the resolution of the possible conflict does not help serialization of the two branches.

Rule 3: (Linking) Unless serialization[7] occurs in the RUF-structure due to the newly generated action, skip the linking operation so as to achieve a goal (a precondition) for the action. This is due to the fact that there is no good candidate to be linked for the precondition in the RUF-structure.

Rule 4: (Plan revision) If the action just generated initiates a new flow in the RUF-structure, check the possibility of serialization between the action and any parallel actions in the RUF-structure. This is to recover the loss of a serialization which could be done if the action had been generated earlier. This allows not only the exploitation of helpful interactions[11] but also the resolution of goal deadlocks due to the goal misordering[10].

3.5 Example

The following example of a blocks world explains conflict detection based on the RUF-structure and conflict resolution based on a global decision.

---

[7]Note that all preconditions of the action are considered in the serialization procedure.

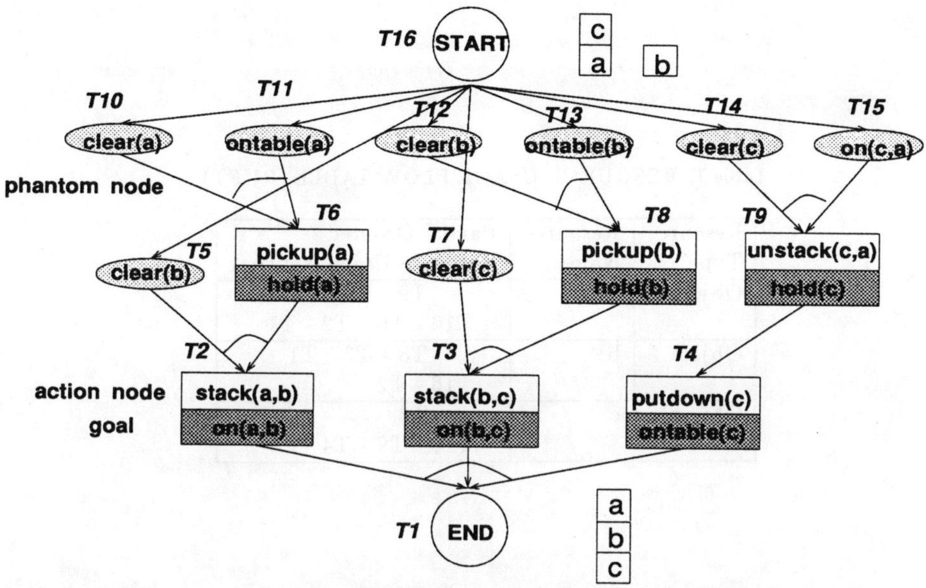

Figure 4. A Tentative Solution Plan Network (TSN)

Example 1:
As depicted in Fig.2, the initial state of the blocks world is given by {on(c,a), ontable(a), ontable(b), clear(b) and clear(c)}, and the goal state is given by {on(a,b), on(b,c), ontable(c) and clear(a)}. The following conventions in Fig.4 are understood: 1) a node number, T1 - T16, is marked on the upperside of each node 2) an action node and a phantom node[8] are indicated by a solid box and an oval, respectively. 3) an action node is composed of two sections: a goal in the lower compartment and an action (to achieve the corresponding goal) in the upper compartment.

Given the initial state and the goal state, the start node, the end node and the three goal nodes ON(a,b), ON(b,c) and ONTABLE(c) are formed in the plan network of Fig.4, and each resource-usage network in Table 1 of the RUFT has two nodes, T1 and T16. Suppose the goal on(a,b) is selected to be achieved. A new action, stack(a,b), should be introduced based on Table 2, and its preconditions become new goals (expansion). Since it can not be achieved by the effect of any existing node (linking fails), action node T2 is generated in the plan network. At this moment, the usages of resources "a" and "b" in the action node are recorded by putting T2 in the entries "a" and "b" of the RUFT such that T2 is precedent to T1 in the corresponding networks. If another goal on(b,c) is achieved by stack(b,c), T3, the usages of resources

[8]It is generated when the corresponding goal is achieved by linking to the start node or an already existing node.

Table 1. RESOURCE-USAGE FLOW TABLE (RUFT)

Resource Type	Resource Name	Partial Ordered Resource-Usage Network
Object	a	T9 T16 - T6 - T2 - T1
Object	b	T8 - T3 - T1 T16 - T2
Object	c	T3 T16 - T9 - T4 - T1

Table 2. ACTION TABLE

action	*precondition*	*postcondition*
stack(x,y)	hold(x)	on(x,y)
	clear(y)	clear(x)
unstack(x,y)	clear(x)	hold(x)
	on(x,y)	clear(y)
pickup(x)	clear(x)	hold(x)
	ontable(x)	
putdown(x)	hold(x)	clear(x)
		ontable(x)

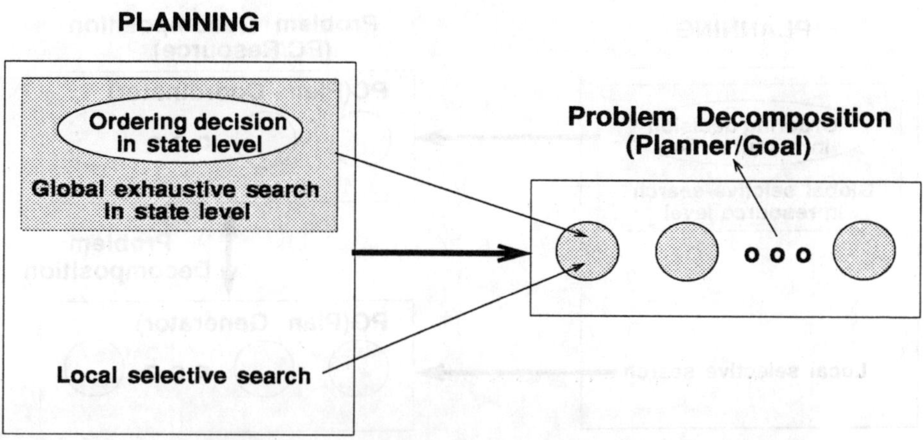

Figure 5. Problem Decomposition in Corkill's Distributed NOAH

"b" and "c" are recorded in a similar way by putting T3 in the corresponding entries such that T3 is precedent to T1 in the networks. In this case, a RUF-structure for "b" will have parallel branches consisting of (T2 - T1) and (T3 - T1), indicating the existence of a conflict in "b". Table 1 represents the RUFT formulated when the plan network shown in Fig. 4 is constructed. Note that only action nodes in parallel branches are considered in the conflict analysis/resolution.

Example 2:
Suppose nodes T4, T9, T14 and T15 have already been generated with START and END nodes in Fig. ?? and T2 is to be expanded to achieve the goal on(a,b). A conflict is detected based on the RUF-structure for "a", parallel branches (T2-T1) and (T9). However, conflict resolution is deferred, since T2 can not be serialized with T9. After T2 is expanded into T5 and T6, T6 is achieved by pickup(a) and (T6-T2-T1) can be serialized with (T9). This shows that the deferred conflict between (T2-T1) and (T9) is resolved based on indirect ordering and a precondition of T6 is achieved by a postcondition of T9.

4 Parallel Architecture

4.1 Problem Decomposition

The distributed NOAH[2] in Fig. 5 adopts a one-level problem decomposition, where the planning process is decomposed in terms of goals. Different from the distributed NOAH, we adopt a two-level problem decomposition in Fig. 6, where the planning process is decomposed into two subprocesses in terms of functionality: 1)

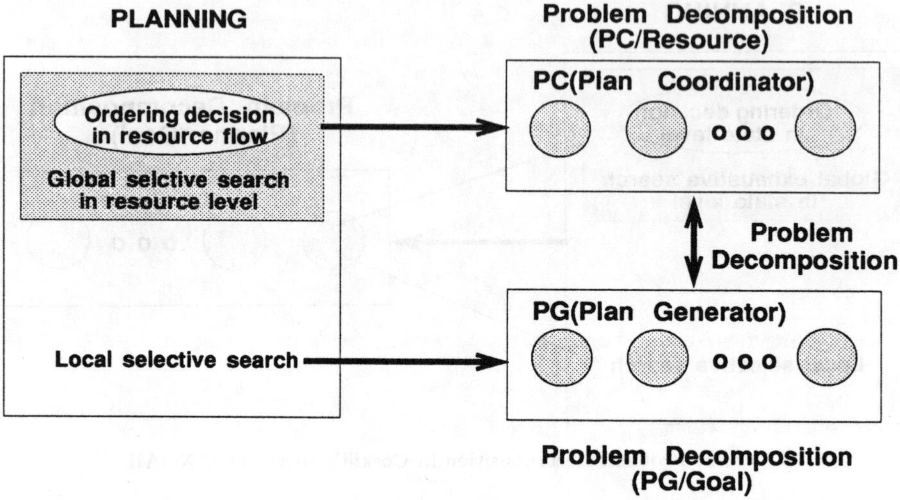

Figure 6. Proposed Problem Decomposition of Nonlinear Planning

Plan Generation, based on local selective search (Expansion) 2) *Plan Coordination* based on global exhaustive search and global decision for ordering actions (Linking and Conflict Detection-Resolution). Furthermore, *Plan Generation* can be decomposed in terms of high-level goals, and *Plan Coordination* in terms of resources. Note that *Plan Coordination* in planning is hardly decomposed unless abstracted at the resource level since it requires global search on the whole plan network.

The proposed decomposition of nonlinear planning allows the distribution of the knowledgebase (action table) and the database (world model and current plan network) during planning. Also, it promotes parallel and distributed processing; low-level parallelism (parallel search or parallel pattern matching) which is inherent in parallel planning can be further exploited based on the proposed scheme.

4.2 Architecture

Based on the above decomposition, we propose an architecture consisting of two modules: Plan Generator(PG) and Plan Coordinator(PC), as shown in Fig. 7.

Plan Generator

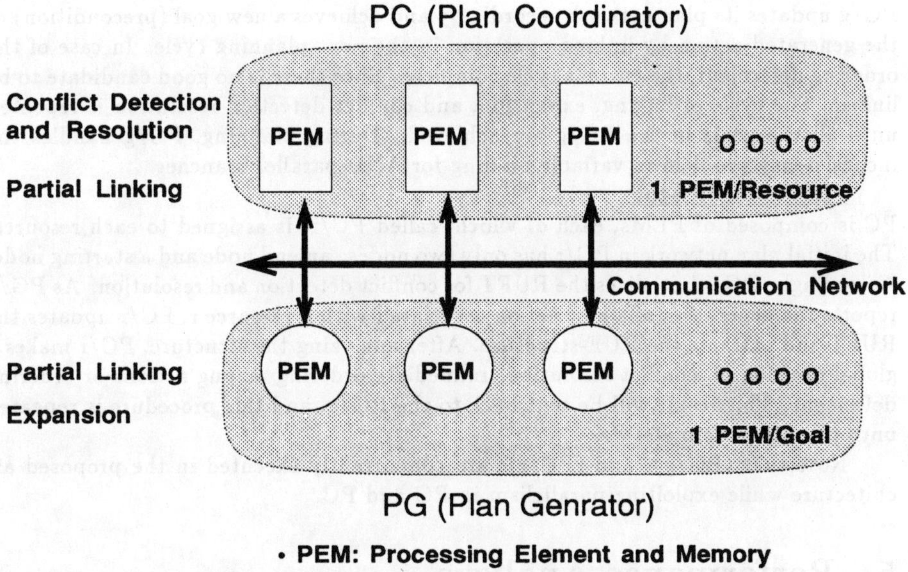

Figure 7. Proposed Parallel Architecture

PG is composed of PEMs (Processing Element and Memory), each of which, called PG/g, is assigned to each high-level goal. For a given goal, the initial plan network in PG/g has only two nodes: an end node and a starting node. When selecting the next goal to be achieved, PG/g checks whether the goal can be satisfied by the starting state (linking). If yes, the current cycle completes. Otherwise, PG/g searchs the action table for an action whose postcondition satisfies the goal (expansion) and generates a new action node in its plan network such that the preconditions of the action become new goals. The generation of the action is reported to some of the PC handling the resources associated with the action. Then, PG/g waits for the PC to send back their results of conflict detection and resolution, which is either an immediate ordering among actions or ordering deferment. In case of an immediate ordering, PG/g updates its plan network accordingly and achieves a new goal (precondition) of the generated action by linking operation in the next planning cycle. In case of the ordering deferment, the current cycle completes since there is no good candidate to be linked. The cycle of linking, expansion, and conflict detection-resolution is repeated until all the goals in the PG/g are achieved. During planning, PG/g handles the inconsistency problem in variable binding for AND parallel branches.

<u>Plan Coordinator</u>

PC is composed of PEMs, each of which, called PC/r, is assigned to each resource. The initial plan network in PC/r has only two nodes: an end node and a starting node. Accordingly, PC/r initializes the RUFT for conflict detection and resolution. As PG/g reports the generation of a new action associated with a resource r, PC/r updates the RUFT and extracts the RUF-structure. After analyzing the structure, PC/r makes a global decision on conflict resolution: immediate ordering among actions or ordering deferment. This result will be sent back to the PG/g, and this procedure is repeated until planning completes.

Note that (PG/g)s and (PC/r)s are concurrently executed in the proposed architecture while exploiting parallelism in PG and PC.

5 Performance Analysis

Suppose a planning task with conjunctive goals is given in Corkill's scheme[2] and the proposed scheme. Assuming that the best choice is selected in each planning operation, let us compare the performance of the two.

Let k denote the number of the conjunctive goals,
n denote the number of the resource types, and
α denote the average number of resources involved in an operator.

Assuming that k processors are available and each goal is assigned to a planner (processor) called (PNR/g), let us compute the execution time of Corkill's scheme. The total number of action nodes in the planners (PNR) is N_a. If we assume that the action nodes are equally distributed across PNR/$g_{g=1,\ldots,k}$, the average number of action nodes in PNR/g becomes N_a/k. While action nodes of N_a/k are generated in PNR/g, they are propagated to all the other subplanners for $\text{PNR/g}_{g=1,\dots k,\ but\ i \neq j}$. Assuming intercommunication between (PNR/g)s takes one unit time, communication overhead due to the propagation is as follows:

$$T_{Communication} = (N_a/k \times (k-1) \times k) \\ = O(k \times N_a)$$

The total execution time of each operation in PNR/g is as follows[9]: $T_{expansion} = O(N_a/k \times N_t)$, and $T_{conflict}$[10] $= O((N_a/k) \times D(N_a))$ where $D(N_a)$ denotes the average depth of the plan network of N_a nodes. As (PNR/g)s operate in parallel while communicating with each other, the total execution time is as follows: $T_{Corkill} \approx O((N_a/k)\log N_a)$.

Assuming that $(k+n)$ processors are available for (PG/g)s and (PC/r)s, let us compute the execution time of the proposed architecture. Plan expansion operation in PG/g takes the same amount of time as that of Corkill's scheme, *i.e.* $T_{expansion} = O(N_a/k \times N_t)$. However, conflict detection and resolution including communication takes different amounts of time. Similar to the Corkill's scheme, the total number of action nodes in PG (plan generator) is N_a. If we assume that the action nodes are equally distributed across $\text{PG/g}_{g=1,\dots,k}$, the average number of action nodes in PG/g becomes N_a/k. While action nodes of N_a/k are generated in PG/g, they are propagated to $\text{PC/r}_{r=1,\dots,\alpha,}$. Assuming intercommunication between PC/r and PG/g takes one unit time, communication overhead due to the propagation is as follows:

$$T_{Communication} = (N_a/k \times \alpha \times k) \\ = O(\alpha \times N_a)$$

The total execution time of operation in PC/r and PG/g is as follows: $T_{PG/g} = O(N_a/k \times N_t)$, and $T_{PC/r}$[11]$= O(\alpha \times N_a/n)$.
As (PC/r)s and (PG/g)s operate in parallel while communicating with each other, the total execution time of the proposed architecture is as follows:
$T_{proposed} = O(N_a/k + \alpha \times N_a/n) \approx O(N_a/k)$ for $\alpha \leq k \approx n \ll N_a$.

Note that the planning speed of the proposed architecture is $O(N_a)$ while that of Corkill's scheme is $O((N_a/k) \times D(N_a))$. The speed-up is achieved mainly due to the avoidance of unnecessary conflict detection based on the RUF-structure, and partly due to the parallel and distributed processing based on the proposed problem partitions. Also, note that the proposed architecture allows limited communication between the PG and the PC due to the decomposition of PC on resource basis.

---

[9] It can be calculated in the same way as in APPENDIX.

[10] This includes the elimination of redundant precondition in NOAH, which is performed after plan expansion. However, in the planning procedure of section 2.1, the corresponding operation is performed before plan expansion under the name of linking.

[11] Conflict detection is performed by comparing two ends of parallel branches in RUFT. Note that this corresponds to a branch level processing while NOAH (NONLIN) corresponds to an action (goal-structure) level processing.

The performance improvement due to the global decision in conflict resolution is not considered in the above comparison. The possible speed up due to the global decision can be best evaluated by simulation.

6 Conclusion

The advantages of the proposed scheme can be stated as follows:

- Conflict detection and resolution based on resource reasoning allows a global decision on conflict resolution, which leads to the avoidance of unnecessary planning operation as well as the assignment of redundant precedence relations among actions in a plan network.
- The two-level problem decomposition allows the exploitation of parallelism in the plan coordinator and the plan generator. Furthermore, it promotes parallel and distributed processing with a distributed knowledgebase/database, *i.e.* low-level parallelism inherent in parallel planning can be further exploited based on the proposed architecture.
- The decomposition of the plan coordinator on a resource-basis allows communication among the plan generator and the plan coordinator to be reduced.

The limitation of the proposed scheme can be stated as follows:

- A load balancing problem may occur in the plan coordinator when part of the resources are heavily involved in a plan, or in plan generator when the variance of the lengths of subplans is large.

References

1. D. Chapman, Planning for Conjunctive Goals, *Artificial Intelligence*,**32** (1987), p.333–377.
2. D. D. Corkill, Hierarchical Planning in a Distributed Environment, *Proceedings of the International Joint Conference on Artificial Intelligence*, (1979), p.168–175.
3. R. E. Fikes and N. J. Nilsson, A New Approach to the Application of Theorem Proving to Problem Solving, *Artificial Intelligence*, **3**, (1971), p.251–268.
4. A. L. Lansky, A Representation of Parallel Activity Based on Events, Structure and Causality, *Technical Report 401,SRI International*, (1986).
5. E. D. Sacerdoti, *A Structure for Plans and Behavior*, (American Elsevier, 1977).
6. G. J. Sussman, *A Computational Model of Skill Acquisition*, (American Elsevier, 1975).

7. A. Tate, Interacting Goals and their Use, *Proceedings of the International Joint Conference on Artificial Intelligence*, (1975), p.215–218.
8. A. Tate, Generating Project Networks, *Proceedings of the International Joint Conference on Artificial Intelligence*, (1977), p.888–893.
9. S. A. Vere, Planning in Time: Windows and Patterns for Activities and Goals, *IEEE Transaction on Pattern Analysis and Machine Intelligence*, **5**, (1983), p.246–247.
10. S. A. Vere, Splicing Plans to Achieve Misordered Goals, *Proceedings of the International Joint Conference on Artificial Intelligence*, (1985), p.1016–1021.
11. D. E. Wilkins, Domain-independent Planning: Representation and Plan Generation, *Artificial Intelligence*, **22**, (1984), p.269–301.

APPENDIX: The Time Complexity in Parallel Planning

The total execution time of parallel planning in a uniprocessor environment (shown in section 2.1) can be calculated as follows:

$$
\begin{aligned}
T_{PP} &= \sum_{i=1}^{N_c} \{t_{expansion(i)} + t_{linking(i)} + t_{conflict(i)}\} \\
&= T_{expansion} + T_{linking} + T_{conflict}
\end{aligned}
$$

where N_c is the number of cycles required for the completion of planning.

The above equation can be further elaborated using the following assumptions:

- No backtracking occurs, *i.e.* the best choice is made in every decision.
- Conflicts between two actions are detected by comparing their effects based on the data structure such as TOME[5] (or GOST[8]).
- A hashing table is used to search an entry of TOME (or GOST).
- Any conflict between two actions in parallel branches is resolved immediately by assigning a direct precedence to the conflicting actions, *i.e.* by promotion in TWEAK[1]. This implies that all conflicts can be resolved without white knight in TWEAK[1]. Note that the time for conflict resolution is less than the time for conflict detection.

Let N_t be the number of action types,
N_a be the number of action nodes generated when planning is completed,
N_{pa} be the average number of preconditions in an action,
N_{sa} be the average number of postconditions in an action, and
N_{cond} be the total number of instantiated condition (predicate) types.

$$
\begin{aligned}
T_{expansion} &= \sum_{j=1}^{N_a} N_t \times (N_{sa}/2) \\
&= 1/2 \times N_a \times N_t \times N_{sa} \\
&= O(N_a \times N_t \times N_{sa})
\end{aligned}
$$

where j, N_t and $(N_{sa}/2)$ respectively indicate the j-th expansion, the number of comparison between the current goal and the postconditions of actions required to find out all possible actions from the action table, and the average number of condition matchings between the current goal and the postconditions of an action.

For linking, an entry of TOME for a condition satisfying the current goal is found using a hashing table and each candidate in the entry is checked to discover whether the possible link with each candidate results in a cycle in the plan network:

$$
\begin{aligned}
T_{linking} &= \sum_{i=1}^{N_c} N_{avc}(i) \times t_{cycle}(i) \\
&= \sum_{i=1}^{N_c} \{(N_a/N_c) \times i \times (N_{sa} + N_{pa})/N_{cond}\} \times D\{(N_a/N_c) \times i\} \\
&\approx N_c \times \{N_a \times (N_{sa} + N_{pa})/N_{cond}\} \times D\{(N_a/N_c) \times i\} \\
&= O(N_c \times N_{avc}(N_a) \times D\{N_a\})
\end{aligned}
$$

where $N_{avc}(N_a)$ denotes $\{N_a \times (N_{sa} + N_{pa})/N_{cond}\}$.

where i, $N_{avc}(i)$, $t_{cycle}(i)$, $(N_a/N_c \times i)$, $D\{N_a\}$ respectively indicate the i-th cycle, the average number of conditions in an entry of TOME at cycle i, the time to check whether the possible link between the two nodes results in a cycle in the plan network (a directed, acyclic graph), the number of action nodes at cycle i assuming the number increases linearly to i, and the depth of the plan network of N_a nodes. $D(N_a)$ varies depending on the number of initial goal states and the (domain-dependent) expansion pattern of the plan network. For example, if the plan network is expanded to be a balanced tree-like acyclic graph, $D(N_a)$ becomes $\log N_a$. Also, the ratio of $N_{avc}(N_a)$ and N_a, *i.e.* the rate of the occurrence of conflicts during planning, varies depending on the application domain.

$$
\begin{aligned}
T_{conflict} &= \textstyle\sum_{j=1}^{N_a}\{j \times (N_{sa} + N_{pa})/N_{cond}\} \times t_{parallel}(i) \times \\
&\quad \{(N_{pa} \times N_{pa}) + (N_{pa} \times N_{sa}) + (N_{sa} \times N_{pa}) + (N_{sa} \times N_{sa})\} \\
&\approx N_a \times \{N_a \times (N_{sa} + N_{pa})/N_{cond}\} \times D\{N_a\} \times (N_{pa} + N_{sa})^2 \\
&\approx \mathrm{O}((N_a) \times N_{avc}(N_a) \times D\{N_a\} \times (N_{pa})^2)
\end{aligned}
$$

where j, $\{j \times (N_{sa} + N_{pa})/N_{cond}\}$, $t_{parallel}(i)$, $(N_{pa} + N_{sa})^2$ respectively indicate the j-th expansion, the average number of conditions in an entry of TOME at j-th conflict detection-resolution, the time[12] to check whether the node to which a candidate condition belongs is in parallel with current node in the plan network, and the number of cross comparison between pre and post conditions of the two action nodes.

Combining the above equations for $T_{expansion}$, $T_{linking}$, $T_{Conflict}$, we have

$$
\begin{aligned}
T_{PP} &\approx \mathrm{O}(N_a \times N_t \times N_{sa}) + \mathrm{O}(N_{avc}(N_a) \times N_c \times D\{N_a\}) + \\
&\quad \mathrm{O}(N_{avc}(N_a) \times (N_{pa})^2 \times N_a \times D\{N_a\}) \\
&\approx \mathrm{O}(N_a \times D\{N_a\}) \\
&\quad \text{for } N_{sa}, N_{pa} \leq N_t, N_{avc}(N_a) \ll N_a \approx N_c
\end{aligned}
$$

T_{PP}[13] indicates that conflict detection and resolution dominate the time complexity in parallel planning, although $T_{conflict}$ does not include the time required to resolve complicated conflicts such as action-deadlocks. Also, note that the actual $T_{expansion}$ and $T_{linking}$ can be less than those calculated above since exhaustive search need not be used if a good heuristic is available.

---

[12] It is same as $t_{cycle}(i)$

[13] $T_{PP} = \mathrm{O}(N_a \times N_a)$ if such a data structure as TOME or GOST is not used. $T_{PP} = \mathrm{O}((N_a)^3)$ if declobbering by white knigt[1] is included.

Applications of Learning and Planning Methods
edited by N. G. Bourbakis

CHAPTER 9

Heuristic Tree Search Using Nonparametric Statistical Inference Methods

Weixiong Zhang † *Nageswara S.V. Rao*
Department of Computer Science
Old Dominion University
Norfolk, VA 23529, USA

ABSTRACT

The computational complexity of widely-used heuristic search algorithm, A^*, could be exponential in the length of the optimal solution path. In several interesting cases, for example in a tree model, we can improve the average-case performance of the search by making use of some global behavior of a suitable statistic derived from the heuristic estimate. These methods assume that the probability distribution of the underlying statistic and also its parameters are *precisely known*. However, such precise knowledge may not be available for some practical situations. In this paper, we propose an approach based on non-parametric statistical inference methods that do not require such precise information about the distribution. We develop search algorithms, *NSA* and *Successive NSA*, and analyze their performance using a uniform *m*-ary search tree *G* with a single goal located at an unknown location at depth *N*. *Successive NSA* has the mean complexity of $O(N(\log N)^2)$, in terms of the average number of nodes expanded when the probability with which the maximum deviation of the heuristic from the actual value exceeds a specified constant, is bounded by another specified constant. The storage required is $O((\log N)^2)$ and the average number of comparisons is $O(N(\log N)^6 \log\log N)$.

1. Introduction

Search is a basic paradigm employed in solving a number of problems in artificial intelligence [3, 5, 7]. Problems, such as puzzles, games, planning, *etc.*, could be naturally cast as search problems. Typically, in many AI problems, the problem space could

† Present address: Computer Science Department, University of California, Los Angeles, CA 90024, USA. This author was partially supported by a fellowship from the College of Sciences, Old Dominion University, Norfolk, VA 23529, USA.

be suitably abstracted as a search graph, where the nodes denote the states and the arcs denote the transitions between the states. A solution to the problem involves finding a path from a start node to a goal node. This process is carried out by expanding the nodes of the search graph. In order to ensure reasonable average run-times, the number of nodes expanded should be kept as low as possible, and heuristics could be employed to achieve this objective.

1.1. *Heuristic Search and Computational Complexity*

A^* search algorithm [1] explores the optimal solution path by iteratively expanding a frontier node n with a minimal value of an evaluation function, $f(n)=g(n)+h(n)$, in the partial search graph, where $g(n)$ is the cost of a minimal cost path from the start node to n, and $h(n)$ is a heuristic estimate of a minimal cost path from n to a goal node.

Pearl [6, 7] and Huyn *et al.* [2] obtained the average computational complexity of A^* using the uniform m-ary tree [5, 7]. They considered a uniform m-ary tree G which contains a unique goal node S_N at an unknown location at depth N. The heuristic estimate, $h(n)$, is assumed to be an *independently and identically distributed* (*i.i.d.*) random variable on $[0, h^*(n)]$, characterized by the distribution $F_{h(n)}(x)=P[h(n)<x]$, where $h^*(n)$ is the cost of a minimal cost path from n to the goal node. The average computational complexity, in terms of the number of nodes expanded by A^*, is $O(e^{cN})$, $c>0$, if the estimate $h(n)$ satisfies the condition

$$P\left[\left|\frac{h^*(n)-h(n)}{\phi(h^*(n))}\right|>\varepsilon\right]>\frac{\alpha}{m}$$

for two constants $\varepsilon>0$, $\alpha>1$, where $\phi(x)$ is a normalizing function. In this case, therefore, A^* does not seem to completely avoid the "combinatorial explosion".

1.2. *Parametric Statistical Heuristic Search*

Zhang and Zhang [11-14] treated the heuristic search as a random sampling process, and introduced the statistical inference techniques into A^* search in obtaining a new search algorithm called the *SA algorithm*. For a node n in a uniform m-ary tree G, *SA* search uses a sequence of statistic $a_k(n)$ to characterize the subtree rooted at n, and assumes that (hypothesis H_0:) $\mu(n)=\mu_0$ when n is on L, the solution path, and (hypothesis H_1:) $\mu(n)=\mu_1>\mu_0$ when n is not on L, where $\mu(n)$ is the mean of $a_k(n)$. Under these assumptions, *SA* recursively explores the subtrees rooted at the children of a node (present state the search agent locates) to extract the statistic which describe these subtrees, and make choices to go to next among these children by using statistical inference. If the hypothesis H_0 is accepted on the child node n_1 of the node n, then the other children nodes are pruned off and the search agent uses n_1 as the present state. This procedure continues until the goal node is used as the present state.

Conceptually, *SA* extracts a statistic $a_k(n)$ from the heuristic information contained in the subtree rooted at n. For a node n in G, let $T_k(n)$ be an expanded tree rooted at n containing k nodes. Let $a_k(n)=\Phi(f(p) \mid p \in T_k(n))$, where Φ is a combination function of $f(p)$. *SA* assumes that $\{a_k(n)\}$ is an *i.i.d.* random variable. In addition, *SA* assumes that all statistic are subjected to normal distributions with known parameters, *i.e.* for a node n on L, the statistic $a(n)$ is subjected to the normal distribution with mean μ_0, otherwise, it is subject to the normal distribution with mean μ_1. By resorting to the sequential probability ratio test (*SPRT*) [8, 10], *SA* can find the goal with probability one with a mean computational complexity of $O(N(\ln N)^2)$.

The major strength of *SA* is the ability of utilizing "global" statistical information in achieving $O(N(\ln N)^2)$ mean complexity. This is significant because the conventional A^* algorithms do not exploit such information, and *SA* guarantees better average performance, in cases where such information is available.

1.3. *Motivation and Contribution*

The performance of *SA* critically hinges upon the fact that the distribution *and* the parameters of the statistic are both *known*. It might be reasonable, in large search problems, to assume that the statistic is subjected to a normal distribution. But, we may not be able to assume that the parameters of the distribution, such as the mean, are *precisely* known. In order to account for the unknown parameters, *SA* uses estimates of these parameters obtained during the search. For example, *SA* uses a minimal sample mean as an estimate of μ_0, and the average of the remaining sample means as an estimate of μ_1 [11]. *SA* assumes that when H_0 (or H_1) is true and accepted (by the test), the solution path is not pruned off. This is not always true if we do not know the precise values for μ_0 and μ_1. As remedies to the difficulties causing by the unknown parameters, *u*-test and *t*-test are suggested to replace *SPRT* [14]. However, by using *u*-test, we assume that the underlying distributions are normal and their variances are known. By using *t*-test, we also assume that the underlying distributions are *t* distributions and their variances are equal [10].

In practice, we usually may not have precise information about the features of the statistic. such as the form of its distribution, its mean and variance. Thus the statistical heuristic search methods that do not rigidly fix the distribution of the statistic (and also do not assume the precise knowledge of the parameters) are more suitable for practical situations. This naturally leads us to the idea of introducing nonparametric statistical inference methods to heuristic search techniques.

However, this idea raises two important questions: (a) What kind of nonparametric tests are to be employed by the algorithm ? (b) What specific nonparametric statistic should be used so that the average complexity of the search can be guaranteed to be approximately equal to that of the parametric statistical search ? In this paper, we discuss these two problems, and develop the algorithm *Successive NSA* based on nonparametric statistical inference methods. In our model we assume that statistic $a(n)$ is subjected to a distribution F, when n is on L, and is subjected to a distribution H, when n is not on L such that $F(x) \geq H(x)$, for all real x. Here F and H need not be known. Our method, being nonparametric, eliminates the errors introduced because of using estimates for the parameters of the distribution. Thus we circumvent the problems encountered by *SA*.

In the analysis of the complexity of *SA* search [11-14], Zhang and Zhang did not take into account the precision of the heuristic estimate. They claimed that the probability of finding the solution with the mean complexity of $O(N(\ln N)^2)$ can be *arbitrarily* fixed by choosing a suitable confidence level (α, β) of a statistical inference, *irrespective* of the heuristic estimate. Their result is counter-intuitive because a precise heuristic (that yields values closer to the actual costs) is more likely to find the solution faster than an imprecise one, even if both yield the same expected value for the statistic. Moreover, the hypothesis H_0 could be accepted when the test is applied to the subtree of minimal estimated mean. Since no restrictions are placed on the heuristic function, any partially expanded subtree can have this minimum mean. On this subtree, which may or may not contain the solution path, H_0 could be accepted with non-zero probability. This probability is not accounted for in computing the probability of pruning the solution path. thus the expression for b (the probability of pruning the solution path) in [11] is probably incorrect.

In the analysis of *NSA*, we also consider the effect of the precision of heuristic estimate on the computational complexity. Our result is that *Successive NSA* finds the solution path with $O(N(\log N)^2)$ mean complexity when the probability with which the maximum deviation of the heuristic from the actual value exceeds a specified constant, is bounded by another specified constant. We achieve this by deriving a new statistic, L_n^*, whose specific properties are tailor-made for our algorithm.

1.4. *Organization of the Paper*

In Section 2, we discuss the search model and present the search algorithm. In Section 3, we introduce a nonparametric statistic L_n^* and a sequential two-sample test based on L_n^*. The computational complexities are derived in Section 4. The conclusions and discussions appear in Section 5. We assume that the readers are familiar with the basic concepts of the parametric and nonparametric statistical inferences [8, 10].

2. Nonparametric Statistical Heuristic Search

2.1. *Search Model*

We consider a uniform *m*-ary tree G (as in [2, 6, 7, 11-14]). G has an initial node S_0 (root), a unique goal node S_N at depth N. Let $L=(S_0,..., S_i,..., S_N)$ be the shortest path from S_0 to S_N. This search model is shown in Fig. 1 at the end of the article. Let $T(n)$ denote the subtree rooted at node n, and $T_k(n)$ denote a subtree (also rooted at n) of $T(n)$ containing k nodes. If node n is located at depth i, we say that $T(n)$ is an i-type subtree, and denote it by $T^i(n)$. We now consider the statistic $a(n)$, that characterizes $T(n)$, such that

$$a(n)=\Psi(f^*(n)), \tag{2.1}$$

where Ψ is a non-decreasing combination function and $f^*(n)$ denotes the exact cost of the solution path constrained to go through n [5]. Consider a node $n \in T^i(n')$, at depth d, $(d>i)$

(1) if n is on the solution path L, then (refer to Figure 1)

$$g^*(n)=d; \quad h^*(n)=N-d; \quad f^*(n)=N;$$

(2) if n is not on the solution path L, then

$$g^*(n)=d; \quad h^*(n)=N+d-2i; \quad f^*(n)=N+2(d-i).$$

Thus we have

$$f^*(n_i) < f^*(n_j); \quad \text{for } n_i \in L, \ n_j \notin L.$$

If Ψ is a nondecreasing function, i.e., $\Psi(x) \ge \Psi(y)$, for $x>y$, we arrive at, for any real t,

$$P[a(n_j)>t] \ge P[a(n_i)>t], \text{ denoted by } F_{a(n_i)} \ge F_{a(n_j)}, \quad \text{for } n_i \in L, \ n_j \notin L. \tag{2.2}$$

Above inequalities mean that $a(n_j)$ is *stochastically greater* [7, pp.176] than $a(n_i)$. We now specify our search model as follows:

Search Model Hypothesis: *If $n \in L$, the statistic $a(n)$ is subjected to an unknown distribution F_1. Otherwise, i.e. $n \notin L$, $a(n)$ is subjected to another independent and unknown distribution F_2, such that $F_1 \geq F_2$. Furthermore, for two nodes n_1 and n_2 in G, if n_1 is neither an ancestor nor a descendant of n_2, then $a(n_1)$ and $a(n_2)$ are independent random variables.*

Note that under this hypothesis, F_1 and F_2 do not have to be any particular distributions. Unlike in [11], F_1 and F_2 do not even have to be of the same type. In general, for two nodes n_1 and n_2 (n_1 is neither an ancestor nor a descendant of n_2), it is not required that $a(n_1)$ and $a(n_2)$ should be subjected to the same type of distribution. Further, we do not presuppose any knowledge about the parameters of these distributions.

For any two nodes n_1 and n_2 in G, if the distribution of $a(n_1)$ and $a(n_2)$ are $F_{a(n_1)}$ and $F_{a(n_2)}$ respectively. If $F_{a(n_1)}$ and $F_{a(n_2)}$ are not identical, denoted by $F_{a(n_1)} \neq F_{a(n_2)}$, then the subtrees $T(n_1)$ and $T(n_2)$ are *probabilistically distinguishable*. Otherwise, $T(n_1)$ and $T(n_2)$ are *probabilistically indistinguishable*.

When the node n is expanded, $a(n)$ is in general not known. So we look deeper into the subtree $T(n)$, and obtain a set of estimated values for $a(n)$. These values are called the *observations* on $a(n)$, and are denoted by $a_k(n)$, $k=1,2,....$ Note that our definition of $a_k(n)$ is in general different from that of [11-14]. We specify two ways of obtaining the observation of $a(n)$, for any node n.

Sampling Principle 1 (depth-first-least-value sampling): *In subtree $T(n)$, take the first sample $a_1(n)=f(n)$. Expand n and take the second sample $a_2(n)=f(n')=\min\{f(n_k) \mid n_k$ is a son of $n\}$. Expand n' next. Similarly, obtain $a_3(n)$, and so on.*

Sampling Principle 2 (breadth-first-least-value sampling): *Expand the subtree $T(n)$ with A search. From all expanded nodes in depth i, let $a_i(n)=\min\{f(n_k) \mid n_k$ is an expanded node in depth i of subtree $T(n)\}$.*

Let $\hat{\mu}(n)$ the mean of the observations $a_k(n)$, $k=1, 2,....$ Under our search model and the above sample principles, we note that

$$P[\hat{\mu}(n_i) \leq \hat{\mu}(n_j)] \geq P[\hat{\mu}(n_i) > \hat{\mu}(n_j)]; \quad n_i \in L, \quad n_j \notin L. \tag{2.3}$$

With no loss of generality, assume that for nodes $n_1, n_2, ..., n_m$ which have the same parent, we have

$$\hat{\mu}(n_1) \leq \hat{\mu}(n_2) \leq \ldots \leq \hat{\mu}(n_m). \tag{2.4}$$

Applying the sequential two-sample test (see next section) with the statistic L_n^* to $Z_j=(a(n_1), a(n_j)), j=2,\ldots,m$, we test the following **statistical hypothesis**

$$\begin{cases} H_0: a(n_1) \text{ and } a(n_j) \text{ are } \textit{probabilistically indistinguishable}, \text{ i.e., } F_{a(n_1)}=F_{a(n_j)}. \\ H_1: a(n_j) \text{ is } \textit{stochastically greater} \text{ than } a(n_1), \text{ i.e., } F_{a(n_1)} \geq F_{a(n_j)}. \end{cases} \tag{2.5}$$

2.2. *NSA Algorithm*

The procedure of algorithm *NSA* is basically the same as that of algorithm *SA*, except that *NSA* uses nonparametric statistical inference methods. We list the algorithm *NSA* as follows:

Algorithm NSA(S_0):

begin

1. $Id \leftarrow 1$; expand S_0 and obtain m 1-type subtrees given by $U \leftarrow \{T^1(n_j) \mid j=1,\ldots,m\}$;
2. **while** ($Id<N$) **do**
   Begin
3. obtain the next sample for each subtree $T^{Id}(n_j) \in U$ by a sampling principle;
4. **if** (a new expanded node is S_N) **then**
5. report the path from S_0 to S_N and exit successfully;
   endif;
6. **for** (every $a(n_j)$ such that statistical hypothesis H_1 is accepted on $(a(n_1), a(n_j))$ or the number of samples on $(a(n_1), a(n_j))$ is d_{Id}) **do**
7. $U \leftarrow U - \{T^{Id}(n_j)\}$;
   endfor;
8. **if** (there is only one subtree $T^{Id}(n_k)$ in U) **then**
9. $Id \leftarrow Id+1$;
10. $U \leftarrow \{T^{Id}(n_i) \mid n_i \text{ is a son of } n_k\}$;
   endif;
   endwhile
11. exit unsuccessfully;

end

Informally, *NSA* progresses level by level in the search tree. At each level *Id*, there are m subtrees of type $T^{Id}(.)$, and *NSA* chooses one of them (line 8). In the next step, the children of this node are considered. This process is continued until either the goal is found (lines 4-5) or we reach the depth $Id=N$ (line 11). The process of choosing one of subtrees at level Id is carried out as follows: We perform the test of (2.5) on each $(a(n_1),a(n_j))$ and obtain the sample size of $(a(n_1),a(n_j))$ (line 6). We delete $T(n_j)$ (line 7) if H_1 is accepted or the sample size surpasses d_{Id}, where, d_{Id} is a finite number whose the exact value will be derived in the Section 4. We terminate the sampling process after we have taken d_{Id} samples at level Id (line 6), If all but one subtree at this level are deleted, then we go to the next level (lines 8-10). If more than one subtrees remain, then we obtain next sample for each of the remaining subtrees (line 3).

The *NSA* algorithm always terminates in finite time, but may or may not find the goal. It is possible that the subtree containing the goal has been pruned off during the search. In order to guarantee that the goal will be found, we apply the *NSA* algorithm on the parts of search tree G that have been pruned off earlier. We call this algorithm *Successive NSA*, along the lines of [11].

3. Nonparametric Statistical Inference: Statistic L_n^*

The performance of the search algorithm presented in the previous section crucially depends on the properties of nonparametric statistical inference methods employed in the search. In this section, we discuss a two-sample nonparametric statistical test and its properties. We then derive a special statistic L_n^* that is employed in our search algorithm. We show in the next section that the employment of L_n^* guarantees that the average search cost is $O(N(\log N)^2)$.

Let $Z=(X, Y)$, where X and Y are independent random variables with unknown distribution functions F_X and F_Y, respectively. The hypothesis

$$H_0: F_X=F_Y; \qquad H_1: F_X \geq F_Y \tag{3.1}$$

is to be tested sequentially using a sequence of independently and identically distributed observations $Z_j=(X_j, Y_j)$, $j=1,2,\ldots$, on Z. Under the hypothesis H_0, X and Y will have the same distribution. While under the hypothesis H_1, Y is *stochastically greater* [7, pp.176] than X, i.e. $F_X \geq F_Y$. At the n-th stage, we have a sample $(X_1,\ldots, X_n)$ from F_X and an independent sample $(Y_1,\ldots, Y_n)$ from F_Y. Let $V_1,\ldots, V_n$ be the ranks of $Y_1,\ldots, Y_n$ respectively in the sorted version of the combined sample $(X_1,\ldots, X_n, Y_1,\ldots, Y_n)$. Now we propose a statistic given by

$$L_n^* = \sum_{i=1}^{n} \left[\frac{2}{2n+1} V_i - \tau \right], \quad 1<\tau<\frac{3}{2}. \tag{3.2}$$

We employ this statistic in a two-sample test [9] to test the hypothesis (3.1) as follows. For a fixed significance level (α, β), α>0, β>0, α+β<1, let

$$A=\frac{1-\beta}{\alpha};\quad B=\frac{\beta}{1-\alpha}. \tag{3.3}$$

Consider a sequential two-sample test based on the statistic L_n^*, with the following stopping rules:

$$\begin{cases} \text{Hypothesis } H_0 \text{ is accepted, if } L_n^*\le\log B\,; \\ \text{Hypothesis } H_1 \text{ is accepted, if } L_n^*\ge\log A\,; \\ \text{Otherwise, obtain the next observation } Z_{n+1}. \end{cases} \tag{3.4}$$

following the terminology of [9], a *stopping time* is the random sample size R when either H_0 or H_1 is accepted. The type *I* (*II*) error probability is the probability that H_0 (H_1) is rejected when H_0 (H_1) is true. We estimate the error probabilities and the bound on the stopping time of this test in the following Lemma.

Lemma 3.1. *The statistic L_n^* is a limiting log probability ratio. Using the above sequential two-sample test with stopping rules in (3.4), the type I and type II error probabilities are respectively given by*

$$P_1=Prob_{H_0}\left[L_n^*\ge\log A\right]<\alpha;\quad P_2=Prob_{H_1}\left[L_n^*\le\log B\right]<\beta. \tag{3.5}$$

The stopping time R of the sequential test is exponentially bounded, i.e., for all $0<\rho<1$, *there exists $k>0$, such that*

$$Prob\,[R>n]\le k\rho^n=ke^{-cn},\quad c=-ln\,\rho>0,\ n=1,2,... \tag{3.6}$$

Proof. We first consider a different test and then relate its treatment to the test of (3.4). Using a sequence of *i.i.d.* observations $Z_j=(X_j,Y_j)$, $j=1,2,...,$ on Z, we test

$$H_0\colon F_X=F_Y;\qquad \hat{H}_1\colon F_Y=F_{X,\Delta} \tag{3.7}$$

where $F_{X,\Delta}=F_X+\Delta\left[F_X^2-\tau F_X\right]$ and Δ is a small constant. If F_X is absolutely continuous with density f, then $F_{X,\Delta}$ is also absolutely continuous, with density $g=f\left[1+\Delta(2F_X-\tau)\right]$, such that

$$\frac{g}{f}=1+\Delta\left[2F_X-\tau\right]. \tag{3.8}$$

Let $p_\Delta(v_1,...,v_n)$ be the probability that $V_j=v_j$, $j=1,2,...,n$, under the alternative $F_Y=F_{X,\Delta}$. At the n-th stage, the probability ratio $R_n(\Delta)=\frac{p_\Delta(v_1,...,v_n)}{p_0(v_1,...,v_n)}$, can be evaluated as

$$R_n(\Delta)=E\left[\prod_{j=1}^{n}\frac{g\,(U^{(v_j)})}{f\,(U^{(v_j)})}\right]. \tag{3.9}$$

in which $U^{(1)},...,U^{(2n)}$ is the order statistic of a sample $U_1,...,U_{2n}$ from F_X, and the expectation is to be taken under F_X. Using $E\left[F_X(U^{(i)})\right]=\frac{i}{2n+1}$ and substituting (3.8) into (3.9), we have

$$R_n(\Delta)=E\left[\prod_{j=1}^{n}\left\{1+\Delta(2F_X(U^{(v_j)})-\tau)\right\}\right]=E\left[1+\Delta\sum_{j=1}^{n}\left\{2F_X(U^{(v_j)})-\tau\right\}+O\,(\Delta^2)\right]$$

$$=1+\Delta\sum_{j=1}^{n}\left\{\frac{2}{2n+1}V_j-\tau\right\}+O\,(\Delta^2).$$

By denoting $\delta=\Delta\sum_{j=1}^{n}\left\{\frac{2}{2n+1}V_j-\tau\right\}+O\,(\Delta^2)$, when Δ is a small quantity, we see that

$$\log R_n(\Delta)=\log(1+\delta)=\delta-\frac{1}{2}\delta^2+\frac{1}{3}\delta^3+...=\delta+O\,(\delta^2)$$

$$=\Delta\sum_{j=1}^{n}\left\{\frac{2}{2n+1}V_j-\tau\right\}+O\,(\Delta^2).$$

We obtain the limiting log probability ratio $L_n^*=\lim_{\Delta\to 0}\frac{\log R_n(\Delta)}{\Delta}=\sum_{i=1}^{n}\left\{\frac{2}{2n+1}V_i-\tau\right\}$. Thus the sequential test based on L_n^* is a sequential probability ratio test (*SPRT*) [8]. Next, if we take A and B according to (3.3), and the stopping rules according to (3.4), then the type I and type II error probabilities of the test are

$$P_1=Prob_{H_0}\left[L_n^*\geq\log A\right]<\alpha;\quad P_2=Prob_{\hat{H}_1}\left[L_n^*\leq\log B\right]<\beta.$$

We are now in a position to show that the stopping time $\hat{R}$ of above test is exponentially bounded. By Theorem 2.3 of [9], we only need to prove that $\lim_{n\to\infty}\frac{L_n^*}{n}\neq 0$. Following the discussion in the section 4.1 of [9], we know that $\lim_{n\to\infty}\frac{L_n^*}{n}=E\left[F_X-F_Y\right]-\tau$. Then $\hat{R}$ is exponentially bounded unless $Prob\left[F_X=F_Y\right]=\tau$, which equivalent to $Prob[X=Y]=\tau$ [9, Lemma 4.4]. This is impossible because $\tau>1$. Hence $\hat{R}$ is exponentially bounded.

For F_Y, such that $F_X \geq F_Y$, consider $\Delta=\max_u\left\{\frac{F_X(u)-F_Y(u)}{\tau F_X(u)-F_X^2(u)}\right\}$. This implies $F_X \geq F_Y \geq F_{X,\Delta}$. Note that H_0 is same in (3.1) and (3.7). $\hat{H}_1$ implies H_1, because if Y is distributed according to $F_{X,\Delta}$, then it is indeed stochastically bigger than X. Thus $Prob\left[H_1\right] \geq Prob\left[\hat{H}_1\right]$. Now, we have

$$Prob\left[L_n^* \leq \log B\right]=Prob\left[H_1\right] Prob_{H_1}\left[L_n^* \leq \log B\right]=Prob\left[\hat{H}_1\right] Prob_{\hat{H}_1}\left[L_n^* \leq \log B\right].$$

This implies

$$Prob_{H_1}\left[L_n^* \leq \log B\right]<\beta.$$

Since $\hat{H}_1$ implies H_1, we have $\hat{R}>R$. To see this, consider the sequence corresponding to the test of (3.7). Using this sequence, once $\hat{H}_1$ becomes true, so does H_1. However, H_1 might become true earlier if the same sequence is used in (3.1). Hence, in conclusion R is exponentially bounded. □

We now state an existing result from [4] which will be used subsequently.

Lemma 3.2. [4, pp.669] *In rank SPRT, for $\alpha>0$, $\beta>0$, $\alpha+\beta<1$, let $A=\frac{1-\beta}{\alpha}$ and $B=\frac{\beta}{1-\alpha}$, then*

$$P_1=Prob_{H_0}[H_0 \text{ is rejected}] \leq \alpha; \quad P_2=Prob_{H_1}[H_1 \text{ is rejected}] \leq \beta.$$

holds. When $\alpha+\beta \to 0$, $A \approx \log\frac{1}{\alpha}$; $B \approx \log\beta$. Suppose there are finite constants $I_0<0<I_1$, such that for all $\zeta>0$,

$$M_i(\zeta)=sup\left\{m>1: \left|\frac{\log Q_m}{m}-I_i\right| \geq \zeta\right\}, \quad i=0,1.$$

has a finite expectation under $H_i (i=0,1)$, where Q_m is the probability ratio, then as $\alpha+\beta \to 0$

$$E_{H_0}(R) \approx \frac{\log\beta}{I_0}; \; E_{H_1}(R) \approx \frac{\log\frac{1}{\alpha}}{I_1}. \; \square$$

We now show that the test in (3.1) based on L_n^* satisfies the conditions stated in Lemma 3.2.

Lemma 3.3. *For the sequential two-sample test of (3.1) with the statistic L_n^*, Lemma 3.2 holds.*

Proof. Since L_n^* is a limiting log probability ratio, $L_n^* = \log Q_n$. It suffices to show that for the sequential two-sample test of (3.1) there exist finite constants $I_0 < 0 < I_1$, such that for all $\zeta > 0$

$$M_i(\zeta) = sup\left\{ n > 1 : \left| \frac{L_n^*}{n} - I_i \right| \geq \zeta \right\}, \quad i = 0, 1 \tag{3.10}$$

has a finite expectation under $H_i (i = 0, 1)$. We prove this Lemma by contradiction. Consider the combined sample $(X_1, ..., X_n, Y_1, ..., Y_n)$. Let S_i and S'_i be the ordered ranks of Y_i and X_i in the combined sample, respectively. Under the hypothesis H_0, X_i and Y_i are from the same distribution, thus when n approaches infinity, we observe

$$\lim_{n\to\infty} \sum_{i=1}^{n} V_i = \lim_{n\to\infty} \sum_{i=1}^{n} V'_i = \lim_{n\to\infty} \frac{1}{2}(1+2+...2n).$$

This in turn implies

$$\lim_{n\to\infty} L_n^* = \lim_{n\to\infty} \left[\frac{2}{2n+1} \sum_{i=1}^{n} V_i - \tau n \right] = \lim_{n\to\infty} (1-\tau)n.$$

Thus,

$$\lim_{n\to\infty} \frac{L_n^*}{n} = 1 - \tau \tag{3.11}$$

Let $I_0 = 1 - \tau$. Because $\tau > 1$, we know that $I_0 < 0$. Assume that under H_0, $M_0(\zeta)$ does not have finite expectation. Then at least there exists a $\zeta' > 0$ such that $n' = M_0(\zeta')$ and

$$\lim_{n'\to\infty} \left| \frac{L_{n'}^*}{n'} - I_0 \right| = \left| \lim_{n'\to\infty} \frac{L_{n'}^*}{n'} - I_0 \right| \geq \zeta'. \tag{3.12}$$

Combining (3.11) and (3.12), we obtain $\zeta \le 0$, which is a contradiction. Thus under H_0, the conclusion holds. In the same way we can prove that under H_I the conclusion is valid. □

In the next section, we analyze the computational complexity of algorithm *NSA* when L_n^* is employed.

4. Computational Complexity of NSA Algorithm

In this section, we first establish a value for d_{Id}. Then we derive a bound on the stopping time R. By using these values, we estimate the average complexity of the *NSA* algorithm.

Definition 4.1. *At depth i of the search tree, the sequential two-sample test, with statistic L_n^*, uses the confidence level (α_i, α_i). Then we define d_i as follows:*

$$d_i = max\left\{\frac{1}{I_1}, -\frac{1}{I_0}\right\}\frac{\log\alpha_i}{k}\ln\frac{\alpha_i}{k}, \tag{4.1}$$

where k is as in (3.6), I_0 and I_1 are given in Lemma 3.3.

Now, we establish a Lemma that characterizes some important properties of sample size when L_n^* is employed by *NSA*.

Lemma 4.1. *Let R be the stopping time of the two sample sequential test on the statistical hypothesis (2.5), then at depth i of the search tree,*

$$Prob_{H_0}\left[R \ge d_i\right] \le \alpha_i; \quad Prob_{H_1}\left[R \ge d_i\right] \le \alpha_i. \tag{4.2}$$

where d_i follows the definition 4.1. Furthermore, we have

$$Prob_{\overline{H}_0 \cap \overline{H}_1}\left[R \ge d_i\right] \le \alpha_i. \tag{4.3}$$

Proof. Consider $Prob_{H_0}\left[R \ge d_i\right]$ first. Following Lemma 3.1, we have $Prob_{H_0}\left[R > n\right] \le ke^{-c_i n} = \int_n^\infty c_i\, k\, e^{-c_i x} dx$. Then $E_{H_0}(R) \le \int_0^\infty x\, c_i\, k\, e^{-c_i x} dx = \frac{k}{c_i}$. By Lemma 3.3, $E_{H_0}(R) \approx \frac{\log\alpha_i}{I_0}$. Thus $\frac{k}{c_i} \approx \frac{\log\alpha_i}{I_0}$, $c_i \approx \frac{kI_0}{\log\alpha_i} > 0$. From (4.1) and with no loss of generality, we have $-c_i d_i = \ln\frac{\alpha_i}{k}$. Finally, we see that $Prob_{H_0}\left[R > d_i\right] \le ke^{-c_i d_i} = \alpha_i$. Similarly, we prove $Prob_{H_1}\left[R > d_i\right] \le \alpha_i$.

Now consider $Prob_{\overline{H}_0 \cap \overline{H}_1}\left[R \geq d_i\right]$. Let us imagine the following combined non-parametric statistical inferences, S_1 and S_{-1}, where S_1 comprises of H_0: $F_X = F_Y$, H_1: $F_X \geq F_Y$; S_{-1} comprises of H_0: $F_X = F_Y$, H_{-1}: $F_X \leq F_Y$. We take the hypothesis H_1 if S_{-1} accepts H_0 and S_1 accepts H_1. We take the hypothesis H_0 if both S_1 and S_{-1} accept H_0. We take the hypothesis H_{-1} if S_1 accepts H_0 and S_{-1} accepts H_{-1}. Otherwise we continue the sampling. Using the same statistical parameters in both S_1 and S_{-1}, and proceeding in a similar manner discussed above, we obtain

$$Prob_{\overline{H}_0 \cap \overline{H}_1}\left[R \geq d_i\right] = Prob_{H_{-1}}\left[R \geq d_i\right] \leq \alpha_i. \quad \square$$

Note that the statistic used in our test is constructed from the heuristic evaluation function. Consequently the probability of finding the solution path not only depends on the correctness of the statistical inference, but also on the precision of the heuristic estimate. We introduce the idea of precision of heuristic estimate in the following definition.

Definition 4.2. *For $T^i(n)$, such that $n \in L$, define the maximal and minimal gaps, $\delta^i_{\max}$ and $\delta^i_{\min}$, of the heuristic estimate $h(n')$, where $n' \in T^i(n)$, as follows:*

$$\delta^i_{\max} = \max_{n' \in T^i(n)} \left\{ h^*(n') - h(n') \right\} \quad \text{for all } n \text{ at level } i \tag{4.4}$$

$$\delta^i_{\min} = \min_{n' \in T^i(n)} \left\{ h^*(n') - h(n') \right\} \quad \text{for all } n \text{ at level } i \tag{4.5}$$

We now show our main result in the following theorem.

Theorem 4.1. *Given $\alpha_0 < 1$, let $\alpha = \dfrac{\alpha_0}{2(m-1)K_1}$, $K_1 = \sum_{i=1}^{N} \dfrac{1}{i}$, and let $\alpha_i = \beta_i = \dfrac{\alpha}{i}$. Let $\dfrac{1}{2}\left(3 - \dfrac{\log A_N}{d_N^2}\right) \leq \tau < \dfrac{3}{2}$ for L_n^*. Assume the heuristic estimate is admissible and its maximal and minimal gaps satisfy the condition*

$$Prob_{\overline{H}_0 \cap \overline{H}_1}\left[\delta^i_{\max} - \delta^i_{\min} > 2\log A_i + 1 + \tau\right] \leq \frac{\alpha_i}{d_N}, \quad A_i = \frac{1-\beta_i}{\alpha_i} \tag{4.6}$$

then NSA finds the solution path with probability of at least $1-\alpha_0$. The average computational complexity (the total node expanded) is $O(N(\log N)^2)$.

Proof. In following proof, we expand the nodes with the *depth-first-least-value sampling principle*. This can be similarly extended to the case of *breadth-first-least-value sampling principle*. Using *depth-first-least-value sampling principle*, there are m i-type subtrees at depth i. Assume that we do not make mistake at the previous level. This means that among the present m i-type subtrees, the root of some subtree is located on L. We will incorrectly prune off L in two cases. The first case is that when $n_j (j \neq 1)$ is on L (*i.e.* under the situation $\overline{H}_0 \cap \overline{H}_1$), and the sample size of $Z_j = (a(n_1), a(n_j))$ is less than the threshold d_i, and we choose H_1. The second case is that when $n_j (j \neq 1)$ is on L, the sample size of Z_j exceeds the threshold d_i. Let z be the sample size of Z_j, and let p_z the probability that we accept H_1 when $n_j (j \neq 1)$ is on L and the sample size of Z_j is z. The probability π_i that we lose L, when we make decision on $(T^i(n_1), T^i(n_j))$, is given by

$$\begin{cases} \pi_i < \left[p_1 + (1-p_1)p_2 + \ldots + (1-p_1)(1-p_2)\ldots(1-p_{d_i-2})p_{d_i-1} \right] + Prob_{\overline{H}_0 \cap \overline{H}_1}[z \geq d_i] \\ p_z = Prob_{\overline{H}_0 \cap \overline{H}_1} \left[L_z^* \geq \log A_i \right] \end{cases} \tag{4.7}$$

We compute the probability p_z as follows. When the sample size of $(a(n_1), a(n_j))$ is z, we have the samples $a_1(n_1), a_2(n_1), \ldots, a_z(n_1)$ from $a(n_1)$, and the samples $a_1(n_j), a_2(n_j), \ldots, a_z(n_j)$ from $a(n_j)$. Denote by $n_{i,j}$ a node in $T(n_i)$ at depth j. Then we have the samples from $a(n_1)$ listed as follows (refer to Fig. 1 and recall that n_1 is not on L):

$$\begin{cases} a_1(n_1) = g(n_{1,1}) + h(n_{1,1}) = g(n_{1,1}) + h^*(n_{1,1}) - \delta_1 = N + 2 - \delta_1; \\ \ldots \\ a_k(n_1) = g(n_{1,k}) + h(n_{1,k}) = g(n_{1,k}) + h^*(n_{1,k}) - \delta_k = N + 2k - \delta_k; \\ \ldots \\ a_z(n_1) = g(n_{1,z}) + h(n_{1,z}) = g(n_{1,z}) + h^*(n_{1,z}) - \delta_z = N + 2z - \delta_z; \end{cases} \tag{4.8}$$

where $\delta_k = h^*(n_{1,k}) - h(n_{1,k}) > 0$, since the heuristic is admissible. On subtree $T(n_j)$ (recall that n_j is on L), consider the case that at depth k of subtree $T(n_j)$ $(2 \leq k \leq z)$, the node selected for expansion is not on L, then the samples from $a(n_j)$ can be listed as follows:

$$\begin{cases} a_1(n_j) = g(n_{j,1}) + h(n_{j,1}) = g(n_{j,1}) + h^*(n_{j,1}) - \delta'_1 = N - \delta'_1; \\ \ldots \\ a_{k-1}(n_j) = g(n_{j,k-1}) + h(n_{j,k-1}) = g(n_{j,k-1}) + h^*(n_{j,k-1}) - \delta'_{k-1} = N - \delta'_{k-1}; \\ a_k(n_j) = g(n_{j,k}) + h(n_{j,k}) = g(n_{j,k}) + h^*(n_{j,k}) - \delta_k = N + 2 - \delta'_k; \end{cases} \tag{4.9}$$

$$\left\{\begin{array}{l} \ldots \\ a_z(n_j)=g(n_{j,z})+h(n_{j,z})=g(n_{j,z})+h^*(n_{j,z})-\delta'_z=N+2(z-k+1)-\delta'_z; \end{array}\right.$$

Suppose that the sorted (in an ascending order) samples from $a(n_1)$ and $a(n_j)$ are given by $a_{k_1}(n_1), a_{k_2}(n_1), \ldots, a_{k_z}(n_1)$ and $a_{k'_1}(n_j), a_{k'_2}(n_j), \ldots, a_{k'_z}(n_j)$ respectively. Now let us consider the combined samples of $a(n_1)$ and $a(n_j)$. Look at the sample $a_{k'_z}(n_j)$ in the combined samples. Consider the sample $a_{k_l}(n_1)$ which is the closest in front of $a_{k'_z}(n_j)$ in the combined samples, where k_l means that there are exactly l samples from $a(n_1)$ in front of $a_{k'_z}(n_j)$. We see the combined samples as

$$\ldots,\ a_{k_l}(n_1), \ldots,\ a_{k'_z}(n_j), \ldots$$

or equivalently

$$\ldots,\ N+2k_l-\delta_{k_l}, \ldots,\ N+2k'_z-\delta'_{k'_z}, \ldots \tag{4.10}$$

where $k_l \in \{1,2,\ldots,z\}$ and $k'_z \in \{0,1,2,\ldots,z-1\}$. Now refer to the set $\{k_1,k_2,\ldots,k_z\}$, where each k_i is distinct from each other and takes a value from $\{1,2,\ldots,z\}$. Choosing l k_i's from $\{k_1,k_2,\ldots,k_z\}$, at least one of them should be equal to or greater than l. We denote this mathematically as follows: Let $k_{l'}=\max\{k_j \mid j=1,\ldots,l, 1\le k_j\le z\}$. Then we have $k_{l'}\ge l$. Consequently, from (4.10) we obtain $2l-\delta^i_{\max}\le 2k_{l'}-\delta_{k_{l'}}<2k'_z-\delta'_{k'_z}\le 2(z-1)-\delta^i_{\min}$, or

$$2l<\delta^i_{\max}-\delta^i_{\min}+2(z-1) \tag{4.11}$$

Then we compute the total ranks of $a_k(n_j)$, and the statistic L_z^* as follows,

$$\sum_{k=1}^{z} V_k \le \sum_{k=1}^{z}(l+i)=lz+\frac{z(z+1)}{2}. \tag{4.12}$$

$$L_z^*=\frac{2}{2z+1}\sum_{k=1}^{z} V_k-\tau z\le\frac{2z}{2z+1}l+\frac{z(z+1)}{(2z+1)}-\tau z. \tag{4.13}$$

Next, respectively according to (4.13), (4.11) and $\frac{1}{2}\left[3-\frac{\log A_N}{d_N^2}\right]\le\tau<\frac{3}{2}$, we have

$$\begin{aligned} p_z &= Prob_{\bar{H}_0\cap\bar{H}_1}\left[L_z^*\ge\log A_i\right] \\ &\le Prob_{\bar{H}_0\cap\bar{H}_1}\left[2l\ge\frac{2z+1}{z}\log A_i+(2\tau-1)z+\tau-1\right] \end{aligned}$$

$$\leq Prob_{\overline{H}_0 \cap \overline{H}_1}\left[\delta^i_{\max}-\delta^i_{\min}\geq 2\log A_i+\frac{\log A_i}{z}+(2\tau-3)z+1+\tau\right]$$

$$\leq Prob_{\overline{H}_0 \cap \overline{H}_1}\left[\delta^i_{\max}-\delta^i_{\min}\geq 2\log A_i+1+\tau\right]\leq \frac{\alpha_i}{d_N}.$$

Consequently, we see that

$$p_1+(1-p_1)p_2+...+(1-p_1)(1-p_2)...(1-p_{d_i-2})p_{d_i-1}<\sum_{z=1}^{d_i} p_z<\alpha_i.$$

We have from Lemma 4.1 $Prob_{\overline{H}_0 \cap \overline{H}_1}[z\geq d_i]\leq\alpha_i$. In consequence, the probability that we eliminate the solution path, π_i, is upper bounded by $2\alpha_i$. At level i, the total probability of making an incorrect decision (pruning $T(n)$, when $n\in L$) is given by $\zeta_i=2(m-1)\alpha_i$. Furthermore, the total probability with which we lose the solution path during the search is

$$\zeta_1+(1-\zeta_1)\zeta_2+...+(1-\zeta_1)(1-\zeta_2)...(1-\zeta_{N-1})\zeta_N$$

$$<\sum_{i=1}^{N}\zeta_i<2(m-1)\sum_{i=1}^{N}\alpha_i=2(m-1)\alpha\sum_{i=1}^{N}\frac{1}{i}=\alpha_0.$$

So, NSA finds the solution path with the probability of at least $1-\alpha_0$.

Consider the mean computational complexity. At depth i, the threshold is given by

$$d_i=max\left\{\frac{1}{I_1},-\frac{1}{I_0}\right\}\frac{\log\alpha_i}{k}\ln\frac{\alpha_i}{k}\sim O\left[\log\frac{\alpha}{i}\right]^2\sim O\left((\log i)^2\right).$$

There are m subtrees $T^i(n_j)$ at depth i. The upper bound on the mean computational complexity is given by

$$m\sum_{i=1}^{N}d_i\sim m\sum_{i=1}^{N}(\log i)^2\sim O\left(N(\log N)^2\right).\ \square$$

Under the search model hypothesis, we characterize the global behavior of the statistic $a(n)$. Namely, for $n_1\in L, n_2\notin L$, we have $Prob[a(n_1)\geq t]\leq Prob[a(n_2)\geq t]$, i.e. $a(n_1)$ is more likely to have a lower value compared to $a(n_2)$. However, in terms of local behavior, an estimate for $a(n_1)$ might have a higher value comparing to that for $a(n_2)$. In particular, because of sampling, we may end up with the condition $\hat{\mu}(n_1)>\hat{\mu}(n_2)$. This phenomenon indicates that the underlying heuristic information is not conducive to the search model discussed in Section 2. The condition (4.6) "captures" the probability with which such phenomenon occurs.

Undoubtedly, it is possible that the goal may not be found by *NSA*. In this case, the solution path L has been incorrectly cut-off during the search. When the goal is not found in the first invocation of *NSA*, it is applied to the remaining (pruned) parts of the search tree. The repeated invocation of *NSA* is continued until the goal is found or the whole search tree (up to depth N) is completely explored. This process is called the *Successive NSA* search. The original idea of successive invocation is due to Zhang and Zhang [12].

It is clear that *Successive NSA* will find the goal, if it exists (possibly after exhausting the entire tree of depth N). Applying *NSA* to G for the first time, the goal will be found with a probability of $(1-\alpha_0)$, and the complexity of the invocation is $O(N(\log N)^2)$. Similarly, the probability that the goal is found at the i-th time is given by $p_i=(1-\alpha_0)\alpha_0^{i-1}$. Then the mean complexity of *Successive NSA* is $T=\sum_{i=1}^{N}\alpha_0^{i-1}(1-\alpha_0)iO(N(\log N)^2)=(1-\alpha_0)O(N(\log N)^2)\sum_{i=1}^{N}i\alpha_0^{i-1}$. The quantity $\sum_{i=1}^{N}i\alpha_0^{i-1}$ in the previous equation converges to a constant. To see this, consider $S=\sum_{i=1}^{N}\alpha_0^i=\dfrac{\alpha_0-\alpha_0^{N+1}}{1-\alpha_0}$. By differentiating S respective to α_0, we obtain the value of $\sum_{i=1}^{N}i\alpha_0^{i-1}$ to be $\dfrac{1-(N+1)\alpha_0^N+N\alpha_0^{N+1}}{(1-\alpha_0)^2}$. Thus $T=O(N(\log N)^2)$. We now present the computational complexities of *Successive NSA* algorithm.

Theorem 4.2: *The Successive NSA search finds the goal, if it exists. Its mean computational complexity is $O(N(\log N)^2)$, under condition (4.6). The total storage space used by Successive NSA search is $O((\log N)^2)$. The average number of comparisons is $O(N(\log N)^6\log\log N)$.*

Proof. The first part follows from the above discussion. Consider the storage complexity. At depth i, we only need to keep all nodes in $T^i(n_j)$ $(j=1,2,...,m)$. Thus the space required is md_i. Taking all search stages into account, and noting that d_i is a monotonically increasing function of i, the total storage space is given by

$$\max_{1\le i\le N}\{md_i\} \sim O((\log N)^2).$$

At depth i, by applying the nonparametric statistical inference which uses the rank statistic, the total number of comparisons is comprised of two parts. The first one is used to sort the combined statistic; the second part is used for calculating the ranks. Using any standard sorting algorithm, the number of comparisons, $s_{i,1}$, needed for sorting is given by

$$s_{i,1} \leq \sum_{l=1}^{d_i} \sum_{l'=1}^{d_i} O((l+l')\log(l+l')) \sim O(d_i^3 \log d_i),$$

where l and l' are the number of samples of $a^i(n_1)$ and $a^i(n_k)$ respectively. The number of comparisons $s_{i,2}$ used to calculate the ranks is given by

$$s_{i,2} \leq O(2d_i) \sim O(d_i).$$

Finally, we see that the total number of comparisons is given by

$$\sum_{i=1}^{N} (s_{i,1}+s_{i,2}) \sim \sum_{i=1}^{N} O(d_i^3 \log d_i) \leq O(Nd_N^3 \log d_N) = O(N(\log N)^6 \log\log N). \ \square$$

5. Conclusions and Discussions

Applying statistical inference methods to heuristic search could greatly improve the efficiency of the heuristic search in an average-case. Until now, however, only the utilization of parametric statistical inference methods, such as *SPRT*, sequential u-test and sequential t-test, has been considered. The effectiveness and efficiency of the parametric statistical heuristic search will mainly depend on the assumption about the distribution of a statistic $a(n)$ that characterizes the subtree $T(n)$ rooted at node n. But, only if we have a sound knowledge of problem domain, we will be able to make a good assumption about the statistic generated from the heuristic information. Further, even if we assume the form of the distribution, we still have to estimate the parameters. These estimates could introduce new errors into the search. In our model, we do not suppose that the statistic $a(n)$ is subjected to any particular distribution. For two brother nodes n_1 and n_2, we only concentrate on whether $a(n_1)$ is *stochastically greater* than or *probabilistically indistinguishable* from $a(n_2)$. In this paper, we present nonparametric statistical heuristic search algorithms, *NSA*, and *Successive NSA*. Their computational complexities are discussed using a uniform m-ary search tree G with a single goal located at an unknown location at depth N. In the analysis, we also investigate the relation between the precision of the heuristic estimate and the computational complexity. The mean computational complexity of *Successive NSA* is $O(N(\log N)^2)$, where N is the length of the solution path. The storage space required is $O((\log N)^2)$ and the number of comparisons needed is $O(N(\log N)^6 \log\log N))$.

There are some limitations on the applicability of *NSA* (and the other search methods using statistical inferences, such as *SA*). We assume that two observations on $a(n)$ are *i.i.d.* Since the observations on $a(n)$ are obtained from the node heuristics in the subtree being inferred, in general the condition of *i.i.d.* observations will not be satisfied. In this circumstance, the theoretical analysis may not be completely verified by an empirical study. It is an open problem of using other statistical inferences which does not require the *i.i.d.* properties.

The search spaces of most application domains are graphs. In this paper, we only considered the utilization of nonparametric statistical inference methods in tree search. The further research should consider the search in general graph.

References

[1] P.E. Hart, N.J. Nilsson, and B. Raphael, "A formal basis for the heuristic determination of minimum cost paths", *IEEE Trans. Systems, Science and Cybern.*, **4(2)**, 1968, pp.100-7.

[2] N. Huyn, R. Dechter, and J. Pearl, "Probabilistic analysis of the complexity of A^*", *Artificial Intelligence*, **15(3)**, 1980, pp.241-54.

[3] R.E. Korf, "Search: A survey of recent results", in *Exploring Artificial Intelligence*, (ed) H.E.Shrobe, Morgan Kaufmann Pub., Inc., 1988, pp.197-238.

[4] U. Muller-Funk, "Sequential nonparametric tests", in *Handbook of Statistics*, vol.4, P.R.Krishnaiah and P.K.Sen (des.), Elsevier Science Pub., 1984, pp.657-98.

[5] N.J. Nilsson, *Principles of Artificial Intelligence*, Palo Alto, Calif.: Tioga, 1980.

[6] J. Pearl, "Knowledge versus search: A quantitative analysis using A^*", *Artificial Intelligence*, **20(1)**, 1983, pp.1-13.

[7] J. Pearl, *Heuristics: Intelligent Search Strategies for Computer Problem Solving*, Addison-Wesley Pub. Company, Inc., 1984.

[8] G.B. Wetherill, and K.D. Glazebreek, *Sequential Methods in Statistics*, third edition, Chapman and Hall, 1986.

[9] R.A. Wijsman, "A general theorem with applications on exponentially bounded stopping time, without moment conditions", *The Annals of Statistics*, **5(2)**, 1977, pp.292-315.

[10] S. Zacks, *The Theory of Statistic Inference*, New York: Wiley, 1971.

[11] B. Zhang and L. Zhang, "A new heuristic search technique---algorithm SA", *IEEE Trans. on Pattern Analysis and Machine Intelligence*, **7(1)**, 1985, pp.103-7.

[12] L. Zhang and B. Zhang, "The successive SA^* search and its computational complexity", *Proc. 6th ECAI-84*, pp.249-58.

[13] B. Zhang and L. Zhang, "A new weighted technique in heuristic search", *Proc. 9th IJCAI-85*, pp.1037-9.

[14] B. Zhang and L. Zhang, "Statistical heuristic search", *Journal of Computer Science and Technology*, **2(1)**, 1987, pp.1-11.

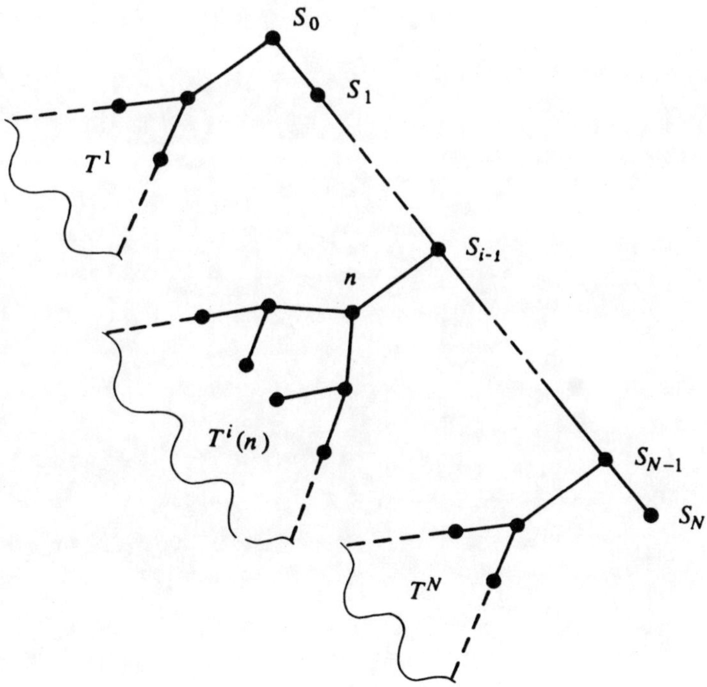

Fig. 1. A uniform m-ary tree model for NSA search.

Applications of Learning and Planning Methods
edited by N. G. Bourbakis

CHAPTER 10

An A* Approach to Robust Plan Recognition for Intelligent Interfaces

Randall J. Calistri-Yeh
Odyssey Research Associates
301A Harris B. Dates Dr.
Ithaca, NY 14850-1313
oravax!calistri@cu-arpa.cs.cornell.edu

Abstract

The traditional problem of plan recognition can be solved in a straightforward manner by a graph-searching algorithm such as A*. A few simple extensions allow the algorithm to handle situations in which the user has an incorrect plan. However, it turns out that recognizing structural errors is computationally much more complex than recognizing plans with simple constraint violations. Four algorithms are presented here, starting with a direct adaptation of A* for recognizing correct plans, and progressing to an optimized algorithm for recognizing violated, ill-formed plans.

1 Introduction

The problem of recognizing an agent's plans has been addressed by many AI researchers over the years (see [8]), but it is still the case that most plan recognition systems assume that the agent will have perfect plans. People are notoriously adept at making mistakes, and plan recognition systems should be robust enough to be able to recognize these mistakes and correct them [4,10,11].

We will begin this paper by expressing a particular version of the plan recognition problem for perfect plans in terms of a classic search algorithm. We will then relax the perfect plan assumption to allow users to make mistakes, and we will investigate how this affects the algorithm and its properties. Finally, we will present a particular implementation of the final algorithm called Pathfinder, and will discuss its theoretical and actual performance.

2 The Basic Plan Recognition Problem

One possible statement of the plan recognition problem, which we will call the "Perfect-Plan Recognition Problem," is as follows:

> *Given a closed world of correct plans, knowledge about the current and past states of the world, and information about a queried step in a plan (possibly described at multiple levels of abstraction), determine how this step relates to the higher level plans and goals of the agent, assuming that the agent's plans do not contain any misconceptions.*

The plan recognition problem has different variations, depending on what types of actions and utterances are observed, and on *how* they are observed. For example, we might be observing a sequence of atomic control adjustments by a pilot, from which we want to infer the plan that the pilot is trying to carry out. Or we might be trying to decide from a novice user's questions to an expert system whether his plan is reasonable. The particular version that we investigate here is especially common in the sort of plan recognition that is needed by intelligent natural language interfaces. For example, in a conversation with an expert system on income taxes, a graduate student might say:

> "I'm trying to fill out my taxes. Where do I get the W2 form for my fellowship?"

This describes a single action — getting a form — at multiple levels: the fact that the student wants to get it, the fact that it is part of a plan to fill in his fellowship information, and the fact that this is part of an even higher plan of doing his taxes.

The first assumption in this problem statement, that there is a "closed world" of correct plans, requires that the observer have complete knowledge about all possible correct plans in the domain. While this is an unrealistic assumption in many cases, we do not address it in this paper. Instead, we focus on the assumption that the agent always has correct plans. This is a much more serious assumption from the perspective of the agent, since it places demands on the abilities of the agent instead of the observer. The closed world assumption means that the observer must be omniscient; the perfect plan assumption means that the agent must be infallible. In later sections of this paper, we will relax the "perfect plan" requirement to eliminate this infallibility assumption.

The plan recognition problem can be viewed as a particular application of the general graph searching problem. The plan terminology can be mapped directly onto graph terminology in the following way. The hierarchy of plans is structured as a directed *AND/OR* graph, where plans are composed of subplans (*AND* nodes) and where there can be alternate ways to execute a plan (*OR* nodes). The input to the problem is this graph (the hierarchy), along with a set of distinguished nodes (steps that the user has mentioned). The task is to search the graph for a path (explanation) that extends from a root of the graph (the agent's top-level goal) to a specific node (the observed action), passing through all of the marked intermediate nodes.

Because of this close relationship between plan recognition and graph searching, classic methods such as the **A*** algorithm for best-first search can be applied directly to the "Perfect-Plan Recognition Problem" to find an optimal explanation for the agent's query. After presenting a version of **A*** for perfect plan recognition, we will show that two modifications need to be made in order to create a more robust algorithm that is tolerant of human errors. Finally, we will present the Pathfinder program, an optimized plan recognition algorithm that handles both violated and ill-formed queries in an efficient manner.

3 PR-1: A* for Perfect Plan Recognition

The **PR-1** algorithm presented here is a direct adaptation of the **A*** algorithm [7,12], as described by Pearl [9, p. 64 & Ch. 3.1], applied to the "Perfect-Plan Recognition Problem". One change that is necessary in order to adapt the normal version of **A*** for perfect plan recognition is the input to the problem. While the original **A*** only had the start and destination nodes (the top-level goal and the actual queried step) as input, we also have other known points along the path (the other steps that the agent has mentioned). Since we know that the solution path must pass through these points, we can change the basic search unit from a single node to a partial path between two consecutive mentioned nodes.

The algorithm works by incrementally extending partial solution paths until one of them reaches the destination node. It maintains two lists: an `OPEN` list, which holds the partial paths that are being considered for expansion, and a `CLOSED` list, which holds partial paths that have already been expanded and might be used in the solution path. We start with a single partial path consisting of the start node, and place it on the `OPEN` list. We then **expand** this partial path by generating all ways that it can be extended to the first mentioned node (these new path segments are called "successors" of the original path).

For each of these successors, we **verify** every step in the new partial path by checking to see that no constraints have been violated. For example, a precondition for one of the steps may not hold in the current state of the world. If there are any violations, we discard this path, since it could not be part of a correct plan (remember, we are currently only doing "perfect plan" recognition). Otherwise, if this partial path has been verified successfully, we compute a heuristic function f. This function estimates our confidence that this partial path will be part of the final answer. Each of the verified successors is then added to the `OPEN` list. Then we remove the original path, which has already been expanded, and place it on `CLOSED`, where we can retrieve it if we need it for the final answer.

Finally, we go back to the `OPEN` list and choose the most likely partial path for

1. `OPEN` ← **`start`**
   `CLOSED` ← ∅.
2. If `OPEN` = ∅ then exit with failure.
3. Remove partial path p with optimal $f(p)$ from `OPEN` and place on `CLOSED`.
4. If the endpoint of p is a destination node, exit with the complete path formed by tracing the pointers back to **`start`**.
5. Otherwise, **Expand**(p):
   5.1. Generate all successors of p (all partial paths from endpoint of p to the next known node), with pointers back to p.
   5.2. For each successor p' of p, **Verify**(p'):
      5.2.1. Check p' for violations.
      5.2.2. If p' contains any violations, remove it.
      5.2.3. Otherwise, estimate $f(p')$ and add p' to `OPEN`.
6. Go to step 2.

Figure 1: The **PR-1** algorithm.

expansion, based on the heuristic function that is stored with the partial path. We continue to repeat the entire process until we find a partial path that reaches a destination node, in which case we have successfully recognized the agent's plan, or until we run out of partial paths to expand, in which case we have failed.

This algorithm is shown in Figure 1. There are several things to note here. First of all, the algorithm is purposely ambiguous about the **Expand** procedure. The successors of p can be generated in any way that is appropriate for the particular domain. Some domains are structured in such a way that a special-purpose method can be used to generate these successors; it may even be possible to precompute the necessary partial paths. In other situations, the most reasonable approach might be to call **A*** recursively with the next mentioned step as a destination.

Another deliberate ambiguity is in the choice of "start" and "destination"; it is equally possible to proceed top-down, from the goal to the mentioned step, or bottom-up, from the mentioned step to the goal. The choice of direction may be influenced by the relative size of the forward and backward branching factor of the plan hierarchy. It is most efficient to search in the direction with the smallest branching factor, since each branch represents a point where the search may need to split. If the domain contains very complex subplans, this signifies a high forward branching factor, and a bottom-up approach might be more efficient. If there are certain very common actions that are found in almost all plans, this signals a high backward branching factor, in which case a top-down approach would make more sense.

Finally, the choice of a partial path with "optimal" $f(p)$ does not distinguish whether the heuristic function f should be maximized or minimized. In fact, the algorithm doesn't specify how to calculate $f(p')$ for the successors at all. The traditional **A*** algorithm defines a minimizing function $f(p') = g(p') + h(n')$. The function $g(p')$ is the earlier cost of expanding the path from the start node to p, plus the new cost of extending p to p'. The function $h(n')$ is an estimate of the cost of expanding p' to a destination node. However, these requirements are not necessary to the algorithm [1,2,6]. It is possible to define f to be either a minimizing function, where f measures the cost of a path, or a maximizing function, where f measures the merit of a path. It is also possible for g and h to be multiplicative instead of additive. In fact, there are only two strict requirements on the heuristic function. In order to guarantee an optimal answer, $f(p')$ must be both *optimistic* and *monotonic*. This means that for a maximizing heuristic, f must always overestimate the merit and must be non-increasing. And for a minimizing heuristic, f must always underestimate the cost and must be non-decreasing.

4 Misconceptions and the Plan Recognition Process

We would now like to extend the basic plan recognition problem to a more robust version that will be able to handle cases in which the user has an incorrect plan. This can be stated as:

> *Given a closed world of correct plans, knowledge about the current and past states of the world, and information about a queried step in a plan (possibly described at multiple levels of abstraction), determine how this step relates to the higher level plans and goals of the agent, when the agent's plans may be flawed in unpredictable ways.*

By the phrase "unpredictable ways", we do not mean that incorrect plans are incapable of being classified into predictable categories. Rather, we mean that it is not possible to predict all specific incorrect plans ahead of time. This is necessary to preclude solutions such as explicitly including a "bug list" of known incorrect plans directly in the plan hierarchy.

There are several ways that an agent can have a flawed plan [4, Ch. 4], but for the purposes of this discussion, it is convenient to group misconception types into two broad classifications: **structural** misconceptions and **constraint** misconceptions. Plans that contain structural misconceptions will be referred to as **ill-formed**, and include cases of substituted, missing, and extra steps. Plans that contain constraint misconceptions will be referred to as **violated**, and encompass violated bindings, temporal violations, and misconceptions involving preconditions. Plans which have no misconceptions at all are called **valid**.

The crucial difference between structural and constraint misconceptions is how the misconception is reflected in the plan hierarchy. For constraint misconceptions, the connections between nodes in the graph are not affected. A path which has only constraint misconceptions can still be located directly in the plan hierarchy. Paths with structural misconceptions, on the other hand, do not exist in the correct plan hierarchy at all. For example, representing an extra step involves adding a new arc to the hierarchy. In order to recognize ill-formed plans, the actual topology of the graph must be altered.

To illustrate this difference, let us once again consider the graduate student requesting a W2 form for his fellowship. In the case where the student has a general fellowship, the correct plan for recording his information does involve getting a form, but it should be a 1099 form instead of a W2 form (an example of a violated binding). This path can still be found in the correct plan hierarchy, because there is an arc connecting the `fill-in-general-fellowship` node and the `get-obj` node, but this particular plan is violated because of the violated binding. On the other hand, if the student has a Brown fellowship, there is no `get-obj` step at all under `fill-in-brown-fellowship`. If he really does believe that he is supposed to get something for his Brown fellowship, this constitutes an extra step, and no amount of searching in the correct hierarchy will find such a path. We would need to add an extra arc in order to recognize this plan, which changes the structure of the hierarchy and makes this an ill-formed plan.

There are two separate modifications to **PR-1** that need to be made in order to handle the two general classes of misconceptions. **PR-2** handles constraint misconceptions, and **PR-3** extends this to handle structural cases as well.

5 PR-2: Allowing Violated Plans

PR-1 can be modified to deal with violated plans fairly easily by changing the **Expand** procedure (Step 5). The key is that plans which only contain constraint misconceptions will already be generated by **PR-1**, and will even be recognized as invalid in Step 5.2. This verification process actually tests for the presence of all of the different types of constraint misconceptions. Bindings are checked to make sure that they are consistent with any pre-stored values. Preconditions are verified to ensure that they hold in the current state of the world. If the user mentioned any preconditions in his query, these are checked to make sure that they actually are preconditions for the current plan. And temporal constraints are checked as well.

The difficulty with **PR-1** is that when a violation is encountered, the improper plan is thrown out in Step 5.2.2. Once agents are allowed to have faulty plans, we can no longer discard a path just because it has a violation, because it could still

1. **1.** OPEN ← start
   CLOSED ← ∅.
2. **2.** If OPEN = ∅ then exit with failure.
3. **3.** Remove partial path p with optimal $f(p)$ from OPEN and place on CLOSED.
4. **4.** If the endpoint of p is a destination node, exit with the complete path formed by tracing the pointers back to start.
5. **5.** Otherwise, **Expand**(p):
   - **5.1.** Generate all successors of p (all partial paths from endpoint of p to the next known node), with pointers back to p.
   - **5.2.** For each successor p' of p, **Verify**(p'):
     - **5.2.1.** Check p' for violations.
     - **5.2.2.** If p' contains any violations, add the appropriate documentation to the path.
     - **5.2.3.** Regardless of violations, estimate $f(p')$ and add p' to OPEN.
6. **6.** Go to step 2.

Figure 2: The **PR-2** algorithm.

be the path that the user is intending to take. So rather than throwing out a path when a constraint misconception has been detected, we will instead add the successor path p' to the OPEN list whether or not it is part of a correct plan. In order to keep track of the fact that we have discovered a misconception in the agent's plan, we will add appropriate tags to the path which document the misconceptions. These misconceptions are also taken into account when calculating the heuristic function f. This gives us the ability to prefer plans with fewer (or less serious) misconceptions in the proper circumstances.

This modification produces the **PR-2** algorithm, as shown in Figure 2. Changes from the **PR-1** algorithm are underlined.

6 PR-3: Allowing Ill-Formed Plans

Ill-formed plans are a bit trickier to handle. While violated plans were found by **PR-1** and then discarded, plans with structural misconceptions were not even generated by **PR-1** in the first place. Since ill-formed plans do not exist in the correct plan hierarchy, **PR-1** will never search for them. So the method for generating successors

1. `OPEN` ← `start`
   `CLOSED` ← ∅.
2. If `OPEN` = ∅ then exit with failure.
3. Remove partial path p with optimal $f(p)$ from `OPEN` and place on `CLOSED`.
4. If the endpoint of p is a destination node, exit with the complete path formed by tracing the pointers back to `start`.
5. Otherwise, **Expand**(p):
   - **5.1.** Generate all successors of p (all partial paths from the endpoint of p to the next known node, plus other ill-formed partial paths), with pointers back to p. If a successor is ill-formed, add the appropriate documentation to the path.
   - **5.2.** For each successor p' of p, **Verify**(p'):
     - **5.2.1.** Check p' for violations.
     - **5.2.2.** If p' contains any violations, add the appropriate documentation to the path.
     - **5.2.3.** Regardless of violations, estimate $f(p')$ and add p' to `OPEN`.
6. Go to step 2.

Figure 3: The **PR-3** algorithm.

in Step 5.1 must be modified to include additional partial paths that do not exist in the plan hierarchy. These paths could result from substituting one step for another, or from adding a step to a path that does not contain it. They will involve changing the structure of the plan hierarchy graph by adding or deleting nodes or arcs in order to form new paths. But in order to keep the algorithm as general as possible, the particular method of generating the ill-formed paths is left up to the implementation.

This results in the **PR-3** algorithm in Figure 3. Constraint misconceptions are handled exactly as in **PR-2**. New changes for dealing with structural misconceptions are underlined. **PR-3** is now able to handle "perfect plan" recognition as well as the complete range of misconceptions presented in [4]. Therefore, **PR-3** qualifies as a "robust" plan recognition algorithm, since it can handle plan-based misconceptions even if they are new and unanticipated.

7 Analysis

It is instructive to analyze each of the three plan recognition algorithms in terms of the following properties:

- **Admissibility**: is the algorithm guaranteed to find the best solution? This can actually be three separate questions, based on whether the user's plan is valid, violated, or ill-formed. For example, an algorithm could be admissible for valid plans but not for violated plans.

- **Complexity**: how long does the algorithm take to find the answer in the worst case? Alternately, how many nodes of the graph does the algorithm expand before finding the correct path?

Note that having an admissible algorithm is *not* the same as having an admissible heuristic function. The only requirement in the latter case is that the heuristic is optimistic and monotonic; it does not have any requirements for applying this heuristic to all appropriate paths. An admissible algorithm must also guarantee that all of the relevant paths will be considered. We will assume that all algorithms always use an admissible heuristic function; the probabilistic heuristics that are actually used in the implementation of the Pathfinder program are admissible.

The A* algorithm

Pearl has shown that if the heuristic function f is admissible, then **A*** is guaranteed to find an optimal solution path if one exists [9, p. 86]. This admissibility result only holds for *valid* plans, since **A*** does not even have a concept of invalid plans.

He has also shown that "a necessary and sufficient condition for maintaining polynomial search complexity is that **A*** be guided by heuristics with logarithmic precision" [9, p. 184]. The complexity here is measured by the number of expanded nodes relative to N, the length of the shortest path acceptable for the answer. For simplicity, assume that this path will be roughly as long as the average depth of the graph (i.e., for the plan hierarchy, assume that the top-level goal is a generic plan and that the mentioned step is an atomic action). This means that **A*** will be polynomial in the depth of the plan hierarchy if and only if $f(p)$ is logarithmically precise. In other words, if $\phi(f)$ describes the growth of the typical error in $f(p)$, then $\phi(N)$ must grow no faster than $c(\log N)^k$ for arbitrary constant k to maintain polynomial complexity. This is a very stringent demand on the precision of the estimation, and it requires choosing the heuristic function carefully. But it still allows **A*** to be a polynomial algorithm under proper circumstances.

As long as f is an admissible heuristic, it is possible to implement **A*** in such a way that it will never visit a node more than once, irrespective of the precision of the heuristic. So for an upper bound, **A*** is guaranteed to be no worse than exponential in the depth of the graph.

The PR-1 algorithm

PR-1 is a direct adaptation of **A***; it generates and selects its nodes in the same manner as **A***, and it has the same requirement that the heuristic function is admissible. So **PR-1** is also admissible for valid plans.

The number of nodes expanded by **PR-1** may actually be *less* than **A***, since **A*** does not have any explicit verification process. Although **PR-1** and **A*** both generate and select nodes in the same way, **PR-1** has the additional step of pruning violated nodes, which can reduce the final number of expanded nodes.

The PR-2 algorithm

The results for **PR-2** are also encouraging. Recall that the change to produce **PR-2** was that paths were not pruned even if they contained violations. The heuristic function takes into account any violations that are detected, and as long as f is still required to be overly optimistic, the answer is still guaranteed to be optimal. This admissibility result now holds for both correct *and* violated plans. And since the original **A*** didn't prune violated paths in the first place (it didn't even have any concept of violation), we are not expanding any more paths than **A***. Assuming that constraints can be checked in an efficient manner, this means that **PR-2** is no worse than **A*** in complexity.

The PR-3 algorithm

Unfortunately, the results for **PR-3** are not as favorable. To produce this algorithm, the successor function was modified to generate additional partial paths. These paths do not even exist in the original plan hierarchy, so there is no way that **A*** would have expanded them. But since the algorithm does not specify how these ill-formed paths are generated, the **Expand** procedure could conceivably produce an exponential (or even larger) number of additional paths at every iteration of the algorithm. And even if the heuristic function is extremely precise, so that we will only expand a node if it lies along the solution path, simply generating the ill-formed paths in the first place may violate the polynomial runtime.

We now have problems with the admissibility claim as well. As long as the heuristic function is optimistic and monotonic, we will still find the optimal path *among those generated by the successor function.* So **PR-3** is still admissible for valid and violated plans. But Step 5.1, which generates the successors of a partial path, does not guarantee that all of the relevant ill-formed paths are generated. The true optimal path may have involved an ill-formed partial path that was never generated because the generation method itself was incomplete. In fact, it may even be the case that a non-exhaustive admissible successor function does not exist at all.

So the properties of **PR-3** are strongly tied to the method by which the successor function generates ill-formed partial paths. There appears to be a fundamental

	PR-1	PR-2	PR-3
Admissible:			
for valid plans	1	1	1
for violated plans	No	1	1
for ill-formed plans	No	No	3
Complexity	2	2	3

1: Admissible iff $f(p)$ is optimistic and monotonic.

2: Polynomial iff $f(p)$ is logarithmically precise.

3: Depends on both $f(p)$ and the successor function; possibly not admissible, probably at least exponential.

Figure 4: Summary of the analysis.

tradeoff between efficiency and admissibility. If too many partial paths are generated, we lose all hope of achieving a reasonable runtime. On the other hand, if we start restricting the successor function, we run the risk of missing the optimal solution. In order to handle ill-formed plans in a sensible manner, the successor function needs to be designed in such a way as to minimize both the number of extraneous paths generated and the chances of missing the actual path that the user chose.

The results of this analysis are summarized in Figure 4.

8 The Pathfinder Program

The general algorithm for robust plan recognition presented in **PR-3** has been optimized and implemented in a program called Pathfinder. Pathfinder is designed to be one component of an integrated natural language system. Its purpose is to act as an interface between the semantic interpreter and an application program such as a command executor or expert system. Rather than just blindly executing the user's commands or retrieving the information that the user requested, Pathfinder analyzes the user's goals and his command or query and attempts to determine if the request the user is making is reasonable.

There were a number of design decisions that were purposely left ambiguous in the presentation of the algorithms which need to be resolved in the Pathfinder program. One of these ambiguities was the selection of the "start" and "destination" nodes. Usually, this issue is decided based on the branching properties of the hierarchy. In the sample plan hierarchy that has been developed for Pathfinder, the average forward branching factor was slightly higher than the backward branching factor. But there were a few very common nodes that had backward branching factors that were twice as high as the largest forward branching factor. So Pathfinder currently runs in a top-down fashion; it starts with the user's top-level goal and works toward a destination of the mentioned step.

Another ambiguity in **PR-1** was the method by which successors were generated. In Pathfinder, well-formed successors are pre-computed and stored in a hash table for efficiency purposes. Each node in the hierarchy has an entry in the hash table which stores all possible partial paths from the root (generic plan) to the node. This information can be used to extract paths between arbitrary nodes very quickly[1]. Ill-formed successors (partial paths involving substituted or extra steps) are generated separately, and will be described later.

Finally, the heuristics that are used to guide the search are implemented using probabilities. For each path that is generated, Pathfinder calculates the probability that the user would actually be following the corresponding plan, and also calculates the probability that the user would actually have the misconceptions that the path entails. This provides an accurate method of choosing among competing explanations for ambiguous plans, as well as a means of measuring our confidence that we have successfully recognized the user's plan. The details of using probabilities for this sort of heuristic are discussed in [4, Ch. 6].

8.1 The Basic Algorithm

The algorithm used in Pathfinder is basically an optimized version of **PR-3**. While **PR-3** can deal properly with incorrect plans, it does not always verify or expand partial paths in an intelligent manner. It turns out that it is much more efficient to split the **Expand** and **Verify** tasks into two separate procedures. In **PR-1**, it made sense to verify partial paths as soon as they were generated, because violated paths were immediately pruned off, reducing the search space. But now paths are kept whether or not they contain any violations. Since the verification process may take up a significant proportion of the time spent on each node, it would now make more sense to postpone verification, and only do it on the partial paths that are likely to be included in the final answer.

In order to separate these two processes, Pathfinder maintains a third list, called the `GENERATED` list. Partial paths are placed on this list if they have been generated but have not yet been verified. The evolution of a partial path progresses from `GENERATED` (after it is created) to `OPEN` (after it is verified) to `CLOSED` (after it is expanded).

At any iteration of the algorithm, the most likely partial path can be selected from either `OPEN` *or* `GENERATED`. If the current partial path is from `OPEN`, it is removed and placed on `CLOSED` as before, but during the **Expand** procedure its generated

[1]Pathfinder is also capable of storing these paths directly, rather than deriving them from the paths to the root.

1. GENERATED ← ∅
   OPEN ← goal
   CLOSED ← ∅.
2. If (OPEN = ∅) and (GENERATED = ∅) then exit with failure.
3. Remove partial path p with optimal $f(p)$ from either OPEN or GENERATED. If p was from OPEN, place on CLOSED.
4. If p was from OPEN and the endpoint of p is the queried step, exit with the complete path formed by tracing the pointers back to goal.
5. Otherwise, if p was from OPEN, **Expand** (p):
   5.1. Generate all successors of p (all partial paths from the endpoint of p to the next mentioned node, plus other ill-formed partial paths), with pointers back to p. If a successor is ill-formed, add the appropriate documentation to the path.
   5.2. For each successor p' of p, estimate $f(p')$ and add p' to GENERATED.
6. Otherwise (p was from GENERATED), **Verify**(p):
   6.1. Check p for violations.
   6.2. If p contains any violations, add the appropriate documentation to the path.
   6.3. Regardless of violations, recalculate $f(p)$ and add p to OPEN.
7. Go to step 2.

Figure 5: The Pathfinder algorithm.

successors are placed on the GENERATED list without being verified. If the current partial path is from GENERATED, it is verified and placed on the OPEN list without being expanded. The Pathfinder algorithm is shown in Figure 5. As before, the changes from **PR-3** are underlined. Although Step 2 (the failure condition) is still included, it will never be used in the current implementation. Because of the way that the successor function works, Pathfinder is always guaranteed to find *some* path to the queried step (in the worst case, just a sequence of extra steps from the goal to the queried step).

One consequence of this new algorithm is that we have to estimate $f(p)$ before p has been verified (in order to choose the most likely unverified partial path from GENERATED). Once p is selected to be verified, $f(p)$ is recalculated to reflect any misconceptions that have been found along the partial path. This involves evaluating $f(p)$ twice for any path that comes from the GENERATED list. So two properties have to hold in order for this method to be feasible: the process of evaluating $f(p)$ (at least the first estimation of $f(p)$) should be inexpensive relative to verifying p itself, and the first estimation should still be accurate. The first property guarantees that the overhead of recalculation is justified for individual partial paths, and the second guarantees that it is justified in the long run by reducing the number of paths that are verified.

8.2 Improving the Efficiency of Ill-Formed Paths

Pathfinder can use two methods to overcome the inherent difficulty involved with ill-formed paths. The first method is to restrict the number of additional paths by using intelligent filtering in the successor function. For instance, Pathfinder will only consider a path with a substituted step if that substituted step is judged by the heuristic function to be "close enough" to the user's mentioned step, and it will only retain that path if it also connects to the next mentioned step. This drastically reduces the number of additional partial paths that the successor function would normally generate. Although this filtering violates the admissibility property of the algorithm in theory, since not all possible successors are being generated, it is a highly reliable method in practice. It will only miss the user's path when the user has at least two incorrect mentioned steps in a row, one of which is a very dissimilar substituted step. In such cases, the user's plan is so seriously flawed that there is very little hope of correctly deciphering his intentions anyway.

The other method for improving Pathfinder's efficiency is to exploit a property of the classification of plan-based misconceptions. It is sometimes possible for a particular misconception to be classified in two different ways, depending on the structure of the plan hierarchy [4, p. 63]. The most common ambiguity is between violated bindings and substituted steps. And it happens that substituted steps are structural misconceptions, which are difficult to detect, while violated bindings are constraint misconceptions, which are much easier to detect. So if it is possible to design the hierarchy in such a way as to minimize the number of potential substituted steps, this can make the algorithm run more efficiently. Without sacrificing the expressiveness and naturalness of the plan hierarchy, the knowledge engineer should attempt to minimize the number of similar steps in the hierarchy. If two steps that accomplish essentially the same task can be combined by introducing an extra parameter, then this should be preferred to keeping them distinct. This will cause the ambiguous misconceptions to tend toward the easier violated bindings instead of the harder substituted steps, and will reduce the number of ill-formed paths that Pathfinder considers.

Finally, remember that just because Pathfinder generates a path does not mean that the path will actually be explored. Assuming that the initial estimates of the probabilities are reasonably accurate, only the most likely paths will be verified.

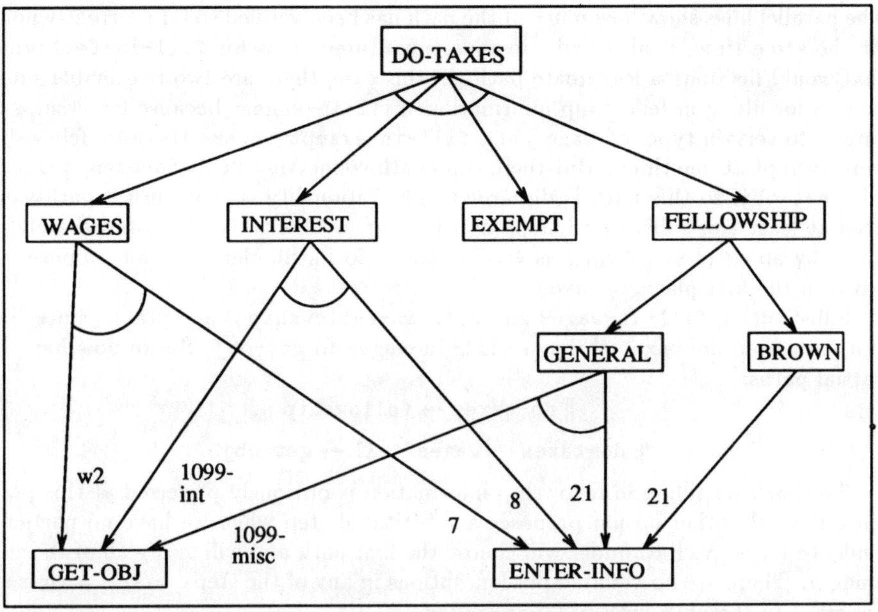

Figure 6: A sample plan hierarchy.

9 A Demonstration

This section simulates the Pathfinder algorithm running on the query from Section 2:

> "I'm trying to fill out my taxes. Where do I get the W2 form for my fellowship?"

It uses the simple plan hierarchy shown in Figure 6. Although the actual program uses probabilities for the heuristics, this demonstration just shows a relative ranking of the various alternatives to simplify matters.

Pathfinder is given a list of three mentioned steps from the parser:

(do-taxes) → ··· → (fill-in-fellowship f235) → ··· → (get-obj form-w2).

It starts out with the top-level goal of do-taxes and finds all ways to extend this to fill-in-fellowship. In this particular hierarchy, there is only one path that connects the two mentioned steps:

|| do-taxes → fellowship.

The parallel lines show how much of the path has been verified so far (currently none). At the same time, it also finds possible substituted steps for `fill-in-fellowship` that would lie along a legitimate path. In this case, there are two reasonable substitutions for filling in fellowship information: `fill-in-wages` (because fellowships are similar to certain types of wages) and `fill-in-exempt-income` (because fellowships were exempt at one time). But there is no path connecting `fill-in-exempt-income` with `get-obj`, so this path is discarded. The rationality is that such a path would need at least two substituted steps in a row, or a substituted step immediately followed by an extra step, which is very unlikely. So Pathfinder does not propose such paths in the first place, to save time.

Substituting `fill-in-wages` and `fill-in-fellowship` is acceptable, since there is a path from `do-taxes` through `fill-in-wages` to `get-obj`. So we now have two partial paths:

$$\| \texttt{do-taxes} \rightarrow \texttt{fellowship}$$

$$\| \texttt{do-taxes} \rightarrow \texttt{wages/fell} \rightarrow \texttt{get-obj}.$$

The path for filling in fellowship information is obviously preferred at this point, since the only other option proposes a substituted step when we have no particular evidence for it. So Pathfinder will choose the first path and will verify all of the steps along it. There are no violated misconceptions in any of the steps, so this path is still preferred to the path with `fill-in-wages`.

Now Pathfinder expands `fill-in-fellowship` down to `get-obj`. There is only one way to do this in the hierarchy, by passing through `fill-in-general-fellowship`. Since this is a very likely path, and since we still have not encountered any misconceptions along this path, Pathfinder will still prefer it to the explanation with a substituted step. So we have:

$$\texttt{do-taxes} \rightarrow \texttt{fellowship} \;\|\!\rightarrow \texttt{general} \rightarrow \texttt{get-obj}$$

$$\| \texttt{do-taxes} \rightarrow \texttt{wages/fell} \rightarrow \texttt{get-obj}.$$

Since we still have not encountered any misconceptions, this path is still preferred to the one with `fill-in-wages`. So Pathfinder will verify the rest of this path. Now it encounters a misconception: since the student's fellowship is from Brown University and is not a general fellowship, there is a precondition in the `fill-in-general-fellowship` plan that does not hold. The most likely explanation here is that the student was not aware that Brown fellowships didn't apply here, so this shows up as a missing precondition. There is also a violated binding on the form that he is supposed to get; the student mentioned a W2 form, but the correct binding under `fill-in-general-fellowship` is a 1099 form. This lowers the likelihood of this path, and since it involves a misconception, Pathfinder tries to find another possible path connecting `fill-in-fellowship` and `get-obj` that avoids this misconception. There is another possibility; the user might be trying to go through `fill-in-brown-fellowship`, which would avoid the misconceptions, but which would require an extra step. Assuming that the missing precondition explanation is still preferred, this gives:

$$\texttt{do-taxes} \rightarrow \texttt{fellowship} \rightarrow \texttt{general*} \rightarrow \texttt{get-obj*} \parallel$$

$$\texttt{do-taxes} \rightarrow \texttt{fellowship} \parallel \rightarrow \texttt{brown} \rightarrow \texttt{get-obj+}$$

$$\parallel \texttt{do-taxes} \rightarrow \texttt{wages/fell} \rightarrow \texttt{get-obj}.$$

The violated binding and missing precondition are marked with a "*", and the extra step is marked with a "+".

When Pathfinder is actually run on this example, it does in fact prefer the explanation that passes through `fill-in-general-fellowship`. It postulates that the user is missing the precondition about fellowship types, and that he has a violated binding on the form that he is supposed to get. This explanation is only slightly preferred over the other two, so the response generator would probably want to phrase the response in such a way as to correct all three possible misconceptions[2].

The results of this simulation are summarized in Figure 7.

10 Complexity Analysis of Pathfinder

This section provides a best-case and worst-case analysis of the Pathfinder program. The analyses of the three other algorithms in Section 7 measured the total number of nodes expanded. But since the expansion process is broken into two pieces in Pathfinder, and since the **Verify** procedure is more costly than generating the successors, we will measure the performance of Pathfinder by the total number of nodes that are verified by the algorithm. The following notation is used:

G = graph (plan hierarchy)
T = tree corresponding to G with duplication of nodes
m = number of mentioned nodes (not counting the root)
n_i = mentioned node in G (and in T)
b = backward branching factor (ambiguity) of G
f = forward branching factor of G (and of T)
d = depth of G (and of T).

The analysis can be made much simpler with the following assumptions:

- There is one unique root, n_0.
- The search starts at n_0.

[2]The current probabilities that are assigned to the three explanations are 0.48 for the missing precondition and violated binding explanation, 0.41 for the extra step explanation, and 0.11 for the substituted step explanation.

	NOTES	GENERATED	OPEN	CLOSED
0	Start		T	
1	Expand T	TF $T{}^{W}\!/_{F}O$		T
2	Verify TF	$T{}^{W}\!/_{F}O$	TF	T
3	Expand TF	FGO FBO^{+} $T{}^{W}\!/_{F}O$		T TF
4	Verify FGO	FBO^{+} $T{}^{W}\!/_{F}O$	$FG^{*}O^{*}$	T TF
5	Exit $TFG^{*}O$	FBO^{+} $T{}^{W}\!/_{F}O$		T TF $FG^{*}O^{*}$

T = do-taxes   G = fill-in-general-fellowship
W = fill-in-wages   B = fill-in-brown-fellowship
F = fill-in-fellowship   O = get-obj

${}^{X}\!/_{Y}$ = substituted step
X^{+} = extra step
X^{*} = constraint misconception

Figure 7: Simulation of Pathfinder on the fellowship example.

- The queried step n_m is at a leaf.
- All leaves are at the same depth (namely, d).
- The branching factors b and f are constant for the entire graph.
- Mentioned nodes are evenly spaced along the path (at a distance of d/m).

In the typical graph searching problems, all important attributes are located at the nodes of the graph. The only values that might be found on the arcs are weights that are used by the heuristic function. In contrast, the plan hierarchy also has attribute information stored on the arcs. For example, binding constraints for a step need to be stored on the arcs because this information may change depending on which plan invokes the step. So instead of performing the analysis on the graph G, it makes more sense to refer to the tree T, which is formed from G by making a separate copy of a node for every incoming arc. This increases the number of nodes, and reduces the backward branching factor b to 1. But it does not affect the forward branching factor f, nor does it affect the depth d.

10.1 Worst-Case Analysis of Pathfinder

In the type of plan recognition that Pathfinder does, the search is constrained to pass through each mentioned node. This is very helpful because the search problem can now be decomposed in such a way that an optimal solution to the entire problem can be constructed from optimal solutions to the subproblems. In other words, the optimal path from the root to n_m is constructed from optimal subpaths between each of the consecutive mentioned nodes.

For the moment, assume that there are no ill-formed misconceptions; all mentioned steps can be found along a path that already exists in the hierarchy. Since Pathfinder has all of the partial paths pre-computed in a hash table, it can construct all well-formed paths between two consecutive mentioned steps directly without actually visiting any of the nodes. It will only visit a node during the verification process, and this is done when a partial path is selected by the **Verify** procedure.

In the worst case, all partial paths between n_i and n_{i+1} will be verified before any are extended to n_{i+2}. Then all will be extended to n_{i+2}, and all of these new extensions will be verified, until one of the extensions finally reaches n_m (the queried step). It may be the very last extension that we verify that ends up being the optimal one, so even at the bottom level we may need to visit all of the nodes in all of the partial paths.

Each partial path contains $\frac{d}{m} + 1$ nodes. But except for the very first path, the root of each partial path will have already been verified on the previous iteration, so it doesn't need to be verified again. This gives $\frac{d}{m}$ nodes that need to be verified along each partial path. There are $f^{\frac{d}{m}}$ partial paths of length $\frac{d}{m}$ starting at n_i, so since each partial path is verified separately, an upper bound on the number of verified nodes between n_i and n_{i+1} is $\frac{d}{m}(f^{\frac{d}{m}})$. This upper bound is only realized in the degenerate case where every partial path out of n_i passes through n_{i+1}.

Note that the combinatorial explosion is completely contained in the subtrees. No matter how many paths there are between two mentioned steps, only one of them (the one judged to be the most optimal) will be expanded, and it will be connected to the optimal path between the next two mentioned steps. So the worst case total cost for Pathfinder (not including ill-formed paths) is simply the number of nodes verified between consecutive mentioned steps, times the number of mentioned steps (m), or

$$1 + d\left(f^{\frac{d}{m}}\right)$$

verified nodes. The leading 1 is the cost of visiting the root of T.

This makes sense, since when $m = 1$ (only the queried step is mentioned) this gives $1 + d(f^d)$ expanded nodes. This corresponds to a complete traversal of every path in the tree from the root to the leaves. And when $m = d$ (every step except the root is mentioned), this gives a worst case of $1 + fd$ expanded nodes. This can be interpreted that at every level of the tree, we will select one node, generate all children of that node, and move to the next level.

Now let us consider the possibility of ill-formed paths. Missing steps are only detectable in special cases, and they take no more time than searching for a correct queried step in the worst case. Extra steps are much cheaper than correct steps; since there is no partial path associated with an extra step, the only node that needs to be verified is the step itself. So they will not occur in a worst-case scenario, and the only additional possibility that we need to consider is the case of substituted steps.

Recall that substituted steps are proposed by finding appropriate substitutions for n_i and making sure that those substitutions lie along an existing path from n_{i-1} to n_{i+1}. In the worst case, every possible path from n_{i-1} to n_{i+1} that does not go through n_i itself can contain a suitable substitution. So the only paths that will *not* be generated during the search for substitutions are the ones that contain n_i, which will already have been generated during the well-formed phase. This means that we are effectively doubling the length of the partial paths; at each mentioned step n_{i-1} instead of finding all partial paths to n_i, we will find all partial paths to n_{i+1} whether they pass through n_i or not.

This gives a total of

$$1 + 2d\left(f^{\frac{2d}{m}}\right)$$

nodes that Pathfinder will have to verify in the worst case, including ill-formed paths.

Note that this upper bound will never be reached in practice. It involves degenerate trees where all leaves are the same, it requires that the heuristic function provide no information at all about alternative paths, and it requires that all substituted steps are equally likely. The number of partial paths between two mentioned steps will be much smaller in any realistic hierarchy, and as long as the heuristic function tells us *anything* useful, many of the paths will never be verified (or even fully extended) in the first place.

It is obvious from this worst case analysis that the mentioned steps play a critical role in the efficiency of Pathfinder. The additional information provided by these mentioned steps is used to constrain the exponential search to isolated partitions of the hierarchy. As m increases and more steps are mentioned, the exponent of the formula decreases, resulting in fewer expanded nodes. As a comparison, the traditional **A*** algorithm does not make use of any intermediate nodes at all. So in the worst case it will verify every node in the tree T, giving a complexity of

$$\frac{f^{d+1}+1}{f-1}$$

verified nodes. So as long as $m \geq 2$, Pathfinder will do at least as well as **A*** in the worst case (and will typically do much better), and will be able to recognize invalid plans that **A*** cannot handle. Pathfinder's advantage comes from the fact that it can make use of the additional information supplied by the mentioned steps to constrain the search space. The more steps that are mentioned, the better Pathfinder does.

10.2 Best-Case Analysis of Pathfinder

In the best case, Pathfinder's heuristic will have perfect knowledge and it will only verify nodes that lie along the final path. Also, in the best case there will be no appropriate steps that can be substituted for anything along the final path, so ill-formed paths can be ignored. So between n_i and n_{i+1} we will only verify the one partial path that will form the final answer; then we will find all extensions to n_{i+2} but again will only verify steps along one of the paths. Note that in the ideal case we do not need to visit any nodes to select the optimal partial path, since an initial estimate of the heuristic can be pre-computed and stored in the hash table along with the path itself. If this initial estimate is perfectly accurate, its value will not decrease during verification, so the first path that Pathfinder selects to verify will remain the optimal choice. So including the root node, this gives us

$$1+d$$

verified nodes for the best case complexity of Pathfinder.

11 Runtime Performance of Pathfinder

So far, Pathfinder has been run on 39 real-life examples taken from actual dialogues containing plan-based misconceptions [3]. It has also been run on 19 variations on these examples (for instance, changing certain facts about the world, or changing the query slightly), for a total of 58 examples.

[E] In EMACS you can define what is called a "REGION".... What you do is move the cursor to the beginning or the end of the block, type esc-M, and move [to the other end].... OK?

[U] ... beginning of Region How ?

[E] Type esc-M.... Now end of region. You move the cursor down to the end of the region. OK?

[U] That's fine, how do I mark, esc-W ?

[E] No, you don't have to mark the end of the region. Only one extreme needs to be marked with the esc-M.

Figure 8: Pathfinder misses the Extra Step.

Accuracy Results

Pathfinder performed remarkably well on the 58 examples. In 46 cases (79% of the examples), it returned interpretations that are indistinguishable from what a human observer would conclude. When the answer was unambiguous, Pathfinder returned the correct answer. And when there were multiple reasonable interpretations, it returned exactly the interpretations that a human observer would, and assigned very believable probabilities to the possible plans.

In 7 additional cases (12% of the examples), Pathfinder also came up with certain highly unlikely possibilities that a person might never have considered in the first place. But it still assigned reasonable probabilities to them (on the order of 10^{-5} to 10^{-10}). So these 7 cases should also be judged "successful".

There were 4 examples (7%) where Pathfinder proposed the appropriate interpretations, but did not assign reasonable probabilities to all of them. This is primarily because the methods for calculating the probabilities of misconceptions are only partially implemented. It should be noted that in all 4 of these examples, Pathfinder still correctly prefers the explanation that people find most plausible.

The only example where Pathfinder came up with the wrong explanation is shown in Figure 8. It correctly catches the violated binding in the user's second question (using esc-W instead of esc-M). But it misses the extra step of setting the mark for a second time. This is because Pathfinder's temporal reasoning is rather weak. It prefers to think that there are two completely separate `define-region` events happening, which require two `set-mark` steps, rather than the normal interpretation that there is only one `define-region` event which has an extra step. Stronger temporal reasoning abilities will correct this problem.

So in all but one of the cases (98% of the examples), Pathfinder correctly selects the best explanation for the user's plan. It should be noted that Pathfinder achieved this performance using only partially implemented heuristic information; once the remaining heuristic features are added to the program, the probabilities should improve even more. This is an extremely strong argument in favor of probabilistic plan recognition. It is also clear that this is not just a "blocks-world" program; Pathfinder achieved this accuracy while running on 58 real-world problems from several different domains.

Time Results

In addition to being highly accurate, the Pathfinder program also suggests that robust plan recognition can be automated in an efficient manner with current technology. Although the worst-case analysis of Pathfinder in Section 10 is alarmingly high, Pathfinder does not even come close to this upper bound in real life.

For these benchmark results, Pathfinder was run on a Sun SparcStation with 16 megabytes of memory, using compiled (but non-optimized) lisp code. It used a non-trivial plan hierarchy consisting of all the plans necessary for the 58 running examples. The graph for this hierarchy contains 157 unique nodes. The depth of this graph is 11 nodes, and it has a maximum forward branching factor of 10 and a maximum backward branching factor of 19.

All 58 examples were run completely, generating all possible interpretations rather than stopping as soon as the most likely explanation was reached. 80% of the examples ran in less than 1.0 second, and every example except one ran in under 2.0 seconds[3]. Many of these examples were multiply ambiguous; one generated 12 different interpretations (6 of which were given probabilities lower than 0.01) and still ran in only 1 second. When Pathfinder was rerun, stopping as soon as it generated the most likely path, most of the examples ran at least two to three times faster.

12 Conclusions

We have shown that the **A*** algorithm can be directly applied to the basic plan recognition problem while keeping its desirable properties of admissibility and the possibility of polynomial runtime. Two relatively minor modifications can extend the robustness by allowing the algorithm to recognize situations in which the user has an incorrect plan. Violated plans are those plans that are formed correctly but are inappropriate in the current situation because of a mistake involving preconditions, bindings, temporal constraints, or optimality. These are easy to handle, and an algorithm to identify them can still retain the desirable properties of the **A*** algorithm. Ill-formed plans are those plans whose structure is incorrectly formed, because of a substituted, missing, or extra step. There is a fundamental difference between violated and ill-formed plans; identifying ill-formed plans is inherently much more difficult, and we cannot guarantee admissibility of the algorithm for this class of plans.

[3]The only example that took more than 2 seconds was the one in Fig. 8, which took 4.3 seconds.

Two critical decisions in implementing the algorithm are the selection of a powerful heuristic function and the selection of an appropriate successor function. The first function is the means by which the search space is pruned, and the second function is the means by which the original search space is expanded to include ill-formed paths. An optimal heuristic function would guarantee that the minimum number of partial paths will be selected for expansion, and an optimal generation function would guarantee that the partial paths that *are* selected will generate a minimum number of successors. If either function is too restrictive, the algorithm will no longer be admissible. If they are too permissive, the algorithm becomes intractable.

The Pathfinder program is an optimized implementation of the **PR-3** algorithm that is capable of doing efficient robust plan recognition in the presence of user misconceptions. It guides its search using probabilities, and overcomes the inherent difficulty of ill-formed plans by using pragmatic filtering heuristics. It can detect the entire range of plan-based misconceptions, and can handle traditionally difficult situations such as multiple misconceptions and multiple ambiguous interpretations.

Acknowledgments

Much of the work described in this paper was done while the author was at Brown University. Special thanks go to Eugene Charniak, Tom Dean, and Jeff Vitter. An earlier version of this paper was presented at the 1989 IEEE International Workshop on Tools for Artificial Intelligence, and appeared in [5].

References

[1] A. Bagchi and A. Mahanti. Search algorithms under different kinds of heuristics — a comparative study. *Journal of the ACM*, 30(1):1–21, 1983.

[2] A. Bagchi and A. Mahanti. Three approaches to heuristic search in networks. *Journal of the ACM*, 32(1):1–27, 1985.

[3] R. J. Calistri. An annotated compendium of naturally occurring plan-based misconceptions. Technical Report CS-89-37, Department of Computer Science, Brown University, Providence, Rhode Island, Oct. 1989.

[4] R. J. Calistri. *Classifying and Detecting Plan-Based Misconceptions for Robust Plan Recognition.* Ph.D. dissertation, Department of Computer Science, Brown University, Providence, Rhode Island, May 1990.

[5] R. J. Calistri. A modified A* algorithm for robust plan recognition. *International Journal of Pattern Recognition and Artificial Intelligence*, 4(2):233–244, June 1990.

[6] R. Dechter and J. Pearl. Generalized best-first search strategies and the optimality of A*. *Journal of the ACM*, 32(3):505–536, 1985.

[7] P. Hart, N. Nilsson, and B. Raphael. A formal basis for the heuristic determination of minimum cost paths. *IEEE Transactions on Systems Science and Cybernetics*, SSC-4(2):100–107, 1968.

[8] H. A. Kautz. *A Formal Theory of Plan Recognition.* PhD thesis, University of Rochester, 1987.

[9] J. Pearl. *Heuristics: Intelligent Search Strategies for Computer Problem Solving.* Addison-Wesley Publishing Company, Reading, Massachusetts, 1984.

[10] M. E. Pollack. *Inferring Domain Plans in Question-Answering.* PhD thesis, Department of Computer and Information Science, University of Pennsylvania, Philadelphia, Pennsylvania, 1986.

[11] A. Quilici, M. G. Dyer, and M. Flowers. Recognizing and responding to plan-oriented misconceptions. *Computational Linguistics*, 14(3):38–52, Sept. 1988.

[12] J. R. Slagle and P. Bursky. Experiments with a multipurpose, theorem-proving heuristic program. *Journal of the ACM*, 15(1):85–99, 1968.

Applications of Learning and Planning Methods
edited by N. G. Bourbakis
© 1991 World Scientific Publishing Co., pp. 253–275

CHAPTER 11

Differential A*: An Adaptive Search Method Illustrated with Robot Path Planning for Moving Obstacles & Goals, and an Uncertain Environment

KAREN I. TROVATO
Philips Laboratories - North American Philips Corporation,
345 Scarborough Road
Briarcliff Manor, New York 10510 USA

ABSTRACT

The robot path planning problem is to plan optimal paths for a robot to follow while ensuring it will not hit any obstacles or itself. In a static or perfectly known world, this has been addressed using the configuration space representation and the A* search algorithm. However, when movement, changes, or unexpected obstacles occur in the environment, a new method, Differential A*, can adapt the solution to the current situation. It updates only the fraction of space that is critically affected. This technique can provide significant speed improvements, with the same desired results, compared to complete space regeneration.

1.0 Introduction

The robot path planning problem is to move a robot from one configuration to another without hitting anything, and to move optimally in some sense (least end-effector distance, effort, etc.). One method to solve this problem is to view the robot and environment in configuration space and generate optimal paths using A*[9]. This method is general, for N-dimensional problems and many optimality criteria.

This paper presents "Differential A*", a method that builds on the A*/configuration-space approach to adapt quickly to changes in the space by determining and updating the localized regions affected by those changes rather than regenerating the entire space. This is particularly effective with moving obstacles or goals and in an uncertain environment because only small parts of the space are affected at a time.

Section 2.0 presents the A* search algorithm and its relationship to the configuration space method of path planning. Described in section 3.0 is the connection of A* to wave propagation in configuration space for path planning. In section 4.0, the Differential A* method is outlined, with the focus on path planning. Section 5.0 presents examples of moving obstacles and goals, and planning in an uncertain environment. Timing for the sample cases is provided in Section 6.0 and indicates that great improvements in speed can sometimes be achieved.

2.0 Background: A* Search

The A* algorithm is a search technique that visits nodes in a graph from a goal node to a start node, searching in the expected direction of the starting node. The search is guided by a heuristic specific to the problem domain. A* with an admissible heuristic, produces optimal paths from a starting position to a goal position. It searches the space in between and leaves pointers (arrows) at each node to mark the path. *There may be more than one arrow leaving a node if equal alternative routes exist.* To apply A*, the search problem must be described in a graph (or graph-like) form where there are "nodes", "transitions", and "start" and "goal" nodes. The transitions have costs and are usually directed connections between nodes. The A* algorithm can quickly find the connection (series of transitions in the graph) between the start and goal. The heuristic guides the search and is often the most optimistic estimate of the distance to the goal. A zero heuristic searches the entire space and is useful in robotics if a goal is repetitively used but the starting positions vary.

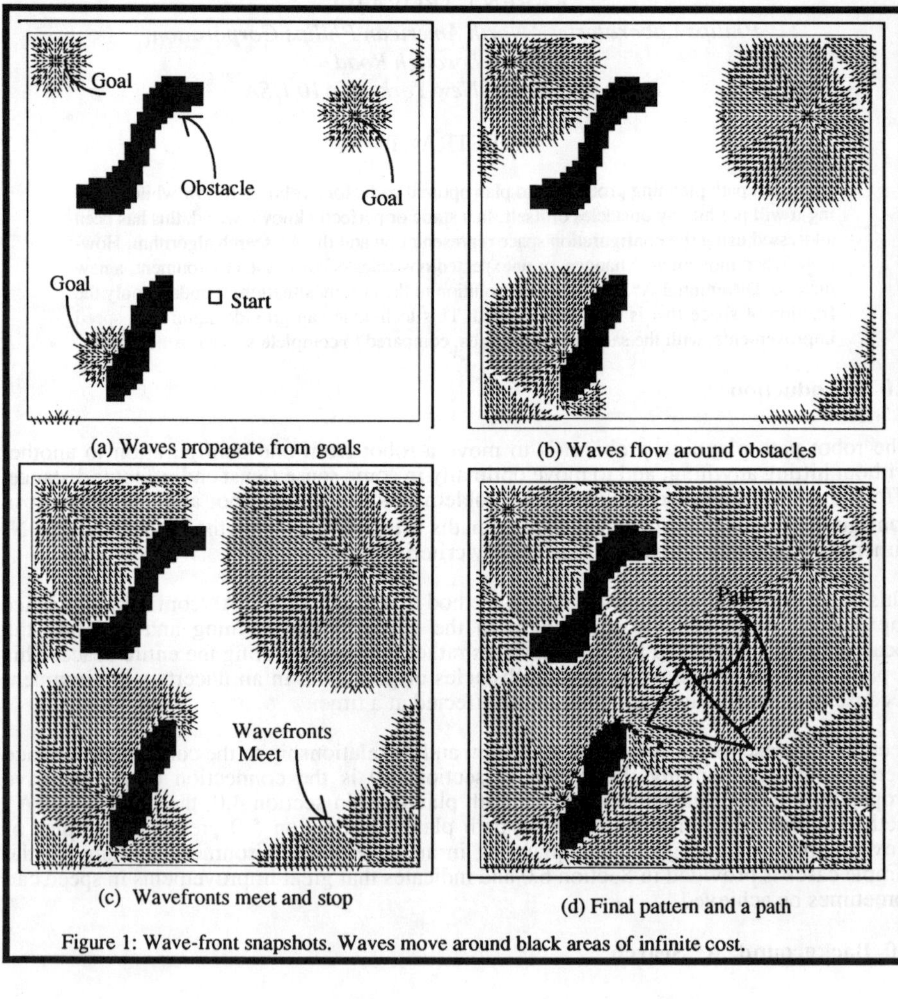

Figure 1: Wave-front snapshots. Waves move around black areas of infinite cost.

The algorithm searches from the goal, in the most probable direction of the start. The search first visits the goal node's neighbor-nodes; these are the nodes connected to the goal via a transition. The neighbors evaluated as most likely to be in the direction of the start are visited next. If a transition is forbidden or obstructed, the cost will be infinite, and the searching will not continue in this direction. This way, the obstructions cause the search to be pruned so that resulting paths point around them. The cycle of visiting and evaluating continues until the start is found.

Figure 1 shows a 64 by 64 grid of interconnected nodes. For clarity, the delineation of nodes is removed. Each node is connected to 16 local neighbors surrounding it, and the topology allows both vertical and horizontal wrap-around. Using A* and a zero heuristic with three goal nodes and two forbidden regions, snapshots of the intermediate searching are shown. The path between a start and a goal is seen in 1(d) as a trace of black nodes. Only the resulting pointers are shown, not the original graph. In examples like this, the spanning of the graph by A* is reminiscent of waves propagating through an area, and is often referred to as wave propagation[4].

3.0 Wave Propagation and the Relationship to Optimal Path Planning

The wave propagation method is useful for finding the path of a point through a space to a goal. Because a robot is not a point, but rather a complex interconnection of joints and bodies, it is necessary to find a transformation from the real world (task space) to another problem space where the robot can be thought of as a point. In the next section such a "configuration space" is identified.

3.1 Configuration Space Representation of the Path Planning Problem

The optimal path planning problem is to compute a path to move a robot, person, etc. from a starting position to the nearest of a set of goal positions "optimally" according to a preferred criterion (minimum motion, shortest end effector distance, least time, etc.) without hitting any obstacles. This apparently difficult problem can be simplified by considering the problem in terms of its configuration space (CS) rather than the actual task space of the moving bodies. The reformulation of the problem in this space is common in robotics [7, 10, 4].

Obstacles and self-intersections (the robot body hitting itself), are represented as forbidden regions; one difficulty using the CS technique is determining them. This is a difficult task, although it has been solved efficiently for specific cases and some progress has been made for arbitrary polygons in two dimensions [1].There are no fast, general solutions for arbitrarily shaped obstacles in a high (particularly >4) dimensional spaces to date. Fortunately, not all problems must be cast in this way. In this paper, obstacles are transformed in real-time by the method of Newman [8]. There are some problem representations, for example the forbidden positions of a vehicle based on its projection, where the obstacle transformation is much less complicated than that of the robot-arm.

In Figure 2 a simple two-link robot and its associated configuration space are shown. The

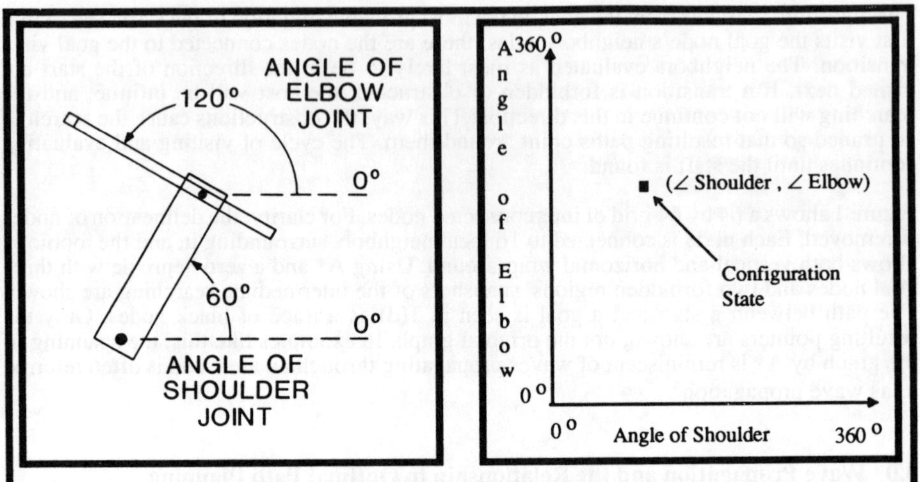

Figure 2: Two-Link Robot and its associated Configuration Space

examples in the remainder of the paper are given for this robot, however the method is not specific to either this space or this robot. Also, there is a link between general graph terminology, in which A* is usually described, and the configuration space terminology of the path planning problem. Chart 1 shows the association of terms.

General Graphs	Config. Space Path Planning
Node	Configuration State or Point
Transition	Transition
Transition Cost	Metric Transition Cost
——	Obstacle State
Connection Graph or Network	Configuration Space
Graph Search	Wave Propagation
Solution Graph	Solution Space or pattern of arrows/pointers

Chart 1: Corresponding terminology for A^* with general graphs and CS Path Planning

3.2 The Path Planning Method

A recent path planning method [10,4] generates the optimal path to follow from a starting state to the nearest goal state in terms of discretized configuration states. The interconnection of neighboring states can be considered a “connected-graph”. We can use A* by simultaneously propagating from all goal states, leaving arrows indicating the optimal transition from each state toward a goal. The method will provide the best global connection sequence from every possible state in the configuration space. Optimal paths will be well defined for every starting state, and will pass through adjacent states (themselves having an optimal path) to the goal. The set of local arrows describing the limited search required to provide the path is called a “solution graph”. It need not span the space.

The connection graph (configuration space in the path planning example) is a collection of states with directed connections weighted by transition costs. Goal states have a zero cost-to-goal and are equally desirable final state(s) of the system.

A* is implemented in the configuration space in the following way: Beginning with the goal states, the neighborhood of local points *(n')* to each configuration state becomes the set of states to be evaluated at each expanded configuration state *n*. The transition between *n'* and *n* can be described by a cost $m_C(n,n')$ that we refer to as the metric. We can establish the lowest cost connection from any given state to the nearest of a set of goals by performing a search of the entire space of states in a least-total-cost first manner.

This method handles various optimality criteria (metrics) to guide the behavior of the robot. A* allows an efficient expansion of the states regardless of the cost metric. Solutions for least joint motion (Figure 3), least end-effector motion (Figure 4), and least effort (Figure 5), have all been computed without any change in the fundamental method. Least joint motion is the shortest distance in configuration space. Least end-effector motion will reduce the distance that the tip of the robot will move, yielding the straightest possible moves. This behavior is also illustrated in figures 9 and 10 where the end-effector-over-shoulder point is clearly exploited because of a transition cost of zero in those configurations. The least effort criterion minimizes the use of high mass robot links to reduce power requirements. For the example in Figure 5, the first link is double the mass of the second link.

3.3 Features of the A Path Planning Method*

The resulting traversal of all reachable configuration states yields gradient arrows directing motion toward the goal. This is particularly useful for robots because the planning may not have taken into account dynamics, or the robot may not be able to follow a predicted path due to mechanical wear. If the entire space is covered and the robot fails to track perfectly, then it can begin from any current state, and try to move optimally toward the goal. In addition, if there are no direction arrows at a starting state, then the path to the goal is blocked.

For each newly added obstacle configuration, a new wave propagation using A* is required. Although this can be performed quickly (< 2 seconds on a Sun 3/260 for the

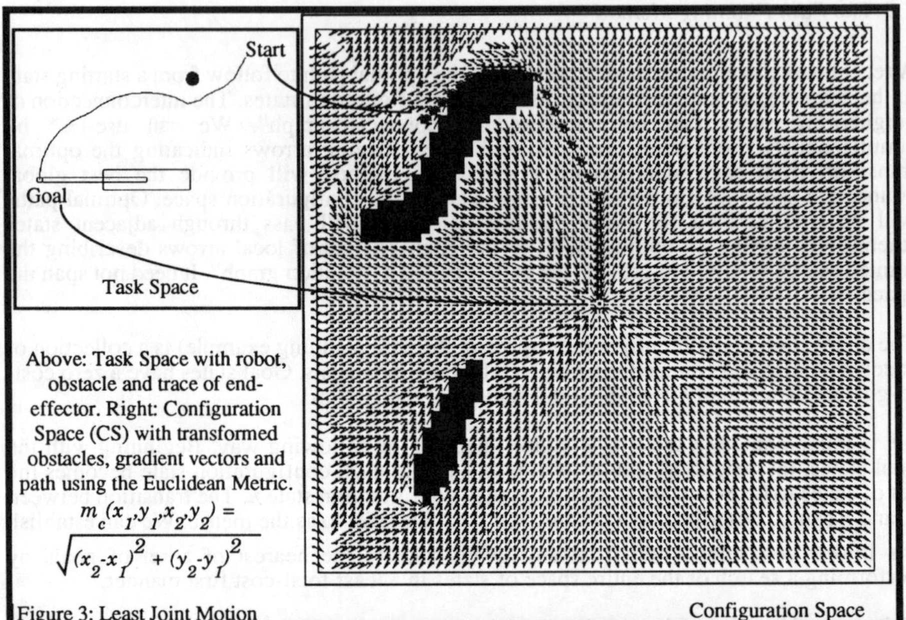

Above: Task Space with robot, obstacle and trace of end-effector. Right: Configuration Space (CS) with transformed obstacles, gradient vectors and path using the Euclidean Metric.

$$m_c(x_1,y_1,x_2,y_2) = \sqrt{(x_2-x_1)^2 + (y_2-y_1)^2}$$

Figure 3: Least Joint Motion

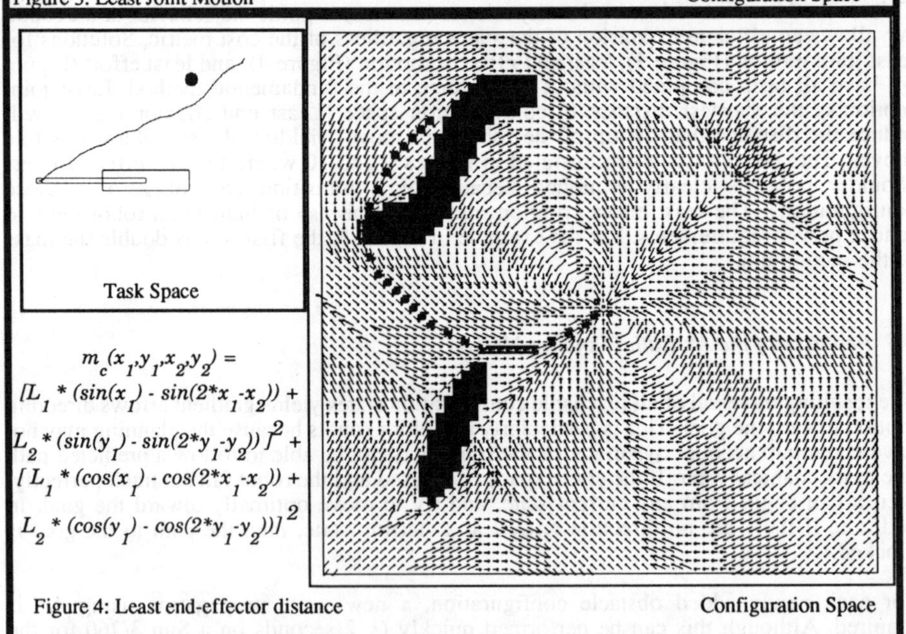

$$m_c(x_1,y_1,x_2,y_2) = [L_1 * (\sin(x_1) - \sin(2*x_1-x_2)) + L_2 * (\sin(y_1) - \sin(2*y_1-y_2))]^2 + [L_1 * (\cos(x_1) - \cos(2*x_1-x_2)) + L_2 * (\cos(y_1) - \cos(2*y_1-y_2))]^2$$

Figure 4: Least end-effector distance

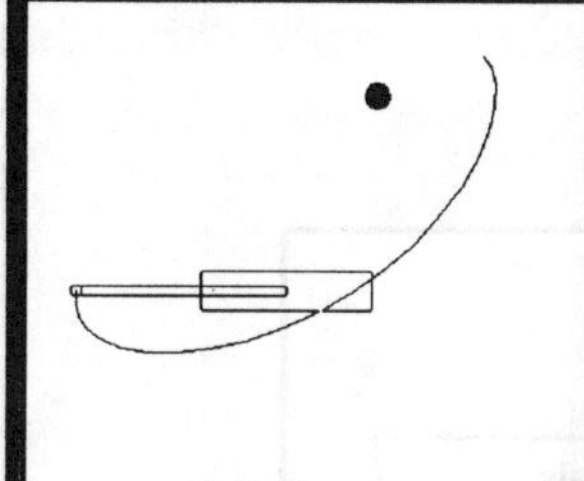

Task Space

Above: Task Space min path corresponding to Config. Space (right). Metric cost for change in joint angle where m_1 and m_2 are mass of link 1 and link 2:

$$m_c(\delta\theta_1,\delta\theta_2) = \sqrt{\left(\frac{m_2}{m_1}\delta\theta_1\right)^2 + \left(\frac{m_2}{m_1}\delta\theta_2\right)^2}$$

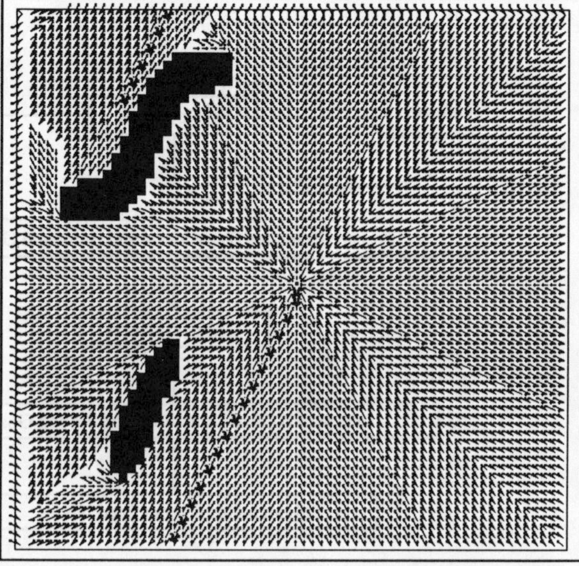

Configuration Space

Figure 5: Least Effort

example in Figure 3) for two degrees of freedom (dof), the fact that the algorithm is exponential in the number of degrees of freedom is motivation to find improvements that are also a function of the dof. One such improvement is specifically targeted at adaptation to changes in the robot environment and uses the Differential A* Method.

4.0 Overview of Differential A*

In the original A* method, any change in the connection graph of nodes forces complete recalculation to obtain the solution graph. Depending on the problem domain, this may be quite often. It is rare that the direction arrows differ completely after recalculation due to a few new or removed nodes. This feature can be exploited to speed the solution. In this sense, the algorithm can adapt the solution to new information. The method identifies the added or removed goals and transitions, and recomputes only the affected areas of the existing solution graph. The affected area is not dependent on the heuristic used, but only on the arrows in the solution graph.

Other techniques are also available to update changes in the configuration space. One method in particular [3], identifies regions that are swept out by the occurrence of obstacles and identifies the shadow areas from the vertices of the obstacle. It is applicable for generating waves using the Constrained Distance Transform [5,6,2]. It also inherits some of the limitations of the CDT, namely, difficulty with variable metrics and unstructured cost transitions.

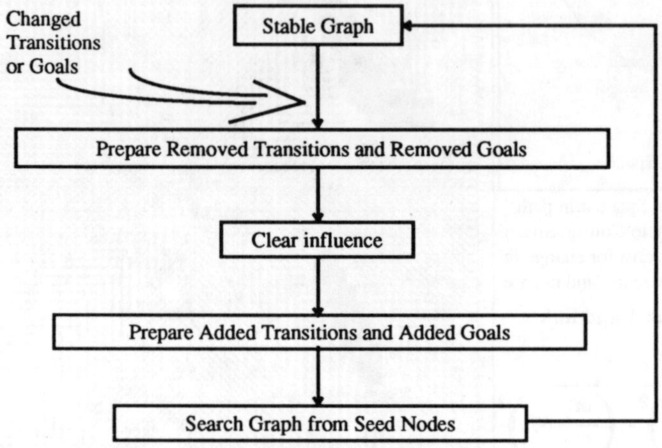

Figure 6: Differential A^* Algorithm

The changes in the connection graph affect only transitions. In path planning, addition of obstacles results in removal of transitions, and removal of obstacles results in introduction of transitions. A change in a transition cost should be classified as an added transition if the cost decreases and as a removed transition if the cost increases. If an entire node is removed or added, then the corresponding transitions must also be removed or added.

Although Differential A* is a method that was originally discovered as a result of performing A* path planning, it is generally applicable to A* problems with changes. Therefore, the algorithm is described in the context of graphs, although the illustrations will be for path planning.

4.1 Differential A Algorithm*

The algorithm consists of five stages and is shown in block diagram form in Figure 6; the corresponding pseudo-code is given in the Appendix. The objective of Differential A* is to adapt the pre-existing stable solution-graph to the new changes and produce a new solution-graph that is identical to one that would have been generated from complete initialization and A*. The essence of the algorithm is to clear and revisit the area influenced by transitions of newly increased cost and only revisit areas affected by transitions of newly decreased cost. The purpose of "clearing" an area is to reset the node values to "uninitialized" so that subsequent wave propagation can correct the area with a new, higher cost. This is necessary since wave propagation is only used to initialize or lower costs during searching.

At the beginning of the algorithm, either an unvisited (initial condition) connection graph *or* a previously-searched solution graph, is proper input for the method. When all searching activity has ceased, a graph *(S)* will be considered stable (unchanging for the moment). Figure 7(a) shows an example of a stable configuration space graph generated with an admissible heuristic which computes the most optimistic cost-to-start. (Remember that the actual cost is propagated from the goal).

The first step of the algorithm is to detect changes in *S* described as added-or-removed transitions and goals. It is also possible to include information about a changed transition cost. If a transition cost increases from the original transition cost, the transition is treated as a "removed transition", otherwise it is treated as an "added transition". The transitions blocked (removed) by the obstacle are situated along the obstacle periphery. If there is a removed goal, then the transitions leading into the goal would become the removed solution space transitions. This case is illustrated later, in Figure 12.

The second step re-initializes the area affected by the removed transitions, and determines the "perimeter" of this "affected area". The area affected by the removed transitions is determined by re-initializing all pointers and costs (from the solution graph) whose paths lead through a removed transition. In practice, any searching method beginning with the removed transitions will be effective. This will reinitialize all directly removed transitions or nodes such as the goal, in addition to reinitializing all of the child nodes (which can be expressed as predecessors in a path). The reinitialized region will be called the affected area. In addition, the perimeter of the affected area will also be identified by this method. Perimeter nodes *P* are the leaves of the search and will become "seed nodes", sources of wave propagation, in a future step.

Removal of transitions involves altering the solution graph so that the transition is invalid, and adding the "parent" (source of the original transition) to a sorted structure such as a sifted heap *(H)*. The least cost node, a, is retrieved from *H*, whereupon all "child" nodes are identified from the connected graph. A child node is one where a path connection exists directly from the child to the parent. All child nodes, that are not forbidden nodes, are added to *H*. An a that is taken from *H* that has no child node is stored in another structure called the perimeter *P*. By repeating this simple procedure until the heap is empty, all affected nodes will be detected, and *P* will be found.

The third step identifies added transitions. If a goal is added, then transitions connecting the goal to its local neighbors will become new transitions in the solution-graph. An obstacle that is removed will also generate new transitions, because the old transitions will no longer be forbidden. These added transitions are simply more seed nodes for wave propagation. (Transitions with lowered cost will also be included here as though they were added transitions.)

The fourth step is to perform wave propagation from the set of nodes in H_S representing the perimeter of the partially cleared solution graph, and the parents of added transitions. Wave propagation using the A* method is initiated from the lowest node of H_S. When the graph search is completed, the resulting interconnection, the new solution graph, will give the local directions at each node to generate globally minimum paths to the nearest goals.

(a) Stable Solution Graph

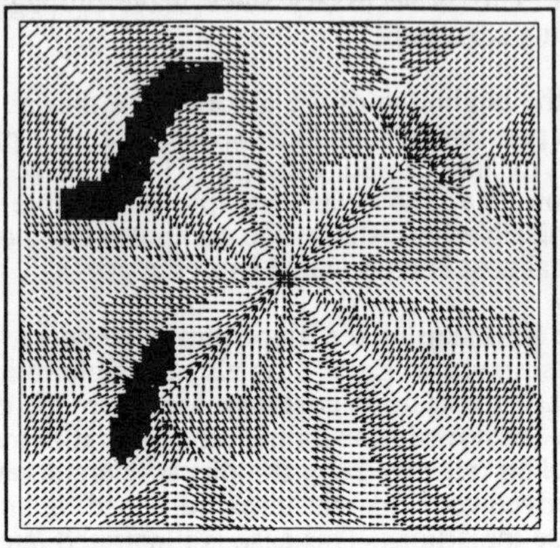

(b) New Obstacle Blocks Some Transitions

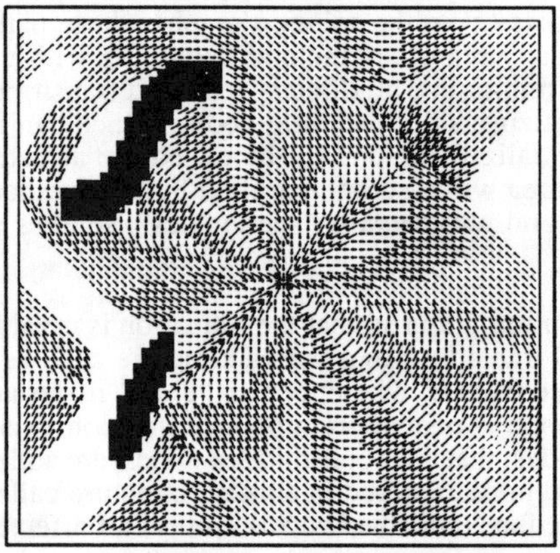

(c) Obstacle's Affected Area Cleared.

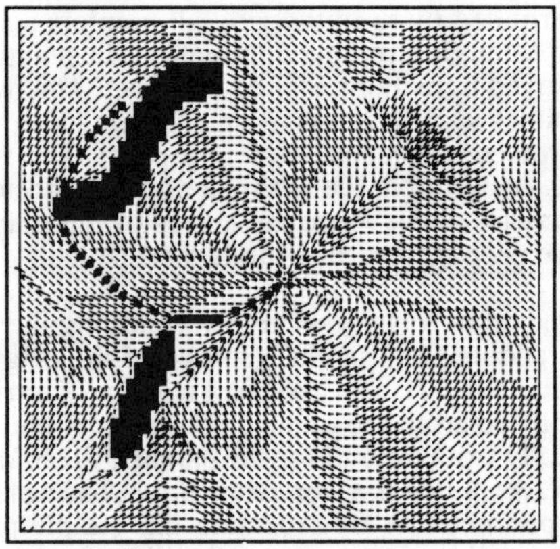

(d) Arrow Pattern Adjusted

Figure 7: Obstacle Appears.

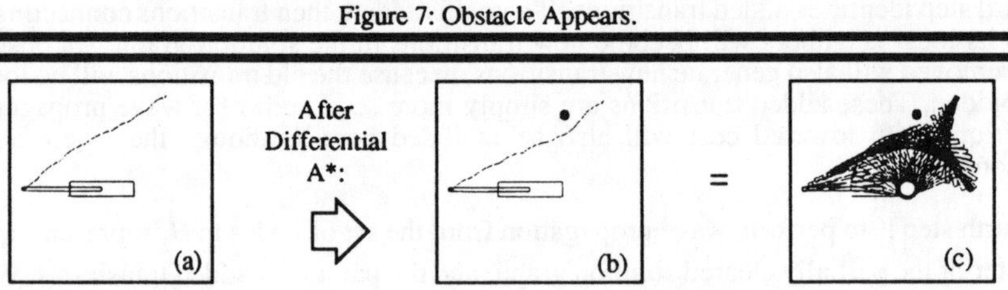

Figure 8: (a) shows Task Space corresponding to CS in 7(a). (b) shows TS of 7(d). (c) shows (b) motion as arm sweep.

5.0 Characteristics of Differential A* in Path Planning

In the case of path planning, it is essential for a general planning algorithm to be able to handle multi-dimensional (and hence large) configuration spaces, since a robot having many degrees of freedom will require it. Often, the optimal path does not need to be completely recomputed when a change affects only a small part of the space.

In a practical robotics problem, a useful artifact of this process is that the robot may be able to carry out an action while the space is being recalculated. This is possible with the Differential A* method if the starting state is not in the affected area (shown as white in Figure 7(c)).

5.1 Examples with Changing Obstacles and Goals

This section shows an example of the Differential A* technique applied to a two-link robot in an environment (as in Figure 2) where: a round black obstacle in the environment suddenly appears (stepwise incorporation in Figure 8), moves (Figure 9), and then disappears completely (Figure 10). Round obstacles are transformed into S shapes in configuration space. The white areas of Figure 9 are the regions affected by the obstacle's sudden presence.

Figure 11 shows seven added goal states. Because the goals are specified in task space, there are generally right and left handed configurations that can be represented symmetrically in configuration space. The addition of new goals is a one step process. Each of the new goals produces reduced cost transitions and becomes a source of wave propagation. Since none of the transitions is deleted or increases in cost, there is no affected area. Removal of one of these goals is shown in Figure 12. This situation will increase the cost of transitions, producing an affected area, in white. It is filled in by the subsequent wave propagation.

Figure 13 shows an example where the minimum end-effector distance from node to start is the admissible heuristic used to guide the search. After an obstacle is introduced, the arrows are cleared behind it, and then repropagation from the existing arrows leads the search to the start. By starting with Fig. 13(a) and Differentially computing, Figure 13(c) required 61% of the time it would have taken to re-compute from scratch.

5.2 Application to Path Planning in an Uncertain Environment

In robotics, often the environment in which the robot moves is changing or unclear. If a robot follows an optimal path based on knowledge of obstacles that was assumed to be correct and complete, yet detects (perhaps with a proximity sensor such as skin or sonar) an unexpected obstacle, it is preferable to be able to incorporate this incremental information quickly so that a more informed path can be followed. Newly-discovered obstructed states in the current environment path may be transformed to a set of forbidden states, which, by using the Differential A* method, can be incorporated quickly into the solution space. In figure 14(a), an obstacle that the robot does not know about is introduced

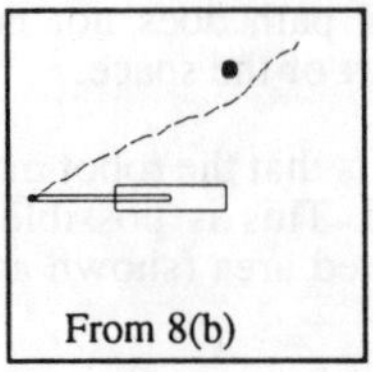

Above: Prior Path from 8(b).
Below: Arm Sweep showing obs. moved into prior path.
Right: CS after clear-influence step.

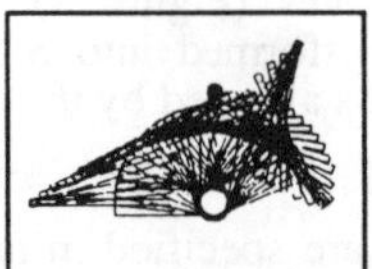

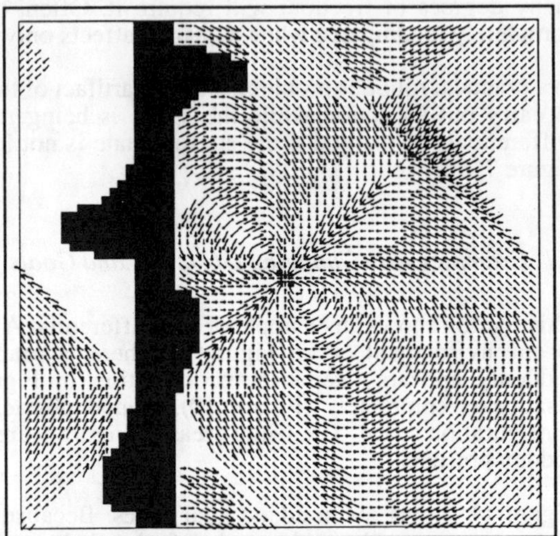

Below: CS after Differential A*.

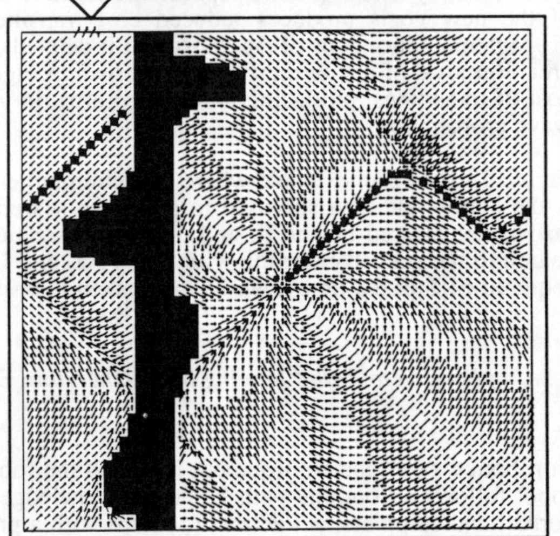

Right: Solution Space determined by Differential A*.
Below: Task Space path corresponding to new CS path.

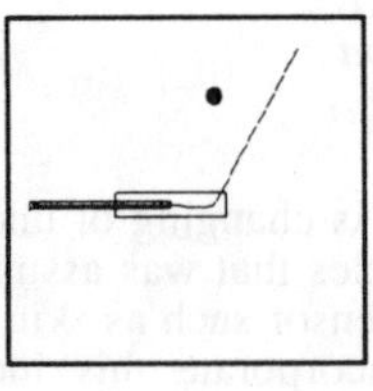

Figure 9: Obstacle Moves.

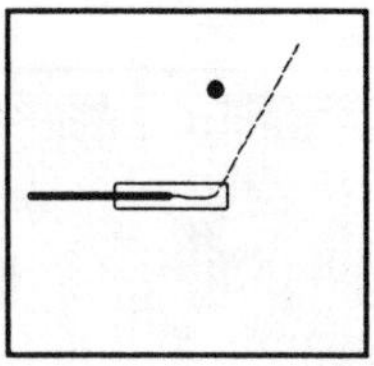

Above: Path with obstacle. When obstacle is removed from reach, the obstacle states disappear from Config Space (right).

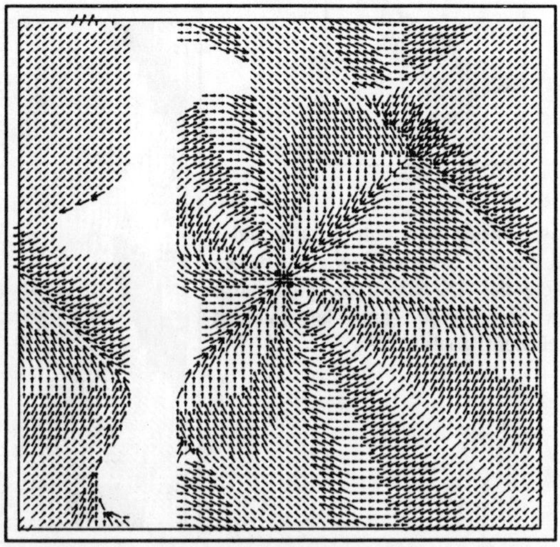

Below: CS after Differential A*.

Below: Task Space path corresponding to new CS path (right).

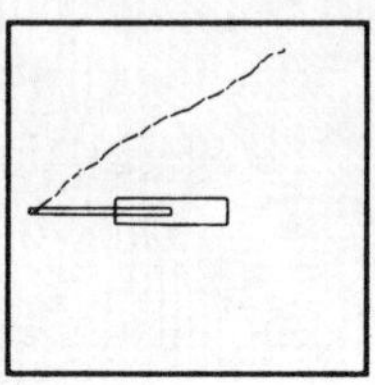

Figure 10: Obstacle Disappears.

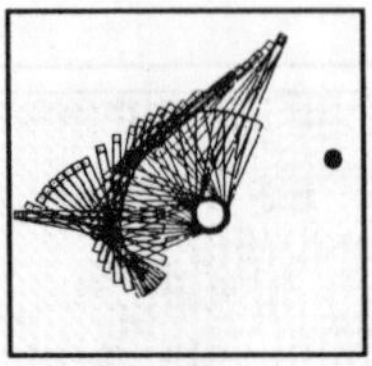

Above: Task Space with obstacle and 1 goal. Right: CS with 7 new goals before Differential A*. Six of the goals arise from three right and left handed (equivalent) goals.

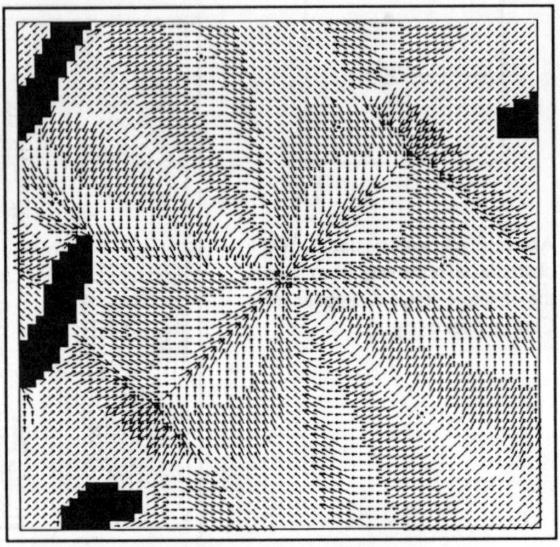

Below: CS after Differential A*.

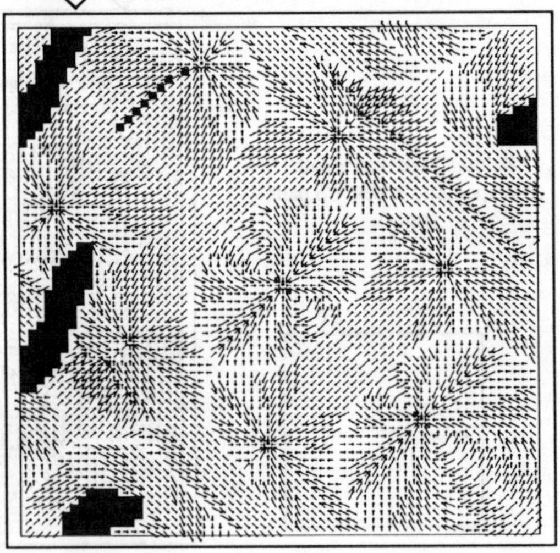

Right: Computed Configuration Space. Below: new path in task space to nearer goal.

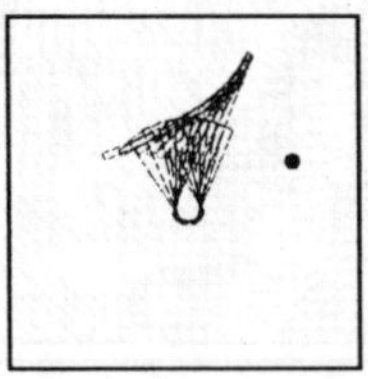

Figure 11: Seven New Goal States in C.S. (from 3 in task space) plus the Center Goal.

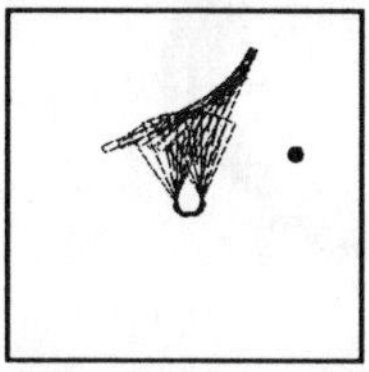

Above: Task Space with path from Fig. 11. Right: CS with center goal removed.

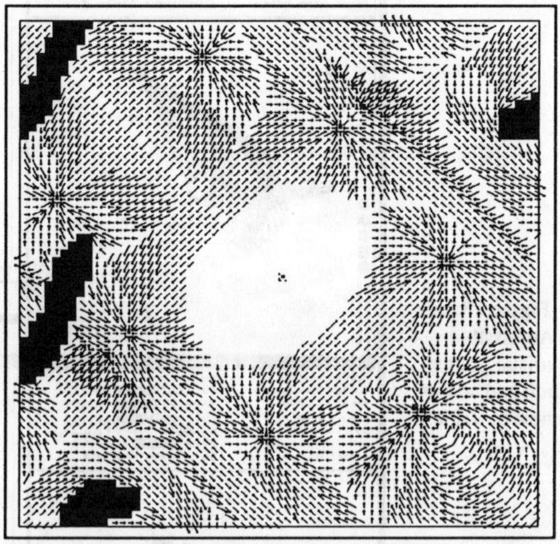

Below: CS after Differential A*.

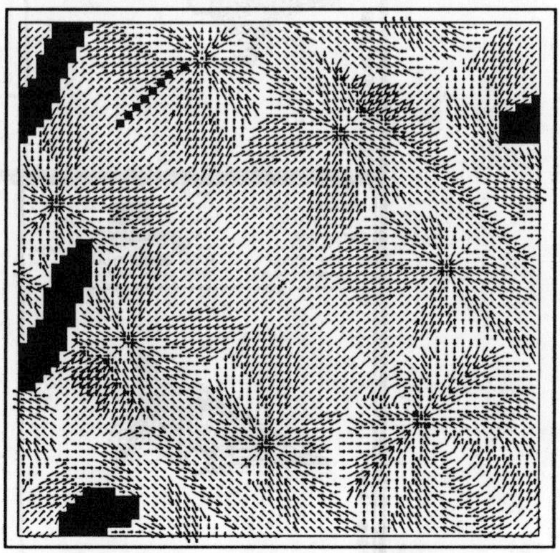

Right: Updated CS.
Below: original path in task space; goal removal caused changes distant from prior path.

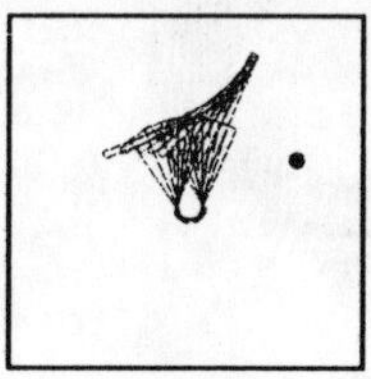

Figure 12: Removed Center Goal. Old path unaffected.

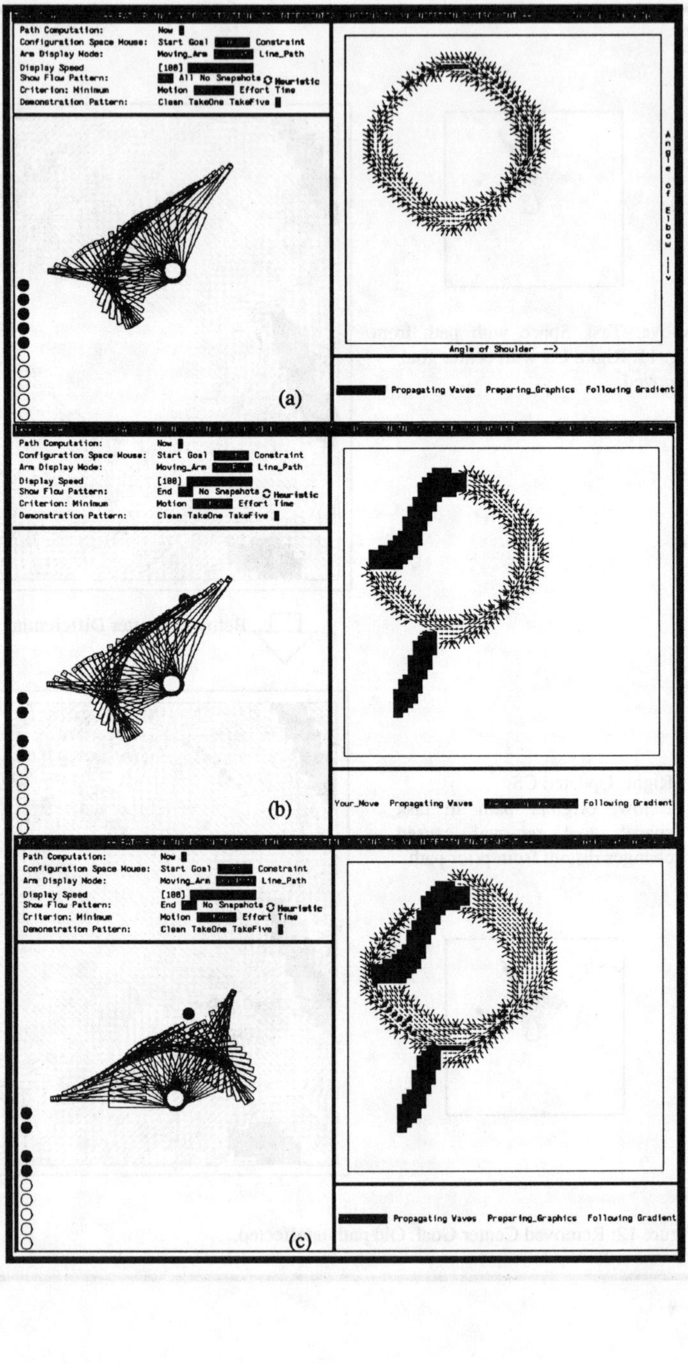

Figure 13: Admissible heuristic guides search and adapts to new obstacle.

Learning About an Obstacle.

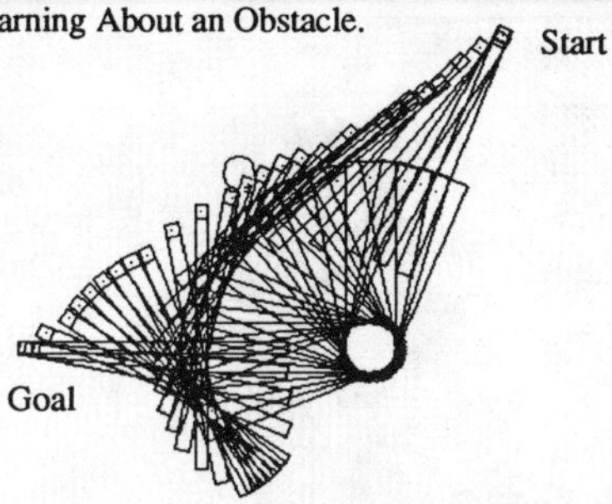

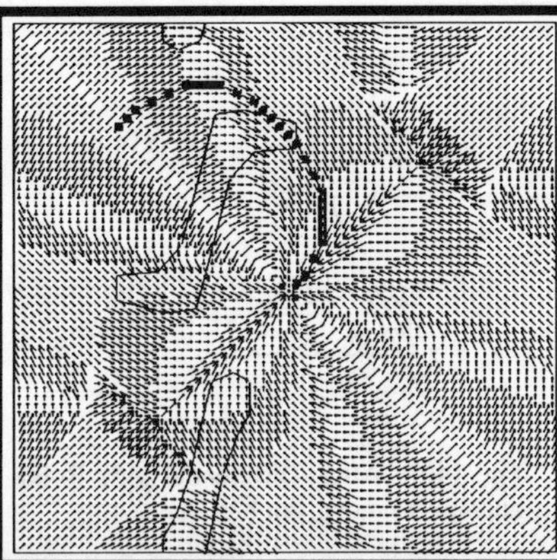

Fig 14(a) Proposed path with an unforeseen obstacle. Right: CS

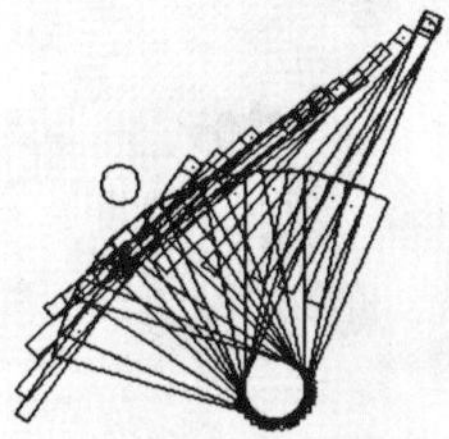

Fig 14(b) Robot first senses the obstacle. Right: Influenced CS Area is cleared.

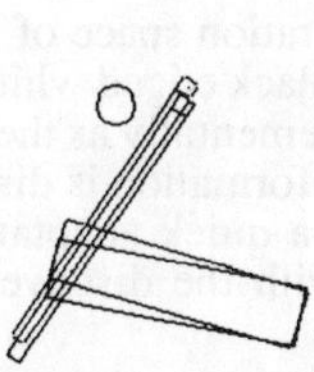

Fig 14(c) After Differential A*, Robot attempts to move, but hits again. Right: Influenced CS Area is cleared.

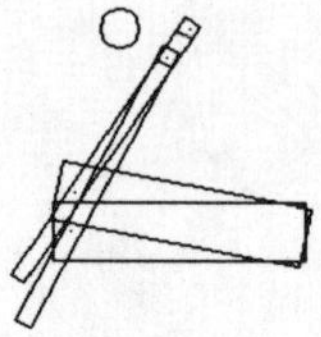

Fig 14(d) After Differential A*, Robot attempts to move, but hits again.

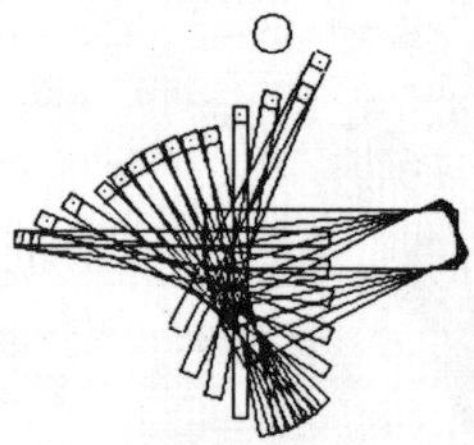

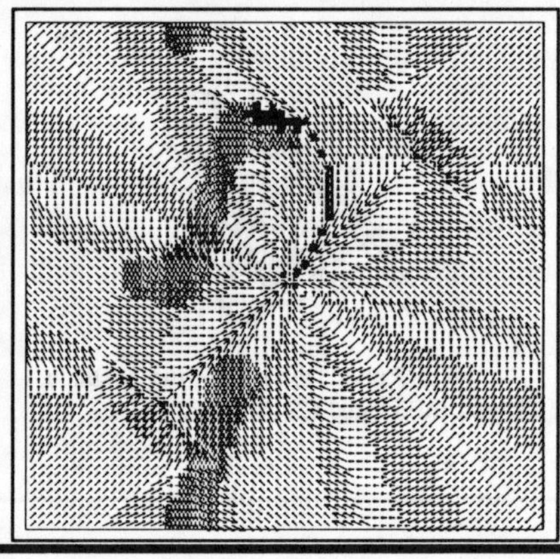

Fig 14(e) After Differential A*, robot is clear of obstacle to goal. Right: Final path piece.

into the environment. The obstacle is depicted in The configuration space of Figure 14 as a black edged obstacle later filled in grey. In task space, it is a black edged white circle. The robot discovers it only when it is hit, and it is incorporated incrementally as the robot "feels around the obstacle". Since only a small amount of obstacle information is discovered and incorporated at a time, the Differential A* method provides a quick adaptation. Figures 14(b-e) shows the configuration space change by change, with the discovered obstacle states represented as small black states.

Further, a blind robot, that is, one with only proximity sensors, might learn about an environment by exploring, so that a self-made configuration space map could be built for future use.

Figure Number	Percent of Time Spent Clearing	Percent of Time Spent Propagating	Total Time (msec.)	Diff-A* Time Required Compared to A* (Diff-A* Time/A* Time)
3: motion metric w/one obs	51	49	1638	50%
4 & 8: dist metric w/one obs	50	50	2119	50%
5: effort metric w/ one obs	42	58	1978	43%
9: move obs	43	57	1779	45%
10: obstacle disappears	0	100	1779	40%
11: New Goals	0	100	3719	90%
12: Remove one goal.	40	60	939	13%
14: Unexpected Obstacle				
b	41	59	438	10%
c	37	63	159	3%
d	40	60	398	9%

Figure 15: Timing for examples.

5.3 Timing

The speed of this method is affected by the location of the change. In general, the closer (in cost) a newly forbidden transition is to the goal, the larger will be the affected area. The calculation to re-initialize and re-propagate this area is obviously proportional to the number of states. This makes it difficult to generalize about the timing. However, since the clearing time is very fast because only arrows are used, and no evaluation function is computed, a significant improvement can be achieved in the average case. For major changes in the configuration space, the problem can be computed somewhat more efficiently by starting the entire wave propagation procedure over.

In the timing chart, Figure 15, speeds are given for the examples in the paper. The percent of time spent clearing the affected area, identifying its perimeter, and repropagating are given. Also, the speed is given for Differential A* compared with A* restarted. Differential A* takes about 50% of the time A* for the adding one obstacle; it takes about 3.5% of the time to incorporate a small amount of the discovered obstacle. The worst possible case (not shown in the examples) occurs in an environment with no obstacles when the one and only goal has moved. This forces the clearing of the entire space followed by repropagation of the entire space. This is roughly on the order of time necessary to repropagate from scratch.

6.0 Memory Requirements

The additional memory requirement for this method beyond A*, is that amount necessary to store a surface in one less dimension than the problem. This is because the clear-influence stage stores the surface of the cleared region.Usually the surface is quite minimal-particularly if heuristics are used. The remainder of the procedure uses the same memory space.

7.0 Conclusion

Differential A* is a general, fast method for updating solution spaces generated by A* when changes to the underlying connections occur. The added; removed, increased or decreased transitions are identified, and the consequences of only those changes are calculated. This is often a small fraction of the entire solution graph that is ordinarily calculated with A*.

The characteristics of A* are also inherited by Differential A*, namely: power to optimally propagate with arbitrary cost transitions and topology, ability to handle multiple dimensions, and timing that is (worst case) exponential in the number of dimensions. The primary advantage of Differential A* is rapid adaptation in environments that have changes affecting only part of the solution-space. Although the examples in this paper fill the entire configuration space (for illustration purposes) this rarely happens in practice; once the start is reached by wave propagation, a path is known, and wave propagation can cease. Differential A* works just as well in these partially filled spaces.

One domain where this is useful is in Path Planning in robotics when unexpected obstacles appear. With this new process, the robot 1) continues movement because the previous solution is usually not affected, 2) encounters obstacles on the fly and incorporates this knowledge, and 3) makes the best possible maneuver based on all knowledge possessed.

Differential A* is applicable for situations that currently use conventional A*, and can often considerably increase the speed of the solution. Good candidate problems are those with transition cost changes, and generally stable goals with multiple starts.

Acknowledgment

This work could come to its full fruition only with the support and enthusiasm of Dr. Leo Dorst - an animated character and my mentor. Special thanks to Dr. Roy Featherstone and Dr. Dave Schaffer for their valuable comments on this manuscript.

References

[1] F. Avnaim, J.D. Boissonnat, and B.Faverjon, A practical exact motion planning algorithm for polygonal objects amidst polygonal obstacles, *Proceedings 1988 IEEE International conference on Robotics and Automation*, (1988) pp. 1656-1661.

[2] G. Borgefors, Distance Transformations in Arbitrary Dimensions. *Computer Vision, Graphics and Image Processing*, **Vol. 27**, (1984) pp. 321-345.

[3] T. Boult, Updating Distance Maps when Objects Move, *1987 SPIE Conference on Mobile Robots*. (1987).

[4] L. Dorst and K. Trovato, Optimal Path Planning by Cost Wave Propagation in Metric Configuration Space, *SPIE Conference on Advances in Intelligent Robotics Systems*, **Vol. 1007** (1988) pp.186-197.

[5] L. Dorst and P. Verbeek, The Constrained Distance Transformation: A Pseudo-Euclidean, Recursive Implementation of the Lee-Algorithm. *Proceedings of the European Signal Processing Conference*, (1986).

[6] C.Y. Lee, An Algorithm for Path Connections and Its Applications, *IRE Transactions on Electronic Computers*, (September, 1961) pp. 346-365.

[7] T. Lozano-Pérez and M. A. Wesley, An Algorithm for Planning Collision-Free Paths Among Polyhedral Obstacles, *Communications of the ACM*, **no. 10** (1979) pp. 560-570.

[8] W. Newman, *High-Speed Robot Control in complex Environments*, Ph.D. Thesis, MIT, (October, 1987).

[9] J. Pearl, *Heuristics: Intelligent Search Strategies for Computer Problem Solving.* (Addison Wesley, Reading, Mass., 1984).

[10] B. Verwer, Heuristic Search in Robot Configuration Space using Variable Metric, *Third Conference on Artificial Intelligence for Space Applications, NASA Conference Publication 2492*, (1987) pp. 153-154.

Appendix: Pseudo-Code for Differential A* Path Planning

Notation:

S	Stable solution graph (may be unspanned or partially spanned)
S_t	S at the beginning of the algorithm.
S_{t+1}	S at the completion of the algorithm.
G	Set of equivalent goal states.
G_{add}	Set of goal states that are in S_{t+1} but not in S_t
G_{rem}	Set of goal states that are in S_t but not in S_{t+1}
CS	Configuration Space graph
W	Wave propagation operator
T	Set of Transitions in S
T_{add}	Transitions in S_{t+1} but not in S_t, or reduced-cost transitions
T_{rem}	Transitions in S_t but not in S_{t+1}, or increased-cost transitions
O_t	Obstacle states at time t
O_{t+1}	Obstacle states at time t+1
O_{add}	Obstacle states that are in S_{t+1} but not in S_t
O_{rem}	Obstacle states in S_t but not in S_{t+1}
n	State or node
n'	Neighbor of n
K(n)	Cost to nearest state in G from state n
D(n)	Set of equivalently optimal path directions (arrows) to nearest state in G
Parent(n)	Set of parent (predecessor) nodes to n or transition in *S*
Child(n)	Set of child (successor) nodes to n or transition in *S*

Algorithm: / ***Comments:***

Given: S, O_{add}, O_{rem}, G_{add}, G_{rem}, T_{add}, T_{rem}.
H, Hs and P are sifted heaps: no redundant entries.

```
For each o element of O_add                          front edge
        if (Parent(o) is not element of O)
                add_to_heap(P,Parent(o))
        if (Child(o) is not element of O)
                add_to_heap(H, Child(o))
For each g element of G_rem                          back edge of obstacle
                add_to_heap(H, g)
For each t element of T_rem                          Increased or Removed
                add_to_heap(H, Parent(t))            transitions.
WHILE H not empty                                    clear influence
    uncostflag=NO
    n <- remove_from_heap(H)
    for each n' of n
        if (n' in O_t+1) or (K(n') = UNCOSTED)
            if(Child(n) = n')
                    uncostflag = YES
            else
                    bag <- n'

    if(uncostflag = NO)                              if n has no children, it is a leaf (member of P)
```

```
            add_to_heap(P,n)
      else
            K(n) = UNCOSTED;
            D(n) = NULL;
            add_to_heap(H,bag)
ENDWHILE

WHILE P not empty                                  put perimeter elements
      p <- remove_from_heap(P)                     into H if they are not
      if (p != UNCOSTED) & (p not in O_t+1)        cleared or obstructions
            add_to_heap(H,p)
ENDWHILE

For each o element of O_rem                        init removed obstacles
      K(o) = UNCOSTED
      D(o) = NULL

For each o element of O_rem                        add to H any
      for each n' of o                             neighbors of obstacle
            if (n' != UNCOSTED) & (n' not in O_t+1) states that are not
                  add_to_heap(H,n')                cleared or obstacles

For each g element of G_add
      K(g) = 0
      D(g)  = NULL;                                added goals
      add_to_heap(H,g)

For each t element of T_add                        H contains all seed nodes.
            add_to_heap(H, Parent(t))

For each element n of H                            Propagate waves with A*
      W(n)

END
```

Applications of Learning and Planning Methods
edited by N. G. Bourbakis

CHAPTER 12

PATH PLANNING UNDER UNCERTAINTY

F.Yegenoglu
School of Information Technology
George Mason University
Fairfax, Virginia 22030-4444, USA

and

Harry E. Stephanou
New York State Center for Advanced Technology in
Automation and Robotics
Rensselaer Polytechnic Institute
Troy, New York 12180-3590, USA

Abstract

This chapter deals with a real-time algorithm for planning collision-free navigation or manipulation paths in dynamic, cluttered environments. A Newton-like iteration is used to determine a sequence of knot points that guide the motion of several robots in a velocity vector field toward their targets while avoiding collisions with a set of moving obstacles. Complex targets and obstacles are modeled with elliptical primitives. The velocity field can be updated on-line, and the path locally replanned in response to environmental changes. The velocity field approach is applied to two main problems: (i) two-robot path planning with curvature constraints, (ii) coordinated redundant and bilateral manipulation.

1. INTRODUCTION

1.1 A velocity fields approach

This paper deals with collision-free kinematic path planning for multirobot systems in complex, dynamic environments. Our approach is based on a local scheme for the real-time generation of a sequence of *knot points* through which the desired path must pass [Yegenoglu, Erkmen and Stephanou, 1988]. The interpolation between knot points can then be performed by using clothoid curves [Kanayama and Miyake, 1986] or some similar scheme. We develop the concept of *velocity fields* to guide the motion of the robot subject to several kinematic constraints. Our scheme uses a local map of the location and identity of obstacles, derived from sensory data.

We assume that this path planning process is preceded by some coarse global path planning scheme, and followed by some trajectory tracking algorithm. We also assume that objects in the environment have been identified and classified [Stephanou and Erkmen, 1988] as obstacles and targets. Obstacles are clustered into groups of objects (e.g. stairs, or trees) sharing common properties. We assume that because of sensing uncertainties [Erkmen, Yegenoglu and Stephanou, 1989], a group of candidate targets (destinations), rather than a single target, has been identified.

The shape of objects (i.e. obstacles and targets) in the environment is described by elliptical primitives. The polynomial functions used to describe these objects are also used to generate a **vector velocity field** consisting of two components: a repulsive field due to the obstacles, and an attractive field due to the targets. Collision free paths are then generated by a Newton-like root finding algorithm, driven by this velocity field, that guides each robot toward a target, and away from the local obstacles. This algorithm finds the roots of a polynomial. Each root is the center of an ellipse representing a target.

1.2 Related work

The various approaches to robot navigation can be broadly clustered into two groups: global, and local. Global techniques use prior knowledge providing a precise and complete description of the environment, and rely on off-line search techniques. Two map representations in which a search can be performed include grid maps [Borenstein and Koren, 1988; Dunlay, 1988; Jun and Shin, 1988], and line segment maps [Fujimura and Samet, 1989; Gex and Campbell 1987; Kanayama, 1988]. For polygonal obstacles, a robot is mapped into a point in configuration space [Lozano-Perez, 1983], thus simplifying the search problem, at the expense of increased dimensionality. Lozano-Perez [1986] uses the configuration of obstacles in the joint space of a manipulator with revolute joints to identify the free space, then an A^* search is performed to define a path to the goal point. These methods can be used with a set of known, stationary obstacles. When a global model of the environment is available, paths can be optimized by using cost functions such as path length [Chen and Vidyasagar, 1988], time [Fujimura and Samet, 1989], probability of dead-ends [Jun and Shin, 1988] and proximity [Kanayama, 1988].

Because of sensing uncertainty, a complete description of unstructured environments is generally not available. Furthermore, the obstacles are not necessarily stationary. In such cases, a local scheme for on-line obstacle avoidance and path planning must be used. Gex [1987] describes a method for mapping the local environment using acoustic sensors. Line segments are used to represent detected obstacle boundaries, and the map is reduced to subgoals and avoidance points. Fujimura and Samet [1989] include time as a dimension of the world model to plan paths among moving obstacles. They assume that the geometry and motion of the obstacles are known in advance.

The two-level hierarchy of [Kant and Zucker, 1988] combines global geometric trajectory planning in dynamic environments with a local avoidance strategy based on velocity servoing. A local method for the coordinated motion of multiple mobile robots or manipulators has been developed in [Tournassoud, 1986]. Obstacle avoidance is performed by using separating hyperplanes between the robot and the obstacles. An extended version of this method was also applied to manipulators with a large number of degrees of freedom [Faverjon and Tournassoud, 1987], using a hierarchical CAD model of the environment and manipulators. Chien and Koivo [1988] propose a hierarchical control architecture for planning paths of multiple manipulators sharing a common workspace in the presence of moving obstacles. Dynamic potential functions were incorporated in a time-optimal manipulator control problem in [Newman and Hogan, 1987]. In [Kheradpir and Thorp, 1988], elliptical/spherical or ellipsoid obstacles are treated as state constraints in a real-time optimal trajectory tracking problem that minimizes the deviation between the acceleration vector of a point robot and a desired acceleration field. A local map of visible obstacles, modeled as convex polygons, is used in [Krogh and Feng, 1989] for the real-time generation of a collision-free path. Current sensory data are used to generate subgoals that take the system dynamics into account. The navigation of a point robot in an uncertain environment using tactile sensing has been discussed in [Lumelsky and Stepanov, 1986].

Our approach to obstacle avoidance is closest to the concept of artificial potential fields [Khatib, 1986]. In the potential fields method, analytic equations of the form $f(X)=0$ are used to describe obstacles. Based on the shortest distance to an obstacle, an artificial repulsive potential field is generated. Similarly, based on the distance to the goal position, an attractive potential field is generated. The total potential energy of the workspace is obtained by the superposition of the potentials for obstacles and targets. The gradient of the total potential energy determines an artificial *force* acting on the robot. The direction of this force points away from an obstacle and toward a target, and its magnitude is inversely proportional to the distance from an obstacle or a target. The dynamic equations of the robot are solved to obtain the kinematic equations that describe the motion of the robot in this field.

Khatib [1986] includes path planning into low-level control by applying the potential field approach to robot manipulators. Potential functions for attraction and repulsion are expressed in Cartesian coordinates, while joint limit avoidance is accomplished by introducing barriers of potential in joint space. Khosla and Volpe [1988] use superquadric potentials, which have spherically symmetric contours at large distances, and closely model the obstacle at short distances, thus minimizing the effects of local minima. Krogh and Thorpe [1986] use potential fields for local planning in conjunction with path relaxation for global path planning. Borenstein and Koren [1988] integrate certainty grids for obstacle representation and potential fields for navigation of mobile robots to accommodate inaccurate sensory data.

Some drawbacks of the artificial potential field methods include: (i) the possibility of reaching a local minimum of the field before the target, (ii) high computational cost because two different functions are used to represent an obstacle and to generate the corresponding potential field, (iii) high computational cost required to compute the shortest Cartesian distance to an obstacle.

1.3 Paper outline

Section 2 of this paper introduces a novel algorithm for real-time path planning among point obstacles in cluttered and dynamic environments. In section 3, the results are generalized to targets and obstacles of more complex shapes by using elliptical primitives. Multirobot navigation with path curvature constraints is considered in section 4. Finally, section 5 deals with path planning for redundant and bilateral planar manipulators.

2. Point obstacle avoidance

In this section, we consider the problem of a point *robot* navigating among point obstacles. We construct polynomial functions, the roots of which represent the target or obstacle locations. Knot points are generated by applying a Newton-like root finding scheme to these functions. The iteration is guided by a (vector) velocity field derived from the polynomial functions.

2.1 Attractor functions

An *attractor function* has the following general form:

$H_A(X) = \{f_1(X), f_2(X), \cdots, f_n(X)\}^T = 0$

where $X \in \Re^n$ is the state vector. Let $\{Z_i \in \Re^n \ ; \ 1 \le i \le m\}$ denote a set of m candidate targets. Each target location is a zero of the nonlinear function $H_A(X)$. The attractor function is generated through the following procedure:

(i) Using the first components z_{1i} of Z_i, construct $f_1(X)$ such that:

$$f_1(X) = x_1 x_2 - [x_1 P_2^{m-1}(x_1) - \{(x_1 - z_{11})(x_1 - z_{12}) \ldots\ldots (x_1 - z_{1n})\}]$$

where z_{ji} is the jth component of the ith zero.

(ii) To account for the jth component of Z_i, construct the function $f_j(X)$ with zeroes at: $\{(z_{1(j-1)}, z_{1j}), (z_{2(j-1)}, z_{2j}), \ldots, (z_{m(j-1)}, z_{mj})\}$

(iii) Let $x_j = P(x_{j-1})$ be a (m-1)Th order polynomial passing through these points. Then $f_j = x_j - P(x_{j-1})$ yields the specified zeroes.

The resulting attractor function is of the form:

$$H_A(X) = \{x_1 x_2 - P_1^m(x_1),\ x_2 - P_2^{m-1}(x_1),\ \cdots,\ x_n - P_n^{m-1}(x_{n-1})\}^T$$

For example, let: m=4 , n=2 ; and $Z_1 = [1,2], Z_2 = [2,4], Z_3 = [3,1], Z_4 = [4,2]$

Then:

$$H_A(X) = \begin{Bmatrix} x_1 x_2 - 0.5x_1^4 + 1.5x_1^3 + 9x_1^2 - 36x_1 + 24 \\ x_2 - 1.5x_1^3 + 11.5x_1^2 - 26x_1 + 14 \end{Bmatrix}$$

2.2 Basins of attraction

In the absence of obstacles, a sequence of knot points in state space is generated by using Newton's method to solve the vector equation $H_A(X) = 0$ by iterating:

$X^{k+1} = X^k + \Phi_A^k$

where $X^k \in \Re^n$ is the state vector, and

$\Phi_A^k = -[\nabla H_A(X^k)]^{-1} H_A(X^k)$

This solution can be viewed as a dynamic process where the roots of the equation are *competing* to attract the robot from its initial state X^0. Each root Z_i has a *basin of attraction* defined by:

$\aleph(Z_i) = \left\{X^0 \in \Re^n \ni \lim_{k\to\infty} X^k = Z_i\right\}$

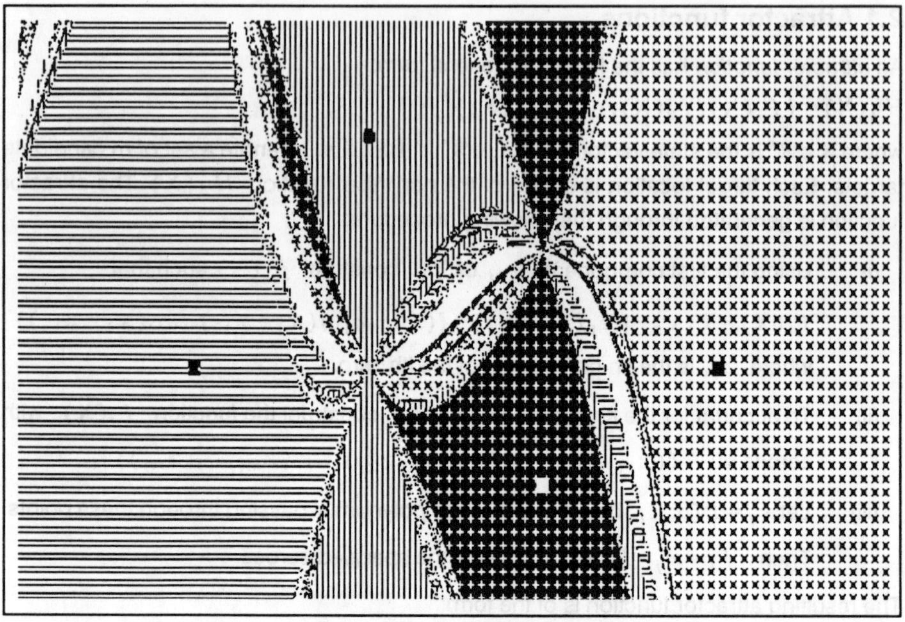

Figure 1 - Basins of attraction

The boundary between attraction basins consists of the roots of $det \nabla H_A(X) = 0$. Fig. 1 shows the basin $\aleph(Z_i)$ (in $\Re^2$) for each of the four attractors of the above example. Note the regions of interference between the basins, near the singularities of $det \nabla H_A(X) = 0$. In these *chaotic* [Peitgen, 1986] regions, two very close points may converge to different attractors.

2.3 Repeller functions

Similarly, obstacles are modeled as roots of the equation $H_{R_i}(X) = 0$, and thought of as repellers. A *repeller function* H_{R_i} is constructed for the ith obstacle group. In the absence of any attractors, the knot points are generated by:

$$X^{k+1} = X^k - \Phi^k_{R_i}$$

where:

$$\Phi^k_{R_i} = -\left[\nabla H_{R_i}(X^k)\right]^{-1} H_{R_i}(X^k)$$

At any point X^k outside the chaotic regions, the *repulsion* (vector) *field* $\Phi^k_{R_i}$ points to a repeller in the same basin as X^k. $\|\Phi^k_{R_i}\|$ is a measure of the distance from the repeller. For three obstacles at {(1.5,3), (2.5,2), (3.5,3.5)}, the repeller function is:

$$H_{R_1}(X) = \begin{Bmatrix} x_1x_2 - 0.25x_1^3 - 1.5x_1^2 + 8.56x_1 - 13.12 \\ x_2 - 1.25x_1^2 + 6x_1 - 9.19 \end{Bmatrix}$$

The magnitude (proportional to the length of the arrows) and direction of $\Phi^k_{R_1}$ are shown in Fig. 2.

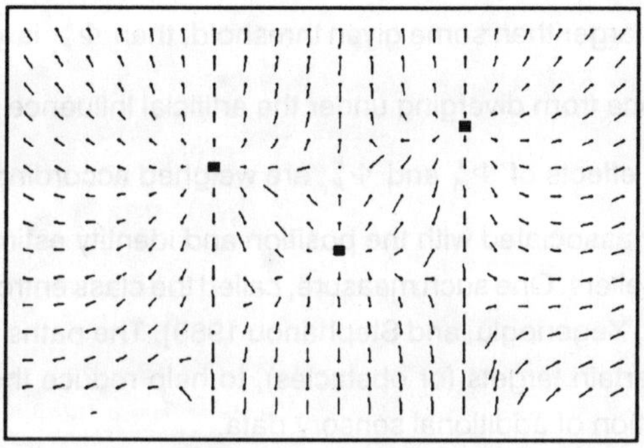

Figure 2 - Repulsion field

2.4 Knot point generation

The functions $H_A(X)$ and $H_{R_i}(X)$ include the necessary information to generate a collision free path to one of the targets. New information about the environment can be incorporated in these two functions. For example, if a candidate target is reclassified and becomes irrelevant to the task, then it is deleted from $H_A(X)$ and added to $H_{R_i}(X)$. Also, the estimated locations of objects are continuously updated.

Φ^k_A and $\Phi^k_{R_i}$ are computed for each point X^k. Φ^k_A directs the motion toward one of the targets, while $\Phi^k_{R_i}$ directs the motion away from the obstacles. The velocity at knot point X^k depends on the relative magnitudes of these vectors. The rule for combining these two vectors must meet the following requirements:

(i) if $|\Phi_A^k|$ is small, then Φ_A^k dominates. If the robot is close to a candidate target and far from an obstacle, then it moves toward the target;

(ii) if $\|\Phi_{R_i}^k\|$ is relatively small, then $\Phi_{R_i}^k$ dominates. If the robot is close to an obstacle and far from a target, then it moves away from the obstacle;

(iii) if both $|\Phi_A^k|$ and $\|\Phi_{R_i}^k\|$ are large, then Φ_A^k is dominant. If the robot is neither close to a target nor to an obstacle, it moves toward the target;

(iv) if $\|\Phi_{R_i}^k\|$ is larger than some given threshold, then $\Phi_{R_i}^k$ is ignored. This prevents the sequence from diverging under the artificial influence of distant obstacles.

(v) the relative effects of Φ_A^k and $\Phi_{R_i}^k$ are weighed according to some measure of uncertainty associated with the position and identity estimates of the attractors and the repellers. One such measure, called the class entropy, has been outlined in [Erkmen, Yegenoglu, and Stephanou 1989]. The paths tend to pass closer to highly uncertain targets (or obstacles), to help reduce the uncertainty through the acquisition of additional sensory data.

(vi) the magnitude of the velocity vector is below some given value

(vii) near the target, the velocity approaches zero

A combination rules that satisfies the above requirements is:

$$V(X) = \frac{\Psi(X)}{|\Psi(X)|}\left\{V_{max}\left(1 - e^{-b_A|\Phi_A(X)|}\right)\right\}$$

where

$$\Psi(X) = \frac{\Phi_A(X)}{\gamma_A\,|\Phi_A(X)|^2} - \sum_{i=1}^{s} \frac{\Phi_{R_i}(X)}{\gamma_{R_i}\,\|\Phi_{R_i}(X)\|^2}\, e^{-b_i|\Phi_{R_i}(X)|}$$

The knot points are obtained by iterating the following sequence:

$$X^{k+1} = X^k + h\,V(X^k)$$

where

h determines the step size

s is the number of obstacle groups

$\| \Phi^{k}_{R_i} \|$ is the *pseudodistance* from the current position to a point obstacle in set i

γ_A is the class entropy for the set of attractors

γ_{R_i} is the class entropy for the ith set of repellers

V_{max} is the maximum allowable velocity

b_A and b_i are spatial repulsion decay rates

2.5 Examples

Fig. 3 shows the combined effect of four attractors at {(1,2), (2,4), (3,1), (4,2), and two groups of obstacles at {(1.5,3), (2.5,2), (3.5,3.5)} and {(1,1), (4,3), (5,2)}. The filled in circles and squares denote the attractors and the two sets of repellers, respectively. The arrows show the direction of the velocity vector $V(X^k)$.

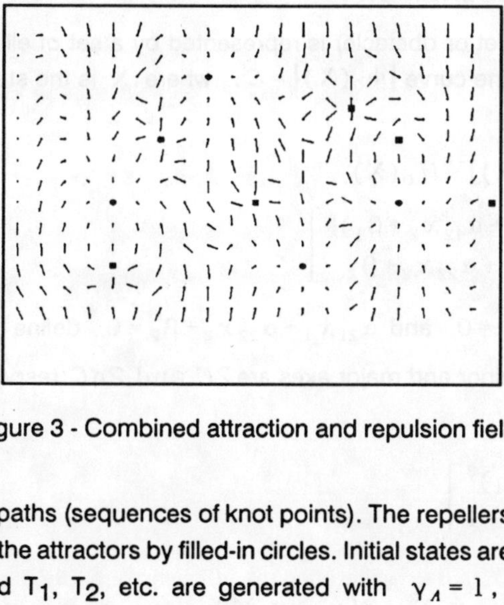

Figure 3 - Combined attraction and repulsion field

Fig. 4 shows several paths (sequences of knot points). The repellers are represented by filled-in squares, and the attractors by filled-in circles. Initial states are indicated by empty circles. Paths labeled T_1, T_2, etc. are generated with $\gamma_A = 1$, $\gamma_{R_1} = 1.25$, and $\gamma_{R_2} = 1.25$, while paths labeled S_1, S_2, etc. are generated by assigning lower values of entropy for the attractors, and higher entropy for the repellers, namely $\gamma_A = 0.5$, $\gamma_{R_1} = 2.0$, and $\gamma_{R_2} = 2.0$. This example illustrates a simple active sensing effect: the robot moves closer to objects whose position/identity estimates are highly uncertain, thus facilitating the acquisition of sensory data.

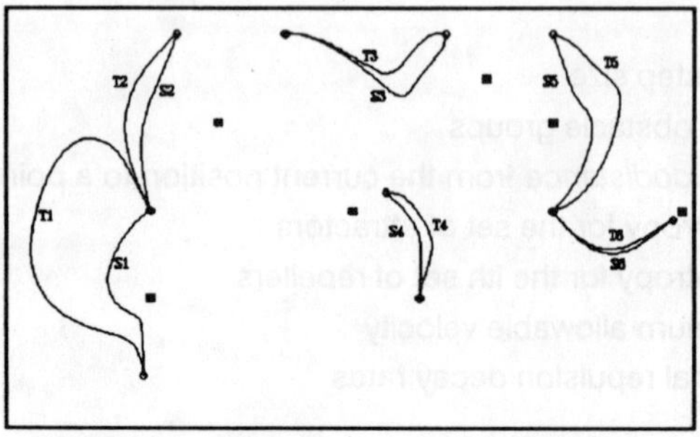

Figure 4 - Effect of sensing uncertainties

3. PATH PLANNING AMONG ELLIPTICAL OBSTACLES

3.1 Object representation

Each 2D object (target or obstacle) is represented by a set of ellipses, i.e. a region of points enclosed by the curve $|\Phi_o(X)| = C$, where X is the state vector, $C > 0$ is a scalar and,

$$\Phi_o(X) = -[\nabla H_o(X)]^{-1} H_o(X)$$

$$H_o(X) = \begin{bmatrix} (\alpha_{11}x_1 + \alpha_{12}x_2 + \beta_1)^{\sigma} \\ \alpha_{21}x_1 + \alpha_{22}x_2 + \beta_2 \end{bmatrix}$$

$\alpha_{11}x_1 + \alpha_{12}x_2 + \beta_1 = 0$ and $\alpha_{21}x_1 + \alpha_{22}x_2 + \beta_2 = 0$ define the axes of an ellipse. The lengths of the minor and major axes are $2C$ and $2\sigma C$ respectively. For example, let:

$$H_o(X) = \begin{bmatrix} (x_2 - x_1)^3 \\ x_2 + x_1 - 1 \end{bmatrix}$$

Then:

$$\nabla H_o(X) = \begin{pmatrix} -3(x_2 - x_1)^2 & 3(x_2 - x_1)^2 \\ 1 & 1 \end{pmatrix}$$

and

$$\Phi_O(X) = \frac{1}{6}\begin{pmatrix} 4x_1 + 2x_2 - 3 \\ 2x_1 + 4x_2 - 3 \end{pmatrix}$$

$$\frac{\|\Phi_O(X)\|^2}{C^2} = \frac{1}{36C^2}[(4x_1 + 2x_2 - 3)^2 + (2x_1 + 4x_2 - 3)^2] = 1$$

$$\frac{\|\Phi_O(X)\|^2}{C^2} = \left[\frac{(x_1 - 0.5)\frac{1}{\sqrt{2}} + (x_2 - 0.5)\frac{-1}{\sqrt{2}}}{3C}\right]^2 + \left[\frac{(x_2 - 0.5)\frac{1}{\sqrt{2}} - (x_1 - 0.5)\frac{-1}{\sqrt{2}}}{C}\right]^2 = 1$$

Therefore, $\|\Phi_O(X)\| = C$ defines an ellipse with center at $(0.5, 0.5)$. The orientation of the major axis is $-45°$. The semi-major and semi-minor axis lengths are $3C$ and C, respectively.

The general expression:

$$\left[\frac{(x_1 - h)\cos(\lambda) + (x_2 - k)\sin(\lambda)}{\mu}\right]^2 + \left[\frac{(x_2 - k)\cos(\lambda) - (x_1 - h)\sin(\lambda)}{\eta}\right]^2 = 1$$

describes an ellipse centered at (h, k), with major axis orientation λ, and axes lengths μ and η.

3.2 Velocity field

In this subsection, we compute the *velocity field* for any given set of 2D targets and obstacles, subject to the following requirements:

(i) the velocity field should be smooth and differentiable to ensure smooth trajectories and avoid high accelerations,

(ii) near obstacle boundaries, the field should point away from or tangential to the obstacle,

(iii) outside the vicinity of an obstacle, the field should point toward a target,

(iv) the field should not have any local minima,

(v) the field should approach zero at the target boundary,

(vi) the field should be additive to handle the case of multiple contacting objects (targets and/or obstacles).

The *pseudodistance* from point X to some object **O** (i.e. an attractor or repeller) is defined as:

$$D_O(X) = \{\|\Phi_O(X)\| - C_O\}\left\{1 + (\sigma - 1)\left(1 - \frac{2\theta(X)}{\pi}\right)\right\}$$

where $\theta(X)$ is the acute angle between the major axis of the ellipse, and the field $\Phi_O(X)$ of the ellipse. The second term in the right-hand side of the above equation causes the velocity field to decay faster along the major axis. This is done by amplifying the pseudodistance at points around the ellipse where Φ_A (Φ_{R_i}) has a large component along the major axis.

This pseudodistance (to an elliptical object) is conceptually similar to the one used in [Khosla and Thorpe, 1988], and considerably less expensive to compute than the Euclidean distance (to the closest point on the boundary of an object), as required in other artificial potential fields method [Khatib, 1986].

The vector velocity field is computed as a linear combination of the attraction and repulsion fields:

$$V(X) = \frac{\Psi(X)}{|\Psi(X)|}\left\{V_{\max}\,(1 - e^{-b_A D_A(X)})\right\}$$

where

$$\Psi(X) = \Psi_A(X) + \Psi_R(X)$$

$$\Psi_A(X) = \frac{\Phi_A(X)}{\gamma_A\, D_A(X)\, |\Phi_A(X)|}$$

$$\Psi_R(X) = -\sum_{i=1}^{s} \frac{\Phi_{R_i}(X)}{\gamma_{R_i}\, D_{R_i}(X)\, \|\Phi_{R_i}(X)\|}\, e^{-b_i D_{R_i}(X)}$$

V_{max} is the maximum allowable velocity,
b_A and b_i are spatial repulsion decay rates.

The repulsion vector $\Phi_{R_i}(X)$ dominates when $D_{R_i}(X)$ is close to zero, i.e. when the robot is close to the ith obstacle boundary. As the robot moves away from the obstacle along the direction determined by $\Phi_{R_i}(X)$, the effect of the repulsion vector becomes insignificant due to the exponential decay, and the robot moves along the direction of the attraction vector. As the robot approaches the target boundary, $D_A(X)$ tends to zero, and $\Psi_A(X)$ dominates. The magnitude of the field depends on the pseudodistance to a target. Far from the target, the magnitude of the field is constant. The velocity tends to zero near the target boundary to approach the target with zero velocity.

Fig. 5 shows the velocity field for two obstacles represented by:

$$H_{R_1} = \begin{Bmatrix} (x_1 - 0.3)^{10} \\ (x_2 - 0.5) \end{Bmatrix} \qquad c_{R_1} = 0.02 \qquad b_1 = 10.0 \qquad \gamma_{R_1} = 1.0$$

$$H_{R_2} = \begin{Bmatrix} (x_2 - 0.7)^{10} \\ (x_1 - 0.5) \end{Bmatrix} \qquad c_{R_2} = 0.02 \qquad b_2 = 10.0 \qquad \gamma_{R_2} = 1.0$$

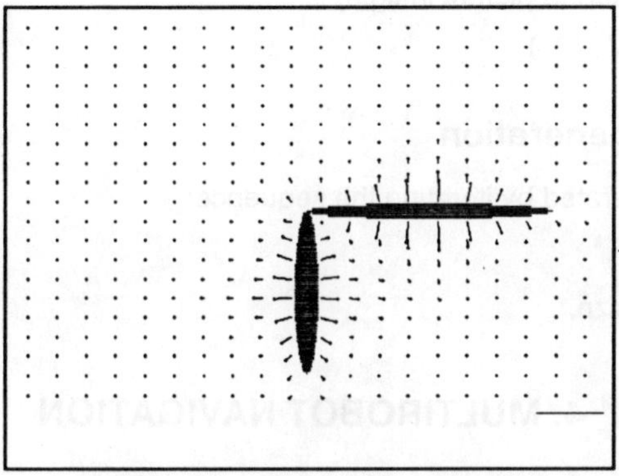

Figure 5 - Repeller velocity field

3.3 Escape velocity

Oscillatory behavior may occur in regions where the attraction and repulsion velocity fields tend to cancel each other. First picture in Fig. 6 shows such a case, caused by the presence of an obstacle between the robot and the target (obstacles are shaded with horizontal lines, while darker shades are used for targets).

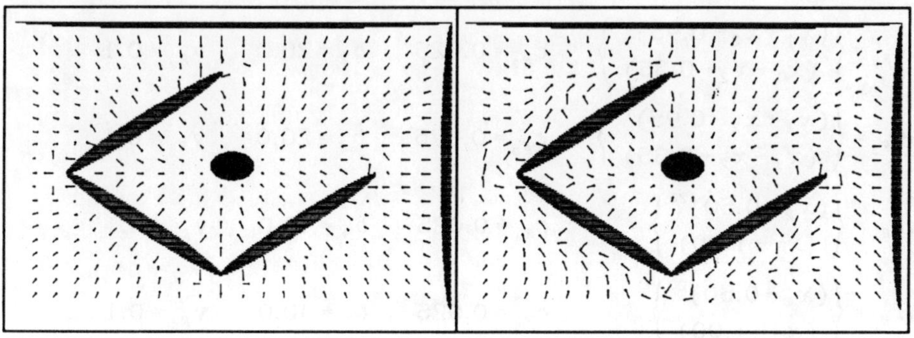

Figure 6 - Velocity field with local minimum, and the result of adding the escape velocity

Oscillatory regions are eliminated by the addition of an *escape velocity* field $V_E(X)$ which is tangential to the obstacle boundary at these regions:

$$V_E(X) = \pm K_E e^{-b_E|\Psi(X)|}\ \xi(X)$$

$\xi(X)$ is a unit vector orthogonal to $\Psi_R(X)$. K_E, b_E are constant coefficients. The *total velocity field* (second picture in fig.6) is:

$$V_T(X) = V(X) + V_E(X)$$

3.4 Knot point generation

Knot points are generated by iterating the sequence:

$$X^{k+1} = X^k + h\ V_T(X^k)$$

where h is the step size.

4. MULTIROBOT NAVIGATION

4.1 Single-robot navigation

Fig. 7 shows a target surrounded by three "walls", and a rectangular robot starting from an initial position where it cannot "see" the target. To avoid collision with any point on the robot, each obstacle is enlarged by the radius of the circumscribing circle to the rectangular robot. Accordingly, the robot is shrunk to a point. The five obstacles and the target are represented by the following functions and parameters:

$$H_{R_1} = \begin{Bmatrix} (x_1 - x_2)^{10} \\ (x_1 + x_2 - 0.60) \end{Bmatrix} \qquad c_{R_1} = 0.025 \qquad b_1 = 20.0 \qquad \gamma_{R_1} = 0.1$$

$$H_{R_2} = \begin{Bmatrix} (x_1 + x_2 - 0.95)^{10} \\ (x_1 - x_2 - 0.35) \end{Bmatrix} \qquad c_{R_2} = 0.025 \qquad b_2 = 20.0 \qquad \gamma_{R_2} = 0.1$$

$$H_{R_3} = \begin{Bmatrix} (x_1 + x_2 - 0.95)^{10} \\ (x_1 - x_2 + 0.35) \end{Bmatrix} \qquad c_{R_3} = 0.025 \qquad b_3 = 20.0 \qquad \gamma_{R_3} = 0.1$$

$$H_{R_4} = \begin{Bmatrix} (x_2 - 0.50)^{20} \\ (x_1 - 1.00) \end{Bmatrix} \qquad c_{R_4} = 0.025 \qquad b_4 = 40.0 \qquad \gamma_{R_4} = 0.1$$

$$H_{R_5} = \begin{Bmatrix} (x_1 - 0.50)^{20} \\ (x_2 - 1.00) \end{Bmatrix} \qquad c_{R_5} = 0.025 \qquad b_5 = 40.0 \qquad \gamma_{R_5} = 0.1$$

$$H_A = \begin{Bmatrix} (x_2 - 0.50)^{20} \\ (x_1 - 0.50) \end{Bmatrix} \qquad c_A = 0.050 \qquad \gamma_A = 1.0$$

The escape velocity parameters are $b_e = 10.0$ and $K_e = 1.0$. The environment is defined in the region $\{0.0 \le x_1 \le 1.0 \text{ and } 0.0 \le x_2 \le 1.0\}$. The "walls" at $x_1 = 1.0$ and $x_2 = 1.0$ are represented by H_{R_4} and H_{R_5} respectively. The robot starts at (0.1, 0.1) and moves toward the target until the resultant attraction and repulsion velocity becomes very small. Then, the robot follows the obstacle boundary along the direction provided by the escape velocity. It finds an opening toward the target, and reaches its surface with zero velocity.

Fig. 8 is the convergence map of the environment. It depicts the length of the path (number of knot points) required to reach the target from each initial point. All sequences converge to the target in a finite number of steps, indicating that there are no dead-ends.

4.2 Two-robot navigation

Fig. 9 shows two robots in a common workspace. The robots start at the opposite ends of a narrow lane and move toward their targets at the other end while avoiding collisions with the walls, obstacles, and with each other. Each robot acts as an elliptical obstacle with time varying location and orientation. In the middle of the lane, the two robots are sufficiently close to repel each other. The shape and size of each robot is determined by $\sigma = 2.0$, and $C = 0.01$ Repulsion velocity parameters are $b = 20.0$ and $\gamma = 1$. The escape velocity parameters are $b_E = 10.0$ and $K_E = 1.0$. The "walls" and the two obstacles on the lane are represented by the following functions and path parameters:

$$H_{R_1} = \begin{Bmatrix} (-x_1 + x_2)^{20} \\ (x_1 + x_2 - 0.80) \end{Bmatrix} \qquad c_{R_1} = 0.025 \qquad b_1 = 20.0 \qquad \gamma_{R_1} = 1.0$$

$$H_{R_2} = \begin{Bmatrix} (-x_1 + x_2)^{20} \\ (x_1 + x_2 - 1.20) \end{Bmatrix} \qquad c_{R_2} = 0.025 \qquad b_2 = 20.0 \qquad \gamma_{R_2} = 1.0$$

$$H_{R_3} = \begin{Bmatrix} (x_2 - 0.25) \\ (x_1 - 0.75) \end{Bmatrix} \qquad c_{R_3} = 0.020 \qquad b_3 = 20.0 \qquad \gamma_{R_3} = 1.0$$

$$H_{R_4} = \begin{Bmatrix} (x_2 - 0.75) \\ (x_1 - 0.25) \end{Bmatrix} \qquad c_{R_4} = 0.020 \qquad b_4 = 20.0 \qquad \gamma_{R_4} = 1.0$$

Robot 1 starts from an initial position (0.1, 0.9), its target is represented by:

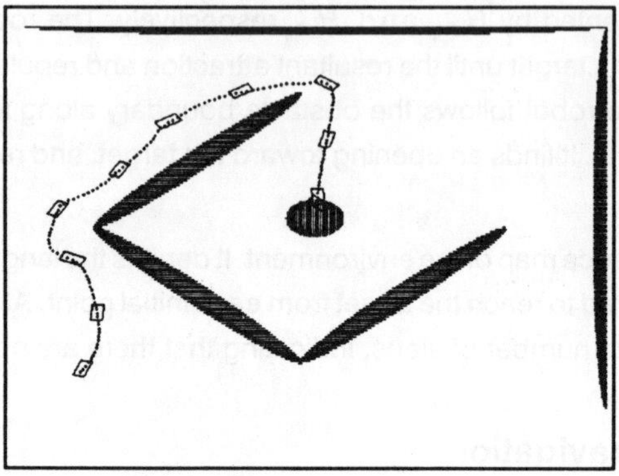

Figure 7 - Navigation in a cluttered environment

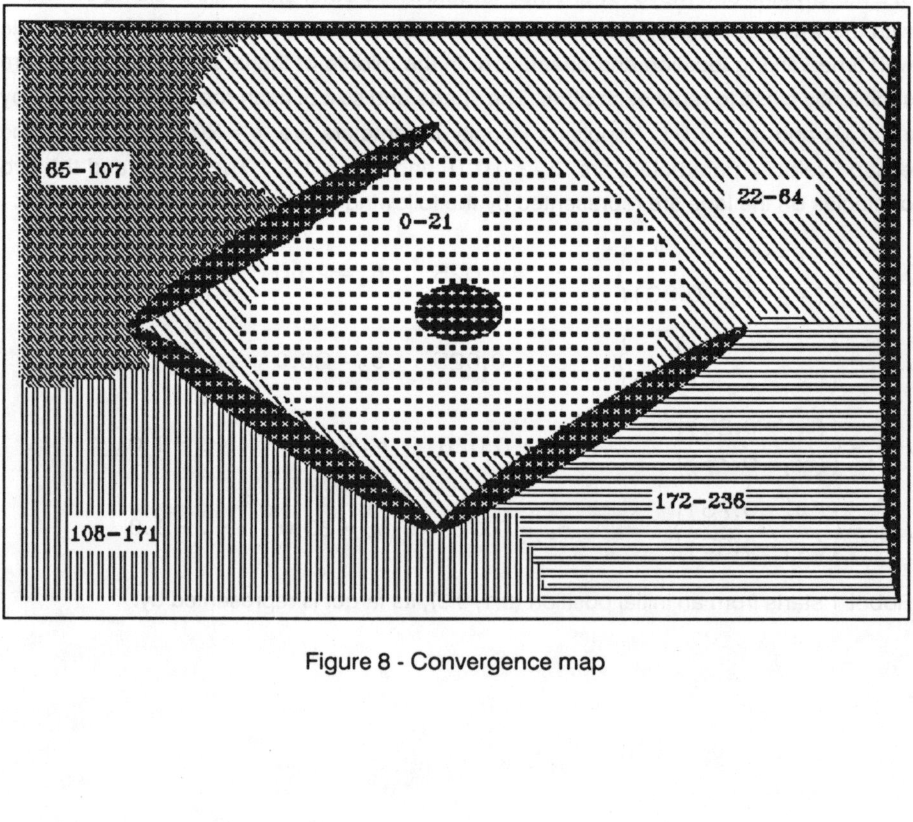

Figure 8 - Convergence map

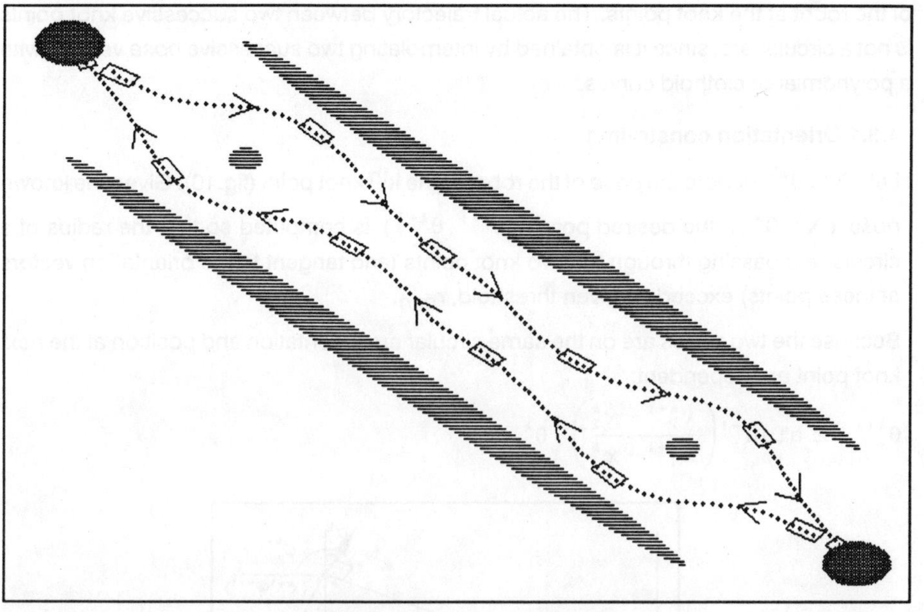

Figure 9 - Two robots navigation

$$H_{A_1} = \begin{Bmatrix} (x_2 - 0.95) \\ (x_1 - 0.05) \end{Bmatrix} \qquad c = 0.04 \qquad \gamma_{A_1} = 1.0$$

Robot 2 starts from an initial position (0.9, 0.1) and moves toward its target represented by:

$$H_{A_2} = \begin{Bmatrix} (x_2 - 0.05) \\ (x_1 - 0.95) \end{Bmatrix} \qquad c = 0.04 \qquad \gamma_{A_2} = 1.0$$

4.3. Path curvature constraints

To accommodate steering range limits for navigation in cluttered environments, we now extend the obstacle avoidance algorithm to include curvature and orientation constraints. Successive knot points lie on circular arcs with radii that are constrained to some minimum value. The tangents to these arcs at the knot points determine the robot's orientation. The circular arcs are only used to satisfy the curvature constraint and to specify the orientation of the robot at the knot points. The actual trajectory between two successive knot points is not a circular arc, since it is obtained by interpolating two successive *pose* vectors with a polynomial or clothoid curves.

4.3.1 Orientation constraints

Let (X^k, θ^k) denote the *pose* of the robot at the kth knot point (fig.10). Given the known pose (X^k, θ^k), the desired pose (X^{k+1}, θ^{k+1}) is computed so that the radius of a circular arc passing through the two knot points (and tangent to the orientation vectors at these points) exceeds a given threshold, r_{min}.

Because the two poses are on the same circular arc, orientation and position at the next knot point are dependent:

$$\theta^{k+1} = 2\,\mathrm{atan}^{-1}\left(\frac{x_2^{k+1} - x_2^k}{x_1^{k+1} - x_1^k}\right) - \theta^k$$

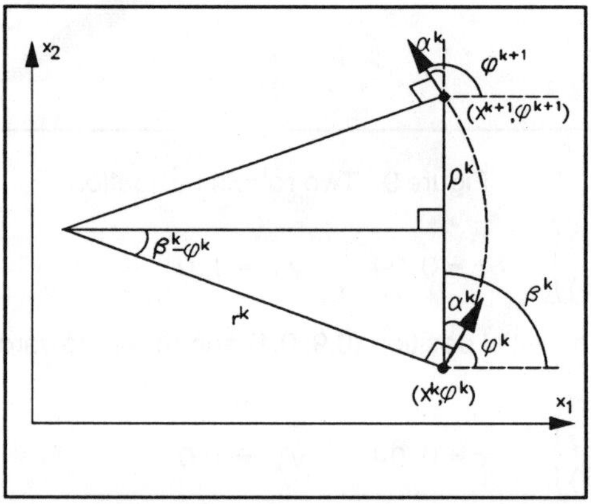

Figure 10 - Curvature and orientation constraints

4.3.2 Curvature constraints

The radius of the circular arc between the two knot points is (fig. 10):

$$r^k \cos[90° - (\beta^k - \theta^k)] = \frac{\rho^k}{2} \Rightarrow r^k = \frac{\rho^k}{2\sin(\beta^k - \theta^k)}$$

where

$$\rho^k = \|X^{k+1} - X^k\| = \|h V_T(X^k)\|$$

The curvature constraint is: $r^k \geq r_{min}$

If $\beta^k - \theta^k > 0$, then:

$$\frac{\rho^k}{2\sin(\beta^k - \theta^k)} > r_{min} \Rightarrow \beta^k \leq \theta^k + \sin^{-1}\left(\frac{\rho^k}{2r_{min}}\right) = \beta^k_{max}$$

If $\beta^k - \theta^k < 0$, then:

$$\beta^k \geq \theta^k - \sin^{-1}\left(\frac{\rho^k}{2r_{min}}\right) = \beta^k_{min}$$

In summary, the curvature constraint is: $\beta^k_{min} \leq \beta^k \leq \beta^k_{min}$

4.3.3 Constrained path generation

X^{k+1} is computed by minimizing:

$$J = \|X^{k+1} - (X^k + h\, V_T(X^k))\|^2$$

subject to the above curvature constraints.

Let β'^k denote the orientation of $V_T(X^k)$. Three cases need to be considered:

(i) $\beta^k_{min} \leq \beta'^k \leq \beta^k_{max}$ Then:

$$X^{k+1} = X^k + \rho^k \begin{bmatrix} \cos\beta'^k \\ \sin\beta'^k \end{bmatrix} \qquad \theta^{k+1} = -\theta^k + 2\beta'^k$$

(ii) $\beta'^k < \beta^k_{min}$ Then:

$$X^{k+1} = X^k + \rho^k \begin{bmatrix} \cos\beta^k_{min} \\ \sin\beta^k_{min} \end{bmatrix} \qquad \theta^{k+1} = -\theta^k + 2\beta^k_{min}$$

(iii) $\beta'^{k} > \beta^{k}_{max}$ Then:

$$X^{k+1} = X^{k} + \rho^{k} \begin{bmatrix} \cos\beta^{k}_{max} \\ \sin\beta^{k}_{max} \end{bmatrix} \qquad \theta^{k+1} = -\theta^{k} + 2\,\beta^{k}_{max}$$

4.3.4 Example

Fig. 11 shows two paths with different curvature constrains. The first path was obtained by using (r_{min}=0.01), the second one shows the effect of increasing r_{min} to 0.03 which results in a smoother path. The environment consists of five obstacles corresponding to walls of a "narrow hallway", and a target located at the end. The obstacle and target functions and path parameters used in this example are:

$$H_{R_1} = \begin{Bmatrix} (x_1 - 0.70)^5 \\ (x_2 - 0.42) \end{Bmatrix} \qquad c_{R_1} = 0.020 \qquad b_1 = 20.0 \qquad \gamma_{R_1} = 0.1$$

$$H_{R_2} = \begin{Bmatrix} (x_1 - 0.32)^{10} \\ (x_2 - 0.42) \end{Bmatrix} \qquad c_{R_2} = 0.020 \qquad b_2 = 20.0 \qquad \gamma_{R_2} = 0.1$$

$$H_{R_3} = \begin{Bmatrix} (x_1 - 0.50)^{20} \\ (x_2 - 0.42) \end{Bmatrix} \qquad c_{R_3} = 0.025 \qquad b_3 = 30.0 \qquad \gamma_{R_3} = 0.1$$

$$H_{R_4} = \begin{Bmatrix} (x_2 - 0.60)^{10} \\ (x_1 - 0.50) \end{Bmatrix} \qquad c_{R_4} = 0.020 \qquad b_4 = 20.0 \qquad \gamma_{R_4} = 0.1$$

$$H_{R_5} = \begin{Bmatrix} (x_2 - 0.60)^{10} \\ (x_1 - 0.62) \end{Bmatrix} \qquad c_{R_5} = 0.020 \qquad b_5 = 20.0 \qquad \gamma_{R_5} = 0.1$$

$$H_{A} = \begin{Bmatrix} (x_1 - 0.55) \\ (x_2 - 0.80) \end{Bmatrix} \qquad c_{A} = 0.020 \qquad \gamma_{A} = 0.3$$

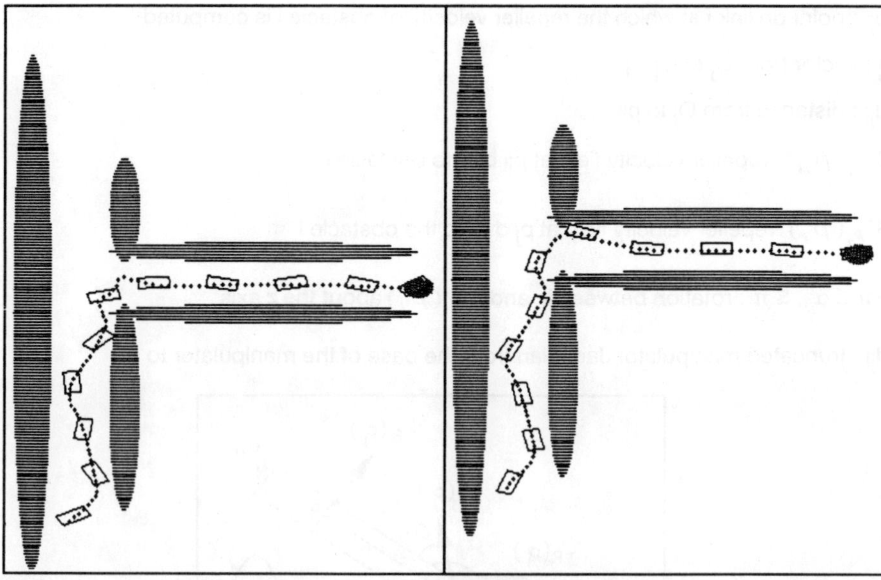

Figure 11 - Effect of curvature constraint

5. REDUNDANT AND MULTIROBOT MANIPULATION

In this section, we apply our path generation algorithm to the collision free motion of a 3-link planar (redundant) manipulator, and that of two 2-link planar manipulators sharing a common workspace in a cluttered environment. To avoid collisions of the links with the obstacles, each obstacle repels each manipulator link, while the targets attract the wrist. Attraction and escape velocities are computed at the wrist. Maximum angular velocity and joint limit constraints are also satisified.

5.1 Attraction, repulsion and escape velocity fields

The following variables are defined with respect to the world coordinate system, and illustrated in fig. 12:

O_j : position of the joint $j = 1,2,..,n$

p_j : midpoint of link j

p_{ji} : point on link j at which the repeller velocity of obstacle i is computed

$\mathbf{l}_j$: vector from O_j to O_{j+1}

d_{ji} : distance from O_j to p_{ji}

$\Psi_{R_i}(p_{ji})$: repeller velocity field at p_{ji} due to obstacle i

$\Psi_{R_i}(p_j)$: repeller velocity field at p_j due to the obstacle i

$-\pi \le \alpha_{ji} \le \pi$: rotation between $\mathbf{l}_j$ and $\Psi_{R_i}(p_j)$ about the *z* axis

$\mathbf{J}_{ji}$: truncated manipulator Jacobian from the base of the manipulator to p_{ji}

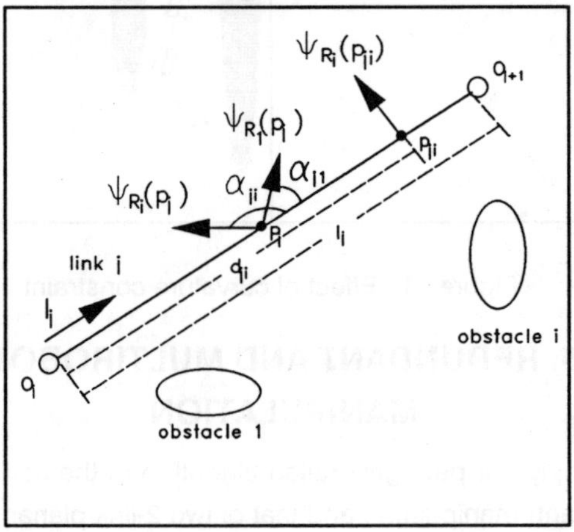

Figure 12 - Planar manipulator variables, "closest" point to an obstacle

To avoid the computation of the shortest Euclidean distance to an object, we consider a point p_{ji} on link *j* that is *sufficiently close* to the obstacle i (fig. 12). If the repeller velocity vector computed at the link midpoint is perpendicular to the link, then the obstacle is *close* to the link midpoint. If this vector points in the direction of $\mathbf{l}_j$, then the obstacle is close to O_j, and if it points in the opposite direction to $\mathbf{l}_j$ then it is close to O_{j+1}.

Therefore, p_{ji} is approximated as :

$$p_{ji} = O_j + \mathbf{l}_j \frac{|\alpha_{ji}|}{\pi}$$

The ith repeller velocity field at p_{ji} is:

$$\Psi_{R_i}(p_{ji}) = \frac{\Phi_{R_i}(p_{ji})}{\gamma_{R_i} D_{R_i}(p_{ji}) \|\Phi_{R_i}(p_{ji})\|} e^{-b_i D_{R_i}(p_{ji})}$$

The attractor velocity field at the wrist O_{n+1} is:

$$\Psi_A(O_{n+1}) = \frac{\Phi_A(O_{n+1})}{\gamma_A D_A(O_{n+1}) |\Phi_A(O_{n+1})|}$$

Let $\Psi_T(O_{n+1}) = \Psi_A(O_{n+1}) + \Psi_R(O_{n+1})$ be the resultant velocity field at the wrist. Then the escape velocity at the wrist is:

$$V_E(O_{n+1}) = \pm K_E e^{-b_E |\Psi_T(O_{n+1})|} \xi(O_{n+1})$$

$\xi(\theta_{n+1})$ is a unit vector perpendicular to $\Psi_R(O_{n+1})$, K_E is a scaling coefficient, and b_E determines the rate at which the magnitude of $V_E(O_{n+1})$ decays as the wrist moves away from the zero velocity region. The sense of $V_E(O_{n+1})$ is initially selected such that the escape velocity points toward the base of the manipulator, so as to reach the target from *below*. If the target cannot be reached from below, then the sense of the escape velocity is reversed, so as to reach the target from *above*.

5.2 Joint velocities

Joint velocities are computed by summing the joint velocities due to the repeller velocity field acting on the links, and the attractor and escape velocity fields acting on the wrist, i.e.

$$\omega_T = \omega_R + \omega_A + \omega_E$$

The angular velocities due to the repeller velocity field are computed by using the truncated Jacobian matrix J_{ji} from the base of the manipulator to p_{ji} at link *j*. The repeller velocity field $\Psi_{R_i}(p_{ji})$ is multiplied by the pseudoinverse of J_{ji} and summed over all *m* obstacles and *n* links:

$$\omega_R = \sum_{i=1}^{m} \sum_{j=1}^{n} \mathbf{J}_{ji}^T [\mathbf{J}_{ji} \mathbf{J}_{ji}^T]^{-1} \Psi_{R_i}(p_{ji})$$

The joint velocities due to the attraction and escape velocities at the wrist are:
$\omega_A = J_T^{-1}\ \Psi_A(O_{n+1})$ and $\omega_E = J_T^{-1}\ V_E(O_{n+1})$
where J_T is the manipulator Jacobian from the base O_1 to the wrist O_{n+1}.

5.3 Joint limits

We now modify the angular velocities to: (i) approach the target with zero velocity, (ii) avoid joint velocity and position limits.

5.3.1 Zero velocity approach

Let Ω_i° be the maximum allowable velocity of the ith joint. Zero velocity at the target boundary can be achieved by redefining a position dependent maximum angular joint velocity as a function of the pseudodistance from the target, i.e.

$$\Omega_i(\Theta) = \Omega_i^\circ(1 - e^{-b_D D_A(X)})$$

where $\Theta = [\theta_1\ \theta_2\ .\ .\ .\theta_n]^T$ is the position vector in joint space.

5.3.2 Joint velocity limits

To satisfy the position dependent maximum joint velocity constraint, the ith joint velocity $\omega_i(\Theta)$ is weighted as follows:

$$\omega'_i(\Theta) = \omega_i(\Theta)\left(1 + \frac{\Omega_i(\Theta) - \omega_i(\Theta)}{\omega_i(\Theta)}\, e^{-b_\omega(\Omega_i(\Theta) - |\omega_i(\Theta)|)}\right)$$

if $|\omega_i(\Theta)| < \Omega_i(\Theta)$, and

$$\omega'_i(\Theta) = \Omega_i(\Theta) \quad \text{if } |\omega_i(\Theta)| \geq \Omega_i(\Theta)$$

5.3.3 Joint position limits

The angular velocities are further scaled to satisfy joint limit constraints, i.e $\theta_{i,\min} < \theta_i < \theta_{i,\max}$. This is accomplished by making the joint velocities go to zero smoothly as the joint limit from either side is reached.

$$\omega''_i(\Theta) = \omega'_i(\Theta)(1 - e^{-b_\theta(\theta_{i,\max} - \theta_i)}) \quad \text{for } \omega'_i > 0$$

$$\omega''_i(\Theta) = \omega'_i(\Theta)(1 - e^{-b_\theta(\theta_i - \theta_{i,\min})}) \quad \text{for } \omega'_i \leq 0$$

5.3.4 Knot point generation

Knot points in joint space are obtained by iterating the sequence:

$$\Theta^{k+1} = \Theta^k + h\,\omega_T(\Theta^k)$$

where Θ is the state vector in joint space, and h is a step size.

5.4 Examples

Two types of experiments were performed, (i) to show how the method is used to move a three link planar manipulator from an initial configuration to a final wrist position avoiding collisions in a cluttered environment, and (ii) to demonstrate the applicability of the method to dynamic environments.

The three obstacles and the target in figs. 13 and 14 were obtained by using the following functions and parameters:

$$H_{R_1} = \begin{Bmatrix} (x_2 - 0.40) \\ (x_1 - 0.60) \end{Bmatrix} \qquad c_{R_1} = 0.04 \qquad b_1 = 20.0 \qquad \gamma_{R_1} = 1.00$$

$$H_{R_2} = \begin{Bmatrix} (x_2 - 0.55) \\ (x_1 - 0.75) \end{Bmatrix} \qquad c_{R_2} = 0.04 \qquad b_2 = 20.0 \qquad \gamma_{R_2} = 1.00$$

$$H_{R_3} = \begin{Bmatrix} (x_2 - 0.70) \\ (x_1 - 0.65) \end{Bmatrix} \qquad c_{R_3} = 0.04 \qquad b_3 = 20.0 \qquad \gamma_{R_3} = 1.00$$

$$H_A = \begin{Bmatrix} (x_2 - 0.35) \\ (x_1 - 0.80) \end{Bmatrix} \qquad c_A = 0.02 \qquad \gamma_A = 0.25$$

Fig. 13 shows the motion of the manipulator from an initial configuration of $\theta_1 = 45°$, $\theta_2 = 30°$, $\theta_3 = -30°$, the base is at (0.5, 0.3) and link lengths are 0.13. The escape velocity parameters are $b_E = 1.0$ and $K_E = 1.0$. Initially the manipulator is at a configuration where there is an obstacle between the wrist and the target. The third link gets closer to the obstacle as the wrist is attracted by the target until the repulsion velocity from the obstacle becomes significant. As the resultant attraction and repulsion velocity at the wrist decreases, the escape velocity dominates and the wrist moves around the obstacle until the path toward the target clears, then approaches the target with zero velocity. Meanwhile, collisions of the other links with the obstacles are avoided.

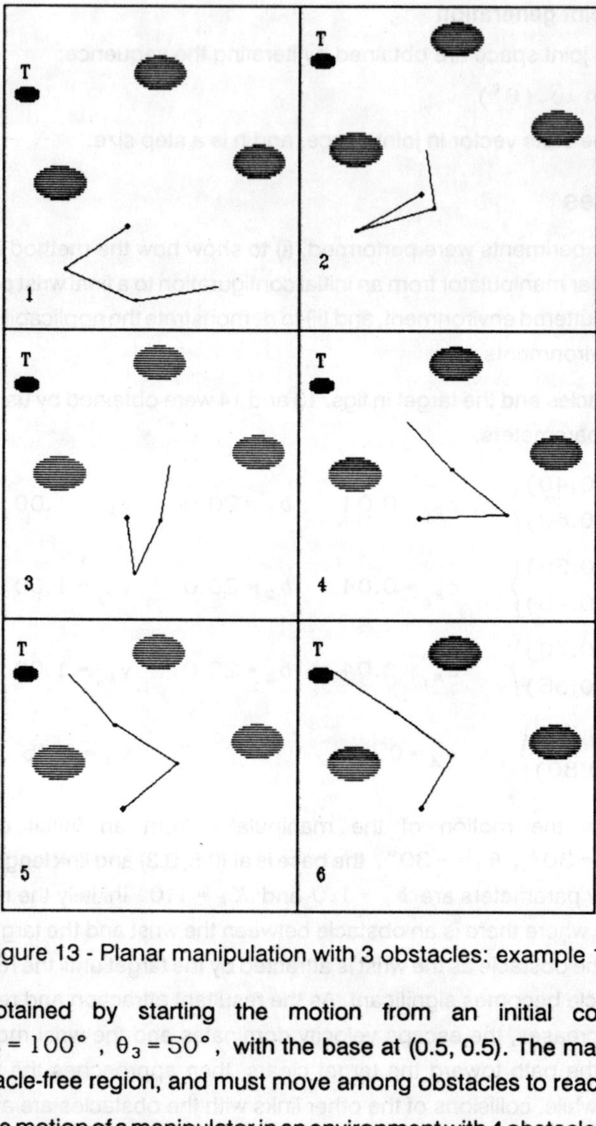

Figure 13 - Planar manipulation with 3 obstacles: example 1

Fig. 14 is obtained by starting the motion from an initial configuration of $\theta_1 = 225°$, $\theta_2 = 100°$, $\theta_3 = 50°$, with the base at (0.5, 0.5). The manipulator is initially in an obstacle-free region, and must move among obstacles to reach the target.

Fig. 15 shows the motion of a manipulator in an environment with 4 obstacles. The obstacle and target functions are:

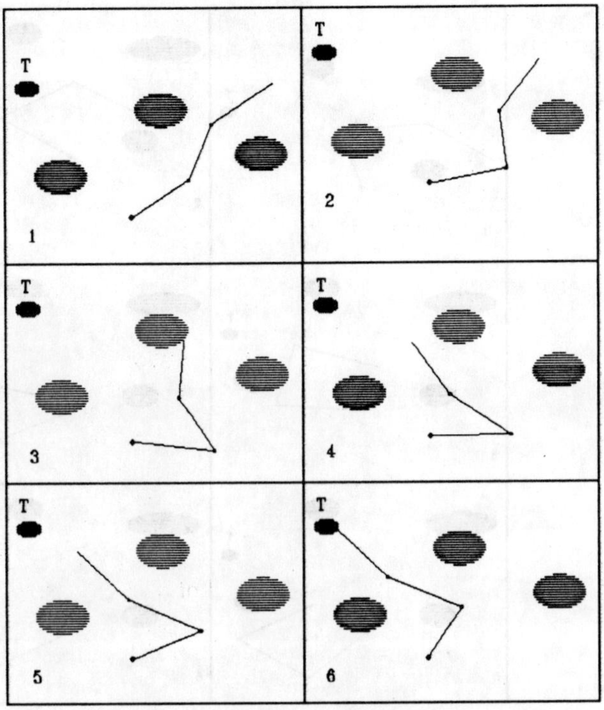

Figure 14 - Planar manipulation with 3 obstacles: example 2

$$H_{R_1} = \begin{Bmatrix} (x_2 - 0.30)^2 \\ (x_1 - 0.65) \end{Bmatrix} \qquad c_{R_1} = 0.04 \qquad b_1 = 30.0 \qquad \gamma_{R_1} = 1.00$$

$$H_{R_2} = \begin{Bmatrix} (x_2 - 0.60) \\ (x_1 - 0.70) \end{Bmatrix} \qquad c_{R_2} = 0.04 \qquad b_2 = 30.0 \qquad \gamma_{R_2} = 1.00$$

$$H_{R_3} = \begin{Bmatrix} (x_2 - 0.35) \\ (x_1 - 0.35) \end{Bmatrix} \qquad c_{R_3} = 0.04 \qquad b_3 = 30.0 \qquad \gamma_{R_3} = 1.00$$

$$H_{R_4} = \begin{Bmatrix} (x_2 - 0.75) \\ (x_1 - 0.40) \end{Bmatrix} \qquad c_{R_4} = 0.04 \qquad b_3 = 30.0 \qquad \gamma_{R_4} = 1.00$$

$$H_{A} = \begin{Bmatrix} (x_2 - 0.80) \\ (x_1 - 0.55) \end{Bmatrix} \qquad c_{A} = 0.02 \qquad \gamma_{A} = 0.25$$

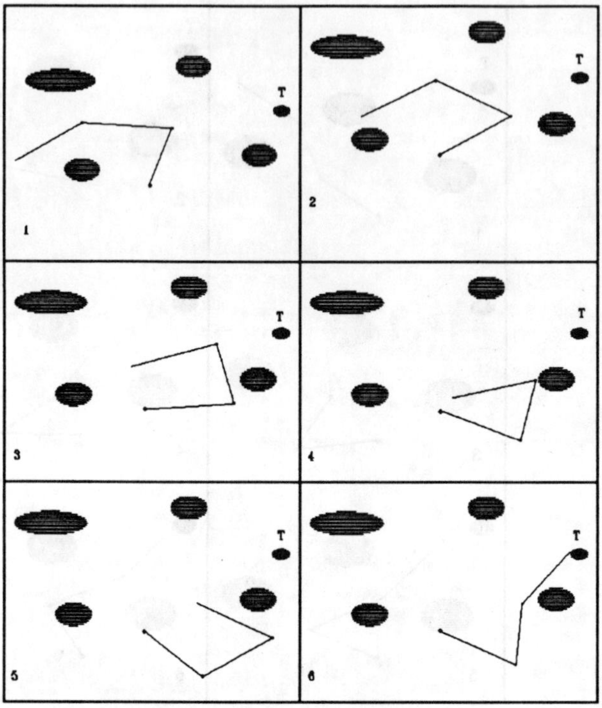

Figure 15 - Planar manipulation with 4 obstacles

The initial configuration of the manipulator is $\theta_1 = 75°$, $\theta_2 = 100°$, $\theta_3 = 45°$. The base is at (0.5, 0.3), and all link lengths are equal to 0.2. Other parameters are the same as in the previous example.

The final example demonstrates path generation for two 2-link manipulators sharing a common workspace. Initially, the two manipulators are at a configuration where their links are obstructing each other's way to the target (fig. 16). The links of each manipulator repel the other's. After they have moved sufficiently apart, the path from the wrist to the target clears, and the manipulators start moving toward their targets while avoiding collisions with each other.

Manipulator 1 is initially at a configuration $\theta_1 = 260°$, $\theta_2 = 10°$, with its base at (0.5, 0.8). The target function and parameters are:

$$H_{A_1} = \begin{Bmatrix} (x_2 - 0.8) \\ (x_1 - 0.7) \end{Bmatrix} \qquad c = 0.02 \qquad \gamma_{A_1} = 0.2$$

Manipulator 2 has an initial configuration of $\theta_1 = 80°$, $\theta_2 = 10°$, with its base at (0.5, 0.2). The target function and parameters are:

$$H_{A_2} = \begin{Bmatrix} (x_2 - 0.2) \\ (x_1 - 0.3) \end{Bmatrix} \qquad c = 0.02 \qquad \gamma_{A_2} = 0.2$$

The escape velocity coefficient, and decay parameter are K_E = 1.0 and b_E = 1.0. The links of each manipulator act as moving elliptical obstacles for the other manipulator. Major and minor axis lengths for the ellipses are 0.25 and 0.05, respectively. Weight coefficients for the ellipses are b = 5.0 and $\gamma = 0.2$.

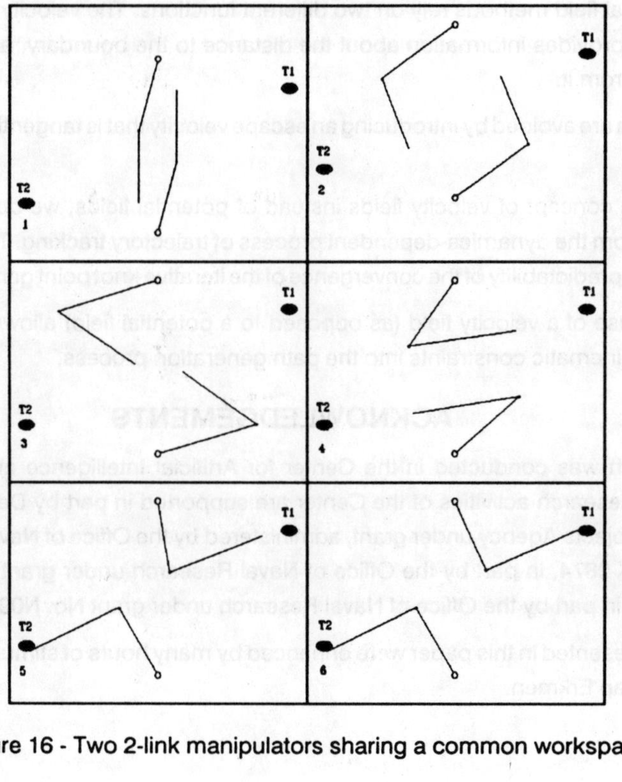

Figure 16 - Two 2-link manipulators sharing a common workspace

6. DISCUSSION

A real-time obstacle avoidance algorithm was developed in this paper, based on the concept of velocity fields. Some properties of this approach include:

Path planning is local, and therefore the method can adapt to changes in the modeled environment without global replanning. As more sensory data are gathered during the motion, the estimated identity (target or obstacle) of an object may change. The estimated location of the object may also change. The path planning algorithm adapts to these changes by making the corresponding changes in the attractor or repeller functions.

The velocity field vectors for each obstacle set provide both distance and direction information, thus avoiding the computation of the shortest Euclidean distance to each obstacle. A single function is used for both object representation and obstacle avoidance. By contrast, potential field methods rely on two different functions. The velocity field vector near an obstacle provides information about the distance to the boundary, and a direction to move away from it.

Local minima are avoided by introducing an escape velocity that is tangential to the obstacle boundary.

By using the concept of velocity fields instead of potential fields, we decouple obstacle avoidance from the dynamics-dependent process of trajectory tracking. This facilitates the analysis and predictability of the convergence of the iterative knot point generation process.

Finally, the use of a velocity field (as opposed to a potential field) allows the easy incorporation of kinematic constraints into the path generation process.

ACKNOWLEDGEMENTS

This research was conducted in the Center for Artificial Intelligence at George Mason University. Research activities of the Center are supported in part by Defense Advanced Research Projects Agency under grant, administered by the Office of Naval Research, No. N00014-87-K-0874, in part by the Office of Naval Research under grant No. N00014-88-K-0226, and in part by the Office of Naval Research under grant No. N00014-88-K-0397.

The ideas presented in this paper were enhanced by many hours of stimulating discussion with Dr. Aydan Erkmen.

REFERENCES

1. J.Borenstein, Y.Koren (1988), "Obstacle avoidance with ultrasonic sensors," *IEEE Journal of Robotics and Automation*, Vol.4, No.2, pp. 213-217.
2. Y.C.Chen, M.Vidyasagar (1988), "Optimal trajectory planning for planar n-link revolute manipulators in the presence of obstacles," *Proc. IEEE Int. Conf. Robotics and Automation*, pp. 202-208.
3. Y.P.Chien, A.J.Koivo, B.H.Lee (1988), "On-line generation of collision free trajectories for multiple robots," *Proc. IEEE Int. Conf. Robotics and Automation*, pp. 209-214.
4. K.Fujimura, H.Samet (1989), "A hierarchical strategy for path planning among moving obstacles," *IEEE Trans. on Robotics and Automation*, Vol.5, No.1, pp. 61-69.
5. R.T.Dunlay (1988), "Obstacle avoidance perception processing for the autonomous land vehicle" *Proc. IEEE Int. Conf. Robotics and Automation*, pp. 912-917.
6. A.Elfes, (1987) "Sonar based real-world mapping and navigation," *IEEE J. of Robotics and Automation*, Vol.3, No.3, pp. 249-265.
7. A.M.Erkmen and H.E.Stephanou (1988), "Information fractals for an evidential classifier," *Proc. IEEE Int. Symposium on Intelligent Control*.
8. A.M.Erkmen, F.Yegenoglu, and H.E.Stephanou (1989), "Entropy-driven on-line control of autonomous robots," to appear, *Int. J. Intelligent and Robotic Systems*.
9. A.M.Erkmen and H.E.Stephanou (1989), "Dexterous grasp planning with incomplete evidence," *NASA Conference on Space Telerobotics*.
10. B.Fajeron, P.Tournassoud (1987), "A local based approach for path planning of manipulators with a high number of degrees of freedom," *Proc. IEEE Int. Conf. Robotics and Automation*, pp. 1152-1159.
11. W.T.Gex, N.L.Campbell (1987), "Local free space mapping and path guidance," *Proc. IEEE Int. Conf. Robotics and Automation*, pp. 424-431.
12. C.Isik and A.Meystel (1988), "Pilot level of a hierarchical controller for an unmanned mobile robot," " *IEEE J. of Robotics and Automation*, Vol.4, No.3, pp. 241-255.
13. S.Jun, K.G.Shin (1988), "A probabilistic approach to collision-free robot path planning," *Proc. IEEE Int. Conf. Robotics and Automation*, pp. 220-225.
14. Y.Kanayama (1988), "Least cost paths with algebraic cost functions," *Proc. IEEE Int. Conf. Robotics and Automation*, pp. 75-80.
15. Y.Kanayama, N.Miyake (1986), "Trajectory generation for mobile robots," *Robotics Research: The Third Int. Symposium*, edited by O.D.Faugeras and G.Giralt, pp. 333-340.
16. K.Kant, S.Zucker (1988), "Planning collision-free trajectories in time-varying environments: A two level hierarchy," *Proc. IEEE Int. Conf. Robotics and Automation*, pp. 1644-1649.
17. O.Khatib (1986), "Real-time obstacle avoidance for manipulators and mobile robots," *International J. Robotics Research*, Vol. 5, No. 1, pp. 90-98.
18. S.Kheradpir and J.S.Thorp (1988), "Real-time control of robot manipulators in the presence of obstacles," *IEEE J. of Robotics and Automation*, Vol. 4, No.6, pp. 687-698.

19. P.Khosla, R.Volpe (1988), "Superquadric artificial potentials for obstacle avoidance and approach," *Proc. IEEE Int. Conf. Robotics and Automation*, pp. 1778-1784.
20. B.H.Krogh, C.E.Thorpe (1986), "Integrated path planning and dynamic steering control for autonomous vehicles," *Proc. IEEE Int. Conf. Robotics and Automation*, pp. 1664-1669.
21. B.H.Krogh and D.Feng (1989), "Dynamic generation of subgoals for autonomous mobile robots using local feedback information," *IEEE Transactions on Automatic Control*, Vol. 34, No.5, pp. 483-493.
22. T.Lozano-Perez (1983), "Spatial planning: a configuration space approach," *IEEE Trans. on Computers*, Vol.C-32, No.2, pp. 108-120.
23. T.Lozano-Perez (1986), "Motion planning for simple manipulators," *Robotics Research: The Third Int. Symposium*, edited by O.D.Faugeras and G. Giralt, pp. 133-140.
24. V.J.Lumelsky (1987), "Algorithmic and complexity issues of robot motion in an uncertain environment," *Journal of Complexity* Vol. 3, pp. 146-182.
25. V.J.Lumelsky and A.A.Stepanov (1986), "Dynamic path planning for a mobile automaton with limited information on the environment," *IEEE Transactions on Automatic Control*, Vol. 31, pp. 1058-1063.
26. A.Meystel (1988), "Intelligent control in robotics," *J. of Robotic Systems*, pp. 269-308.
27. R.S.Michalski and P.H.Winston (1986), *Int. J. Artificial Intelligence*, pp. 121-146.
28. R.S.Michalski, I.Mozetic, J.Hong, and N.Lavrac, (1986), "The multipurpose incremental learning system AQ15 and its testing application to three medical domains," *Proc. 5th National Conference on Artificial Intelligence*, pp. 1041-1045.
29. W.S.Newman, N.Hogan (1987), "High speed robot control and obstacle avoidance using dynamic potential functions," *Proc. IEEE Int. Conf. Robotics and Automation*, pp. 14-24.
30. T.N.Nguyen and H.E.Stephanou (1989), "A model of prehension for multifingered hands," *IEEE Int. Conf. Robotics and Automation*, pp. 446-451.
31. H.O.Peitgen and P.H.Richter (1986), *The Beauty of Fractals*, Springer-Verlag.
32. J.S.Pettinato and H.E.Stephanou (1989), "Manipulability and stability of a tentacle based robot manipulator," *IEEE Int. Conf. Robotics and Automation*, pp. 458-463.
33. H.E.Stephanou, A.M.Erkmen (1988), "Evidential classification of dexterous grasps," *J. Robotic Systems*, Vol. 5, No.4, August, pp. 309-336.
34. P.Tournassoud (1986), "A strategy for obstacle avoidance and its application to multi-robot systems," *Proc. IEEE Int. Conf. Robotics and Automation*, pp. 1224-1229.
36. F.Yegenoglu, A.M.Erkmen, H.E.Stephanou (1988), "One-line path planning under uncertainty," *Proc. IEEE Conf. Decision and Control*, pp. 1075-1079.
37. F.Yegenoglu, and H.E.Stephanou (1989), "Real-time obstacle avoidance in dynamic environments, part I : a velocity fields approach," *IEEE Transactions on Automatic Control.* (submitted)

38. F.Yegenoglu, and H.E.Stephanou (1989), "Real-time obstacle avoidance in dynamic environments, part II : navigation and manipulation algorithms," *IEEE Transactions on Automatic Control.* (submitted)
39. F.Yegenoglu, and H.E.Stephanou (1989), "Collision-Free Path Planning For Multirobot Systems," *Proc. IEEE Conf. Intelligent Control.* (submitted)

Applications of Learning and Planning Methods
edited by N. G. Bourbakis

CHAPTER 13

KNOWLEDGE-BASED ACQUISITION IN REAL-TIME PATH PLANNING IN UNKNOWN SPACE

Nikolaos G. Bourbakis
IBM, San Jose, CA

ABSTRACT

This chapter deals with the representation and acquisition of knowledge extracted during a real-time path planning process in an unknown space. In particular, the extracted knowledge from each current free navigation space is represented by attributed graphs. The graph knowledge representation forms are ″processed and combined″ appropriately by generating a scheme which acquires the extracted knowledge during the navigation. Illustrated examples are provided for simple navigation cases.

keywords : *Knowledge Representation,Knowledge Acqusition, Path Planning in Unknown Space.*

This work is a part of the ODYSSEUS research project.

1. INTRODUCTION

The path planning of autonomous robots is always an interesting topic and a variety of different methodologies has been proposed. In particular, Lozano-Perez [1] discusses a method for finding the shortest path of a moving object, by using the concept of configuration space. Brooks [2] presents the free space by generalized cones and searches a collision free path of a moving object inside the generalized cones. Gonzenes [11] uses a intrinsic tree structure to represent the free space, in order to find the collision free path of a robot, by using graph search methods. Schwartz-Shamir [3] study the peano mover's problems. They define the existence of collision paths of multiple robots by using critical curves. The free space is partitioned by these critical curves into many regions and the solution is searched in the complexity graph corresponding to the regions Barraquand-Latombe [4] propose a distributed representation of the environment model which provides flexibility, parallelism and fast execution. Bourbakis [8] defines the navigation path by using the shapes of the current free navigation spaces selecting low cost corridors in unknown navigation environment.

The path planning methodologies can be classified into three main categories. The first category, *Known Navigation Map*, includes themethods which define a path in a well known navigation space *Incomplete Navigation Map*,includes the methods which deal with the determination of a path under incomplete knowledge of the entire navigation space. The third category, *Unknown Navigation Map*, includes the methods which attempt to define a navigation path using only the knowledge of the current navigation space and without any knowledge about the overall navigation environment.

The path planning approach used in this work belongs to the unknown navigation map category. The only knowledge for the determination of navigation path is the shape of the current free navigation space. The shape of the surrounding free space is defined by the distances of the robot from the existing obstacles. Closed and open corridors are also defined by the use of these distances. Thus, the moving robot has to make a decision and to travel through the appropriate corridor in order to approach or to reach the destination point. The decision making process for the selection of the proper corridor is based on a heuristic "human like" approach. In particular, a *trial and error* methodology is used, which attempts to minimize the current navigation cost in respect with the minimum traveling time and the corridor with the maximum probability to "drive" the robot close to the destination point. When a corridor is selected, the *diamesos* of the corridor triangular surface is defined and the robot travels on it until the changes of the new free space provide a new low cost corridor for traveling.

The knowledge representation and acquisition are two important issues on the development of an intelligent system able to support perception, learning and planning, [5].Moreover, the acquisition of the knowledge requires the development of the appropriate strategies in order to make flexible the overall behavior of the system, [6]. The knowledge, however, can be represented by various forms [12], such as predicate logic, production systems, semantic networks, frame representations, procedural representations, connectionist representations, etc.. Thus it is important to define the appropriate medium for the representation of knowledge with generic features and powerful capabilities. Based on the assumptions above, we have select the attributed graphs as the form for the representation of the knowledge extracted from a 2-D navigation space.

In this chapter, we present a knowledge based acquisition scheme using the synthesis of the shapes extracted from the current free navigation spaces. In particular, the extracted knowledge from each current navigation space is represented by attributed graphs for a generic and flexible representation and process of the accumulated knowledge. The synthesis of the shapes implemented with attributed graphs is the base of the knowledge acquisition scheme (KAS). The KAS scheme repeats the synthesis process between graph forms accumulating and generalizing the knowledge extracted from the navigation space. The global configuration and some operations of a knowledge-baseare also discussed.

2. NOTATIONS AND DEFINITIONS

In this section we provide a number of notations and definitions in order to accommodate the understanding of the following sections.

Notation 1: GNS represents the global navigation space.

Definition 1: A 2-D obstacle, $b_j, j \in Z$, is defined as the two dimensional surface, $Sb_j \in GNS$, where a moving object (*robot R*) cannot go through it, and the visual and sonar rays stop or reflect on it.

*Definition 2:*A robot, R_n, $n \in Z$,(or object) is a moving obstacle in GNS by using its own power to do so.

Notation 2: BS represents the set of all the obstacles in GNS, BS $\subset$GNS.

Notation 3: RS represents the set of all the robots in GNS, RS $\subset$GNS.

Notation 4: Sr_n represents the 2-D surface covered by a robot R_n in GNS.

Notation 5: $E_b = \bigcup Sb_j$ and $E_r = \bigcup Sr_n$ represent the total surfaces covered by the total number of obstacles and the total number of robots in GNS respectively.

Notation 6: SHb_j and SHr_n represent the shape of the b_j obstacle and the R_n robot respectively.

Definition 3: The total free navigation space inside GNS, is defined as FNS = (GNS - E_b - E_r).

Notation 7: t_i, $i \in Z$, represents the current time.

Notation 8: NS_{t_i} represents the navigation space at the time t_i

Definition 4: The current free navigation space $FNS^n_{t_i}$ is defined as the free space perceived by the robot R_n at the time t_i, $FNS^n_{t_i} \subset$ FNS.

Corollary 1: If there are two robots R_{n_1} and R_{n_2} in the same NS_{t_i} with $n_1 \neq n_2$ then $FNS^{n1}_{t_i} \neq FNS^{n1}_{t_i}$ except in the case that both $SHr_{n_1} \equiv SHr_{n_2}$ and the shape SH(NS_{t_i}) is symmetrical between these two robots.

3. KNOWLEDGE EXTRACTION AND REPRESENTATION FROM THE FREE NAVIGATION SPACE

In this section we attempt to extract and represent knowledge from the geometric form of the current free navigation space $FNS^n_{t_i}$ perceived by a moving robot R_n. In order to do it, we extract the shape SH($FNS^n_{t_i}$) = $SH^n_{t_i}$ of the current free navigation space. Then, we define the relationships among the straight line and curve line segments of the shape $SH^n_{t_i}$ and, then we represent the syntactic and semantic information of the shape with a directed graph with attributes.

3.1 Shape Extraction from $FNS^n_{t_i}$

We assume that the moving robot R^n includes the appropriate devices (sonar system and vision camera) to greate the shape $SH^n_{t_i}$ from the current free navigation space $FNS^n_{t_i}$ by using the distances $d^k_{t_i}$, $k \in Z$, of the robot R^n from the surrounding obstacles and boundaries.

$$f : FNS^{n}_{t_i} \rightarrow SH^{n}_{t_i} \in FNS^{n}_{t_i} \qquad (1)$$

such that, for each $d^{k}_{t_i} \in FNS^{n}_{t_i}$ there is a point $B^{n}_{t_i} \in SH^{n}_{t_i}$ with $f(d^{k}_{t_i}) = B^{k}_{t_i}$, $i, k \in Z^{+}$

Now, we connect all the points $(B^{n}_{t_i})$ with short straight line segments $(B^{n-j}_{t_i} B^{n-j+1}_{t_i}) \in SH^{n}_{t_i}$, $j = n, \ldots, 0$. Thus, the shape can be expressed as

$$SH^{n}_{t_i} \equiv \bigcup (B^{n-j}_{t_i} B^{n-j+1}_{t_i}) \qquad \text{for } j = n, \ldots, 0. \qquad (2)$$

Figure 1 illustrates the generation of the shape of a hypothetical free navigation space.

At this point, we code these primitive straight lines $(B^{n-j}_{t_i} B^{n-j+1}_{t_i})$ by using the chain code [8] and the shape takes a new form:

$$SH^{n}_{t_i} \equiv a_1 d'_{j1} a_2 d'_{j2} \ldots a_m d'_{jm} \qquad (3)$$

where, $a_i \in Z$, $i = 1, \ldots, m$, represents the number of the primitive straight line segments with the same direction d'_{ji}.

From the formula (3), we can define the straight lines and curve lines their properties and determine their connection relationships R^{c}_{jk}, $j,k \in Z$. Thus, the shape can be also expressed as

$$SH^{n}_{t_i} \equiv X_1 R^{c}_{12} X_2 R^{c}_{23} \ldots X_{w-1} R^{c}_{w-1w} X_w R^{c}_{w1} X_1 \qquad (4)$$

where, $w \in Z$, and X_j, $j = 1, 2, \ldots, w$ represents either a straight line segment $[L_{w_1}]$, or a curve line segment $[C_{w_2}]$, with $w_1 + w_2 = w$.

Thus, the generalized form of a shape can be:

$$SH^{n}_{t_i} \equiv \int_{j=1}^{w} (X_j R^{c}_{jj+1} X_{j+1}) \qquad (5)$$

where, $X_1 \equiv X_{w+1}$ and $\int$ represents the synthesis operator.

Figure 2 shows a hypothetical shape $SH^{1}_{t_0}$ after the determination of the straight line and curve line segments and their connectivity. The exponent index of the shape shows that this shape was extracted by the robot R_1 at the time t_0.

The shape of the figure 2 can be expressed as:

$$SH^{1}_{t_0} \equiv L_1 R^{c}_{12} C_1 R^{c}_{23} L_2 R^{c}_{34} L_3 R^{c}_{45} L_4 R^{c}_{56} L_5 R^{c}_{67} L_6 R^{c}_{78} L_7 R^{c}_{89} L_8 R^{c}_{910} C_2 R^{c}_{1011} C_3 R^{c}_{111} L_1 \qquad (6)$$

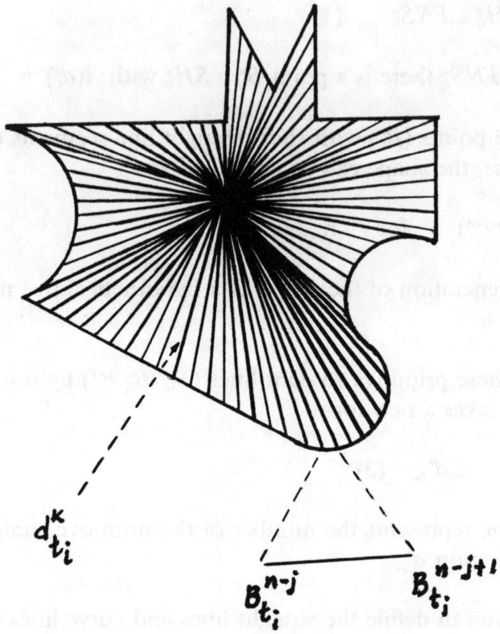

Figure 1: Generation of the shape.

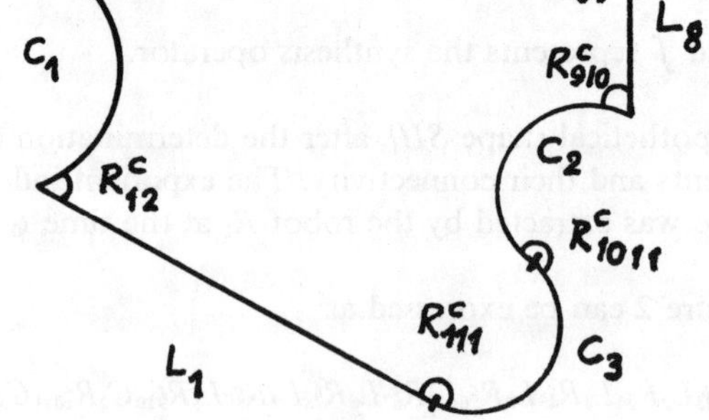

Figure 2: Straihgt line and curve line segments of a shape.

Note that, since every extracted shape is a closed line, then the indeces jk of R_{jk}^{c} represent two consecutive segments.

The complete representation of a shape $SH_{t_i}^{n}$, however, requires the determination of two more factors. (i) The individual properties P_j^x of a segment X_j

$$P_j^x = [starting\ point(sp), size(s), orientation(d'), curvature(cu)] \qquad (7)$$

where, the index j indicate the appropriate segment, and the values of the index x indicate the properties of the particular segment. (ii) The relationships R_{km}^{y} among the segments,

$$R_{km}^{y} = [connectivity(c), parallelism(p), distance(d), symmerty(sy), similarity(si),$$
$$relative\ magnitude(rm), relative\ location(rl)] \qquad (8)$$

where, indeces km point the appropriate segments and the values of the index y indicate the appropriate relationships among the segments.

Thus, given of a current $FNS_{t_i}^{n}$, we define the the segments X_j, their properties P^x and the relationships R^y among the segments for a sufficient description of the current navigation space.

3.2 Abstract Representation of the Shape $SH_{t_i}^{n}$ Using Attributed Graphs

In this subsection, we express the extracted shape, formula (4), and its properties with an abstracted and generic form, the attributed graphs, which accommodate the representation and acquisition of the extracted navigation knowledge. In particular, the new expression of the shape is defined as follows [7, 10]

$$h : SH_{t_i}^{n} \rightarrow G_{t_i}^{n} \qquad (9)$$

with, $h(X_j) = N_j, j \in Z$, where the graph node N_j represents the X_j segment and its properties P^x ; and $h(R_{km}^{y}) = A_{km},\ k,m \in Z$, where the graph arrow A_{km} represents the relationships R_{km}^{y} among the X_j segments. Thus,

$$SH_{t_i}^{n} \equiv G_{t_i}^{n} = [N_j, A_{km}]_{t_i} = N_1 A_{12} N_2 A_{23} N_3 \ldots A_{pp'} N_q \qquad (10)$$

Note that, the graph representation of a shape starts and ends always with a graph node.

In order to make understandable the overall process for the graph representation of an extracted shape $SH_{t_i}^{n}$, we use the shape of the figure 2.

Initially, we present graphically the equivalent graph generated by this particular shape by starting with the connectivity relationship R^c_{jk} among the consecutive segments, as shown in fig. 3. After that, we start a *comparison* process comparing the first segment X_1 with all the rest segments of the graph G^1_{t0}, "marking" (or saving) the existing relations, if any, on the graph. We continue the same process with the second segment X_2 marking the relations on the graph, ect. At the end of the comparison process, we start another process, called *elimination* . The elimination process remove all the non-important relations among the graph segments, such as *a segment X is not parallel with the segment Y*. The explanation is that we "mark" only the parallelism among the segments by implying that the rest segments are not parallel. We also eliminate all the relations that produce redundancy by using *induction* rules, such as

if there are three major relations:

1) $X_iR^p_{ij}X_j$,

2) $X_iR^p_{ik}X_k$ *and*

3) $X_jR^p_{jk}X_k$

then the relation 3) is eliminated since it can be (11)

regenerated by induction between 1) and 2).

Note that, some redundancy remains at the graph representation to accommodate the quick interpretation of the abstract form of the graph.

Figure 3 shows the generated graph after the elimination process. This graph includes the segments properties and the remaining major relationships among the graph segments.

Figure 4 shows the same graph using the nodes N_j and the arrows A_{km}.

Thus, the equivalent graph form G^1_{t0} for the shape SH^1_{t0} is

$$G^1_{t0} =]N_1A_{12}N_2A_{23}N_3A_{34}N_4A_{45}N_5A_{56}N_6A_{67}N_7A_{78}N_8A_{89}N_9A_{910}N_{10}A_{1011}N_{11}A_{111}N_1] \bullet$$
$$\bullet [N_1A_{13}N_3A_{38}N_8] \bullet [N_4A_{47}N_7A_{79}N_9] \qquad (12)$$

where, • represents the graph connection operator.

At this point, we start a search-matching process at the generated graph by using existing "knowledge" stored in a knowledge-base in graph forms. In particular, we retrieve the abstract graph form of a recognizable pattern and we attempt to match the pattern's interrelations with some group of relations existing in a particular "neighborhood" or subgraph ${}^{l}SG^n_{ti} \in G^n_{ti}, j \in Z$, If there is a successful matching then we replace that particular subgraph with a new node ${}^{l}N_{j'}$ abstracting the represen-

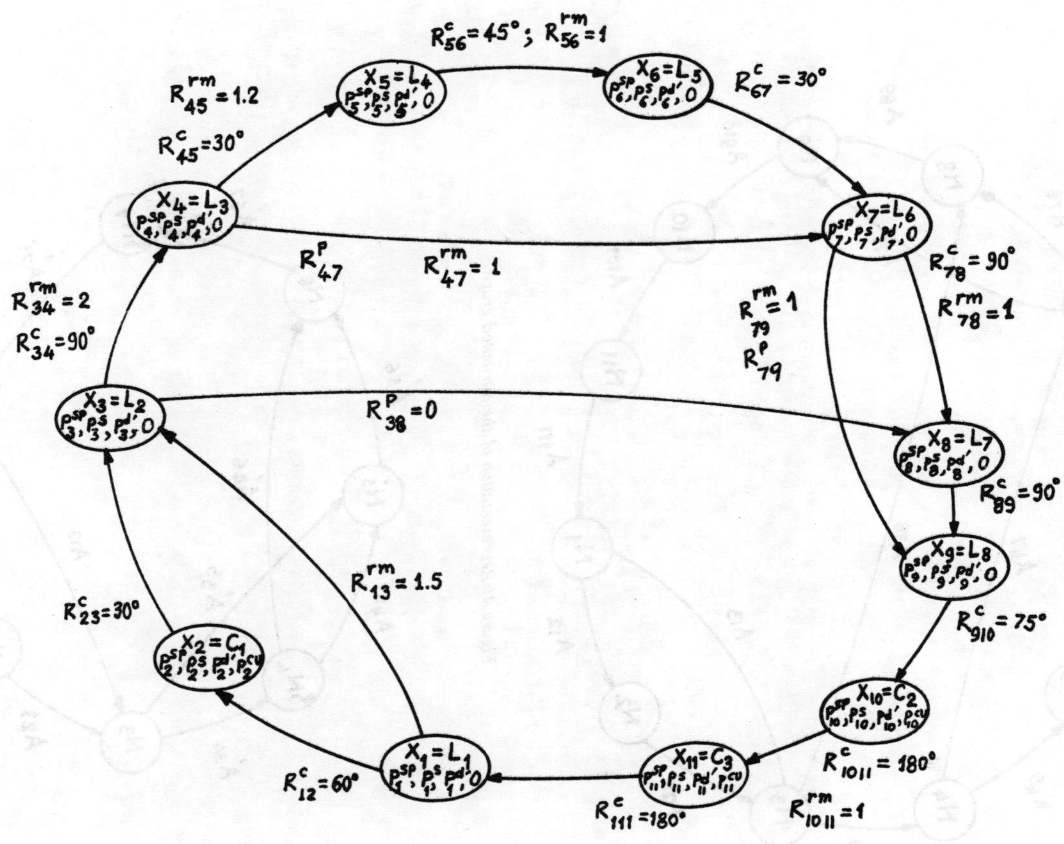

Figure 3: Graphical representation of a shape using graphs.

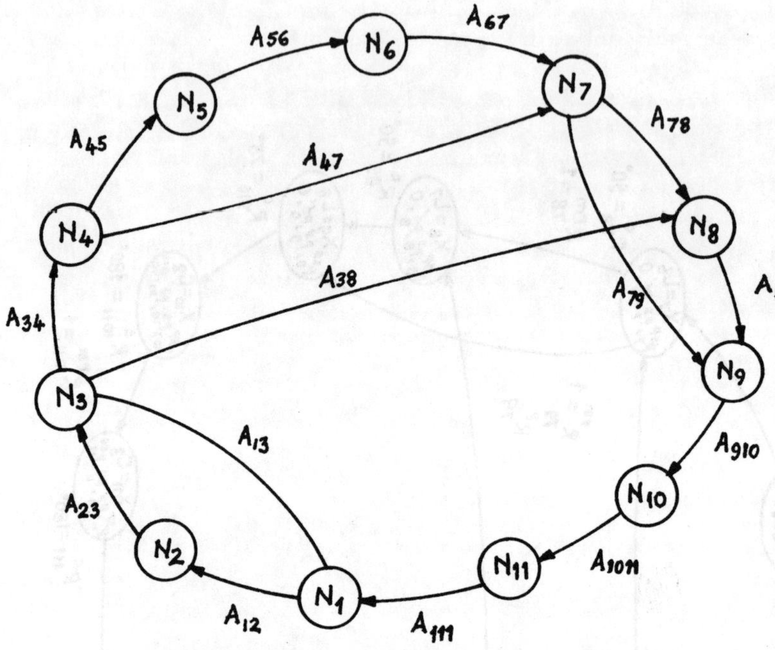

Figure 4: Representation of the attributed graph.

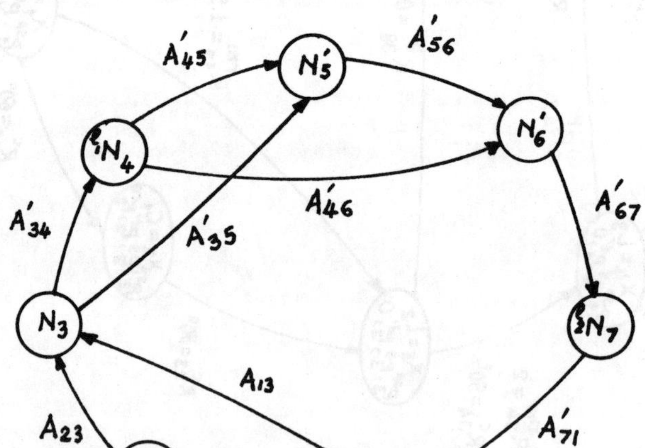

Figure 5: Abstract representation of the graph

tation of the graph. Note that, the replaced subgraph is saved with an appropriate indexing scheme in the knowledge-base (KB) for future use.

Applying the searching-matching process on the graph formula (12). We observe that, in the subgraph ${}^{l1}SG^{1}_{t_0}$, $l1 = N_4N_5N_6N_7$, the relations $R^{\xi}_{45} = R^{\xi}_{67}$, $R^{\xi\pi}_{56} = 1$, $R^{\xi\pi}_{47} = 1$, R^{ξ}_{47} compose a subgraph with similar attributes with the graph form of the capital letter M stored as primitive knowledge in KB. Thus, we can replace the subgraph ${}^{l1}SG_{1t_i} \equiv {}^{l1}N_4$, $l1 = N_4N_5N_6N_7$ by making the appropriate arrangement of the arrows. Moreover, for the same reasoning the subgraph ${}^{N_{10}N_{11}}SG^{1}_{t_0}$ composes similar attributes with the graph form of the capital letter S stored as primitive knowledge in KB. Thus, we replace the subgraph ${}^{l2}SG^{1}_{t_0} \equiv {}^{l2}N_7$, $l2 = N_{10}N_{11}$ by making rearrangement of the arrows and the numbering of the rest nodes. Based on the abstractions above, the new form of the graph, fig. 5, is

$$G^{1}_{t_0} = [N_1A_{12}N_2A_{23}N_3A'_{34}{}^{l1}N_4A'_{45}N'_5A'_{56}N'6A'_{67}{}^{l2}N_7A'_{71}N_1] \bullet$$

$$\bullet [N_1A_{13}N_3A'_{35}N'_5] \bullet [{}^{l1}N_4A'_{46}N'_6] \qquad (13)$$

where,
$N_8 \equiv N'_5$, $N_9 \equiv N'_6$, $A'_{34} \equiv A_{34}$, $A'_{35} \equiv A_{38}$, $A'_{45} \equiv A_{78}$, $A'_{46} \equiv A_{79}$, $A'_{56} \equiv A_{89}$, $A'_{67} \equiv A_{910}$, $A'_{71} \equiv A_{111}$.

The abstract representation of a graph form does not stop with the replacement of some subgraphs with new nodes, which represent the recognizable form of a primitive pattern stored in a KB. The process continues to higher levels of global representation. Thus, if the abstract representation process continues to higher levels, then the form of the graph becomes more abstract (by saving the intermmediate levels of replacements in the knowledge-base). In the highest level of abstraction, the graph form can be represented by a single meaning (or *semantic object (SO)*). Some very simple semantic object representations of a graph could be: "star", "circle", etc. where the extracted shape $SH^{n}_{t_i}$ of the current navigation space has the form of a "circle" , or a star", thus, a higher abstract representation cannot be accomplished than that. If, however, it is not possible the highest abstracted representation to be reached, then the graph is expressed as a combination of intermmediate abstract representation forms and primitive non-recognizable subgraph forms.

Generalizing the representation of the graph form, we have that

$$G^{n}_{t_i} \equiv \int_{lk'm'}([G'^{n'}_{t_{i-j}}]_{l'},[{}^{lk'}N^{v}_{j'}]_{k'},[{}^{lm'}SG^{n}_{t_i}]_{m'}) \qquad (14)$$

where, v = 0,1,2,3,... and v represents the hierarchical level of the abstract representation. In particular, if v = 0 then the graph node is the primitive node of the graph.

Note that, in 3-D navigation the graph form is more complicate and the semantic object forms include many intermmediate levels of abstraction.

4 KNOWLEDGE ACQUISITION USING THE SYNTHESIS OF GRAPHS

In this section, we describe a knowledge acquisition scheme based on the synthesis of successive shapes expressed in graph forms. The general configuration of a knowledge-base, by using graph forms to represent knowledge is also discussed.

4.1 Synthesis of Shapes

The composition of two shapes by generating their synthesis shape plays the key role for the development of an efficient knowledge acquisition scheme, which is based on the repitition of the synthesis process of consecutive pairs of shapes.

The synthesis of two shapes $SH^n_{t_i}$, $SH^n_{t_j}$, $i \neq j$, starts always from the highest representation level of their abstracted graph forms $G^n_{t_i}$, $G^n_{t_j}$ respectively, stored in KB. In order two shapes to be considered for synthesis, it is necessary their equivalent graph forms to satisfy the follwing proposition.

Proposition 1: Two shapes $SH^n_{t_i}$, $SH^n_{t_j}$, $i \neq j$, with $SH^n_{t_i} \in NS^n_t$, $SH^n_{t_j} \in NS^n_t$, where t represents the time between t_i and t_j in the same free navigation space, can be considered as candidates for composition into a new shape $SH^n_{t_{ij}}$ if and only if their graph forms have at least one common node, $G^n_{t_i} \ni N_k \equiv N_m \in G^n_{t_j}$ with the same properties and the same relationships R^p, R^d from some other nodes.

Proof: see illustrative example in this subsection.

Proposition 2: In any graph $G^n_{t_i}$ the condition $A_{km} = A_{mk}$ is true for R^c, R^p, R^d, R^{sy}, R^{si}, but the reverse for R^{rm}, R^{rl}.

Definition 5: The length $L(G^n_{t_i})$ of a graph $G^n_{t_i}$ is defined as the number of primitive nodes N^0_j without repititions of the same values of j.

For instance, if $G^1_{t_2} = N_1 A_{12} N_2 A_{23} N_3 A_{31} N_1$ then the length $L(G^1_{t_2}) = 3$.

Definition 6: Two graphs G_{t_i}, G_{t_j}, $i \neq j$, are similar each other if both they have the same number of nodes and for each node $N_k \in G_{t_i}$ the corresponding equivalent node $N_m \in G_{t_j}$ has the same relations $G_{t_i} \ni R^y_{k1m1} \equiv R^y_{k2m2} \in G_{t_j}$ and $P^{cu}_k \equiv P^{cu}_m$

Definition 7: If two graphs G_{t_i}, G_{t_j} are similar and their corresponding nodes have the same properties , then these graphs are identical.

Starting the synthesis process, we have to search initially the higher level abstract nodes ${}^{l}N^{v}_{k}$, if any, in the first candidate graph $G^{n}_{t_i}$, by examining the value of the index v. If we detect such a node, then we save its characteristics and we proceed for the detection of additional high level nodes, with $v > 0$ by saving their characteristics. At this point, we proceed with the nodes of the second candidate graph $G^{n}_{t_j}$. In particular, we attempt to identify graph nodes with the same v values. If there is at least such a node in the second graph, then we attempt to match its characteristics with the corresponding ones of the node which belongs in the first graph. If there is a successful matching, then these particular nodes will be the starting point for the synthesis of the two graphs. In case that there are not nodes with $v > 0$, then we search the primitive nodes, $v = 0$, which satisfy the proposition 1 in order to use them as the starting point for their synthesis. The synthesis process of the rest nodes for these two candidate graphs is based on the detection of closed subgraphs and determination of their "extended" new subgraph forms.

4.1.1 An Illustrative Example

In this illustrative example, figure 6, we show the synthesis of two shapes by using their graph forms. In particular, lets consider a hypothetical free navigation space FNS_t, fig.6a, in which we select three different positions (1,2,3) for the moving robot R_1. For each position, $m = 1,2,3$, the robot processors generate the corresponding shape $SH^{1}_{t_m}$, see fig. 6b,6c,6d , and the equivalent graphs $G^{1}_{t_m}$:

$$G^{1}_{t_1} = N_1A_{12}N_2A_{23}N_3A_{34}N_4A_{45}N_5A_{56}N_6A_{67}N_7A_{78}N_8A_{89}N_9A_{910}N_{10}A_{101}N_1$$

$$G^{1}_{t_2} = N'_1A'_{12}N'_2A'_{23}N'_3A'_{34}N'_4A'_{45}N'_5A'_{56}N'_6A'_{67}N'_7A'_{78}N'_8A'_{81}N'_1 \qquad (15)$$

$$G^{2}_{t_3} = \dot{N}_1\dot{A}_{12}\dot{N}_2\dot{A}_{23}\dot{N}_3\dot{A}_{34}\dot{N}_4\dot{A}_{45}\dot{N}_5\dot{A}_{56}\dot{N}_6\dot{A}_{67}\dot{N}_7\dot{A}_{78}\dot{N}_8\dot{A}_{89}\dot{N}_9\dot{A}_{910}\dot{N}_{10}\dot{A}_{101}\dot{N}_1$$

For this particular example there are not nodes with $v > 0$, thus our searching-matching process is focus on nodes with $v = 0$. Firstly, we examine if the graphs $G^{1}{}_{t_1}$, $G^{1}_{t_2}$ are candidates for synthesis. From the search-matching process we observe that the nodes $N^4 \equiv N'_3$, $N_8 \equiv N'_7$ are the same and satisfy the proposition 1. Thus, the synthesis of these two graphs is possible by producing a new graph $G^{1}{}_{t_{12}}$:

$$G^{1}_{t_{12}} \equiv G^{1}_{t_1} \odot G^{1}_{t_2} = [N_1A_{12}N_2A_{23}N_3A_{34}\mathbf{N}_4A_{45}N_5A_{56}N_6A_{67}N_7A_{78}\mathbf{N}_8A_{89}N_9A_{910}N_{10}A_{101}N_1] \odot$$
$$\odot [N'_1A'_{12}N'_2A'_{23}\mathbf{N}_4A'_{34}N'_4A'_{45}N'_5A'_{56}N'_6A'_{67}\mathbf{N}_8A'_{78}N'_8A'_{81}N'_1] \qquad (16)$$

where, $\odot$ represents the "synthesis" operator.

The synthesis of these two graphs starts from the "common" node N_4. Thus, with the first common node N_4 as starting point, we attempt to "unite" firstly both the left subgraphs:

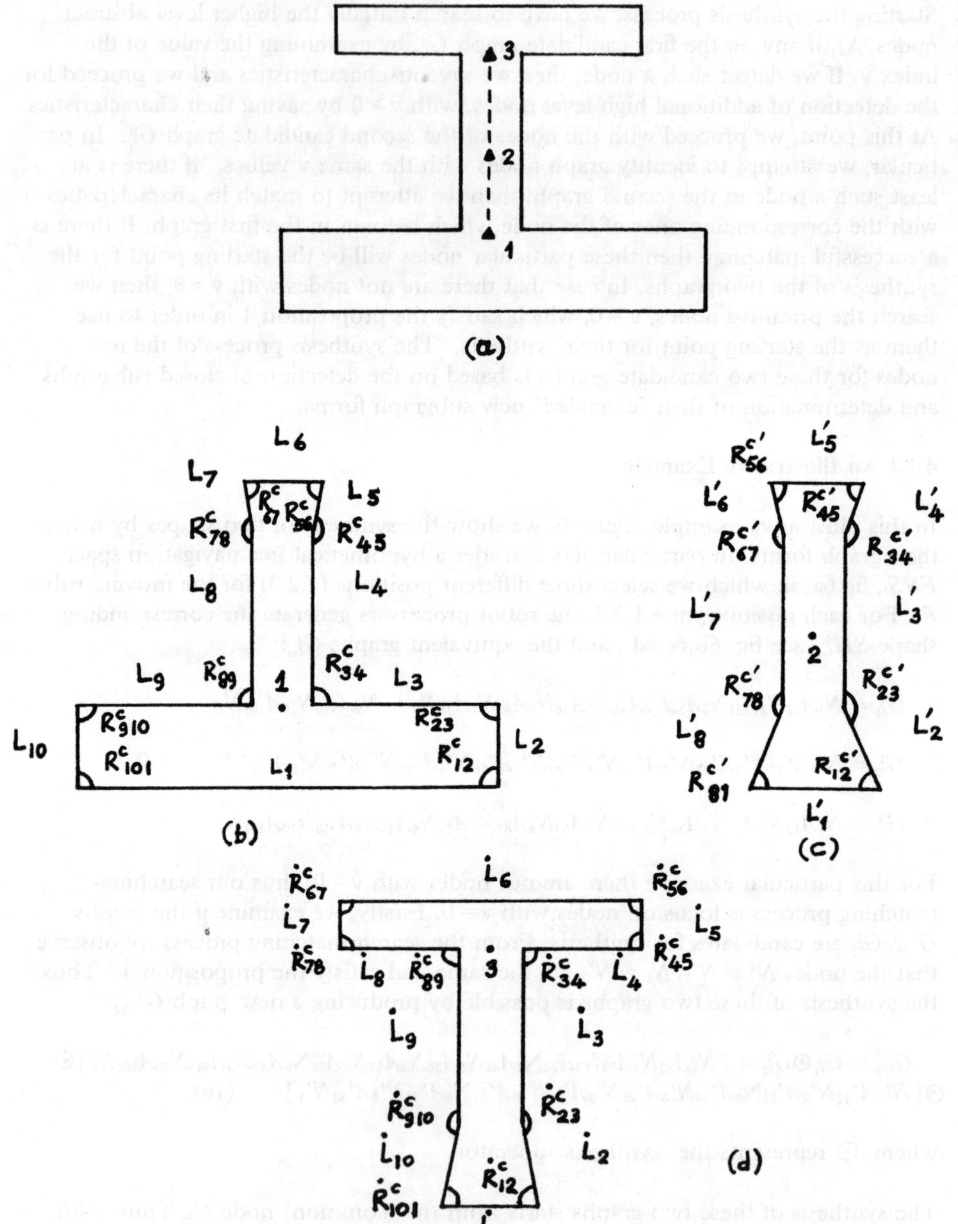

Figure 6: a) The shape of a free space. b) The first shape c) The second shape and d) The third shape.

$$''SG^1_{t_1} = N_1A_{12}N_2A_{23}N_3A_{34}N_4 \qquad (17)$$

$$''SG^1_{t_2} = N'_1A'_{12}N'_2A'_{23}N_4$$

by producing a new "extended" subgraph $''SG^1_{t_1U2}$ which represents their union. We start with the longest left subgraph $''SG^1_{t_1}$ and we attempt to define redundancies by using the nodes and their relationships from the shorter subgraph $''SG^1_{t_2}$.

$$''SG^1_{t_1U2} = [N_4A_{43}N_3A_{32}N_2A_{21}N_1] \cup [N'_1A'_{12}N'2A'_{23}N_4] =$$
$$= N_4A_{43}N_3A_{32}N_2A_{21}\mathrm{N}_1\mathrm{N}'_1A'_{12}N'_2A'_{23}N_4. \qquad (18)$$

The first "redundancy" in the formula (18) is the existence of two consecutive nodes N_1 and N'_1 without a relationship between them. Thus, we compare the nodes N_1, N'_1 by using P_x, R_y :

a) $P^{cf}_{N_1} \equiv P^{cf}_{N'_1}$

b) $P^{cu}_{N_1} \equiv P^{cu}_{N'_1}$

c) $R^d(N_4)_{(N_1)} = R^d(N_4)_{(N'_1)}$ (19)

$R^d(N_8)_{(N_1)} = R^d(N_8)_{(N'_1)}$

d) $R^p_{(N_4N_8)}$

e) $R^{cl}_{N_1} = R^{cl}_{N'_1}$

This means that $N_1 \cap N'_1 \neq \emptyset$

Moreover, the $P^{sp}_{N'_1} \in N_1$ and $P^s_{N_1} > P^s_{N'_1}$

This concludes that $N'_1 \in N_1$

Thus, we can replace the two nodes $N_1N'_1$ with the node N_1.

At this point we attempt an appropriate rearrangement of the relations A'_{12}, A'_{23}. In particular, we can prove easily that $N_3 \cap N'_2 = P^{sp}_{N_4}$. Thus, $N_3A_{32}N_2A_{21}N_1\dot{A}_{12}N'_2\dot{A}_{23}N_3$ is a closed subgraph, where $R^{\dot{f}}_{12} = (180 - R^{f}_{12})$ and $R^{\dot{f}}_{23} = (R^{f}_{23} - R^{f}_{43})$. Since, however, the entire shape of the current navigation space is the only closed graph, and since the node N'_2 has a connection with the node N_1 but not at its end, then the node N'_2 has be removed, thus, new form of the subgraph $''SG^1_{t12}$, which represents the synthesis of these two sub graphs above, is:

$$''SG^1_{t12} = N_4A_{43}N_3A_{32}N_2A_{21}N_1 \qquad (20)$$

A similar process is followed for the the right subgraph of N_4 and at the end of the entire synthesis process the new graph G^1_{t2} is

$$G^1_{t2} = N_1A_{12}N_2A_{23}N_3A_{34}N_4A'_{34}N'4A'_{45}N'_5A'_{56}N'_6A'_{67}N_8A_{89}N_9A_{910}N_{10}A_{101}N_1 \qquad (21)$$

4.2 Knowledge Acquisition

Since the path planning strategy used here is based on a heuristic *human like* scheme, where the chosen path is defined by the *First Best Choice (FBC)* open navigation corridors (OC), the shape of the free navigation space plays the key role for the determination of the path.

Note that, an $FBC(OC^n_{ti})$ provides the current minimum traveling cost $C_o = G(t_{min}, Pr_{max})$, where t_{min} represents the current minimum traveling time through an OC^n_{ti} and Pr_{max} is the current maximum probability of the OC corridor to "drive" the moving robot R_n close to the destination point, [8].

The knowledge acquisition scheme proposed here is based on the "acquisition" of the shapes of the free navigation spaces FNS^n_{ti}, where i= 1,2,,...,K. Acquisition of the shapes SH^n_{ti} means the synthesis of them by producing new ones including the knowledge of their original shapes. It is defined that each current shape SH^n_{ti} determines a part of the traveling path PH^n_{ti}, thus the synthesis of consecutive shapes defines the synthesis of a path. Moreover, the shape of a traveling path can be also expressed as an attributed graph:

$$f{:}PH^n_{ti} \rightarrow GP^n_{ti} \qquad (22)$$

Thus, the knowledge saved in the KB, fig. 7, after K consecutive shapes is:

a) K graph forms G^n_{ti}, i= 1,2,3,...,K, for the K consecutive shapes;

b) 1 graph form $G^n_{t1,2,...,K}$ for the total shape;

c) 1 graph form $GP^n_{t1,2,...,K}$ for the total shape of the travelled path.

In order to manipulate the knowledge stored in the KB, we present a number of operations performed on the graph forms of knowledge.

1) ***Synthesis Operation***

$$G^n_{t12} \equiv G^n_{t1} \odot G^n_{t2}$$

$\odot$ represents the synthesis operator.

2) ***Union Operation*** (see formula 18)

$$G^n_{t1U2} \equiv G^n_{t1} \cup G^n_{t2}$$

$\cup$ represents the union operator.

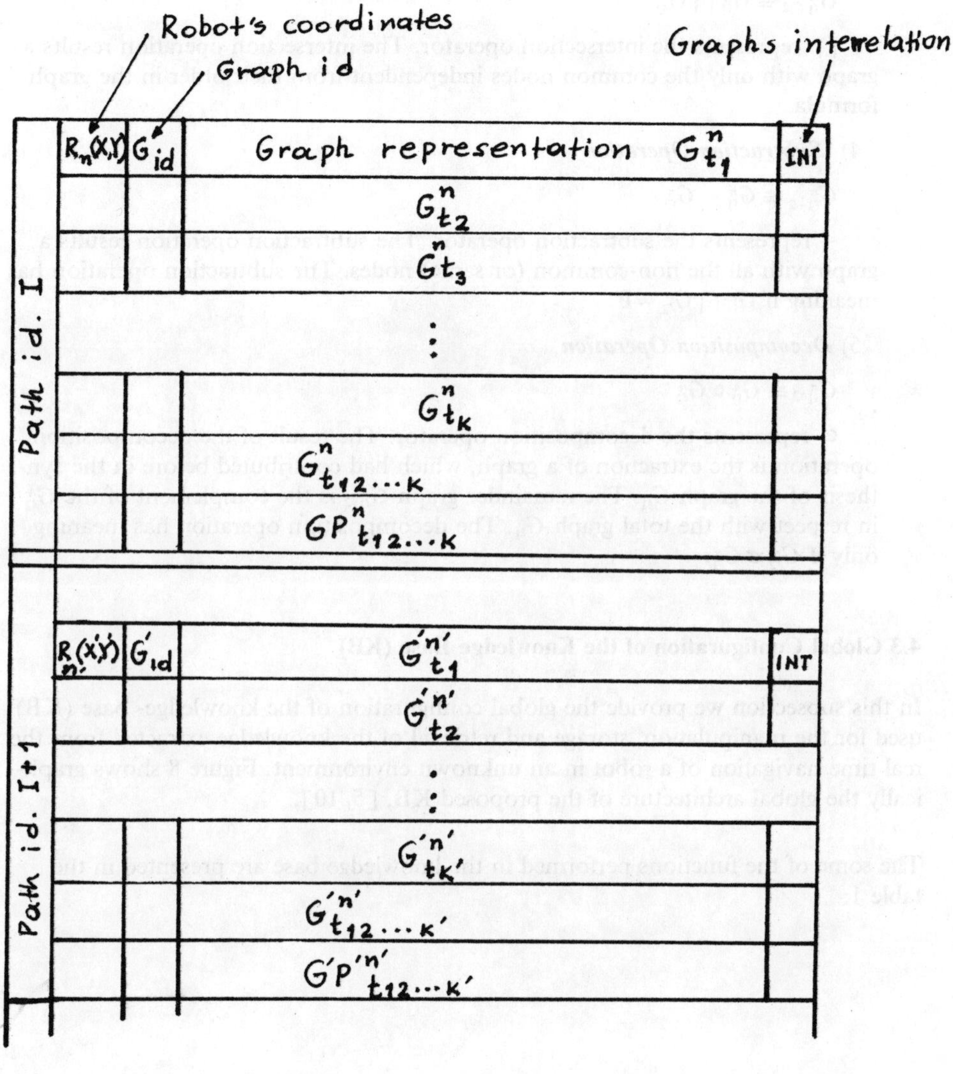

Figure 7: Representation of the graph forms stored in KB.

3) ***Intersection Operation***

$$G^n_{t1\cap 2} \equiv G^n_{t1} \cap G^n_{t2}$$

$\cap$ represents the intersection operator. The intersection operation results a graph with only the common nodes independent from their order in the graph formula.

4) ***Subtraction Operation***

$$G^n_{t1-2} \equiv G^n_{t1} - G^n_{t2}$$

– represents the subtraction operator. The subtraction operation results a graph with all the non-common (or same) nodes. The subtraction operation has meaning if $G^n_{t1} \cap G^n_{t2} \neq \emptyset$.

5) ***Decomposition Operation***

$$G^n_{t1\between 2} \equiv G^n_{t1} \between G^n_{t2}$$

$\between$ represents the decomposition operator. The result of the decomposition operation is the extraction of a graph, which had contributed before in the synthesis of the graph G_{t1}. The remainder graph $G^n_{t1\between 2}$ is the complement of the G^n_{t2} in respect with the total graph G^n_{t1}. The decomposition operation has meaning only if $G^n_{t2} \subset G^n_{t1}$.

4.3 Global Configuration of the Knowledge Base (KB)

In this subsection we provide the global configuration of the knowledge- base (KB) used for the manipulation, storage and retrieval of the knowledge extracted from the real time navigation of a robot in an unknown environment. Figure 8 shows graphically the global architecture of the proposed KB, [5, 10].

The some of the functions performed in the knowledge base are presented in the table 1.

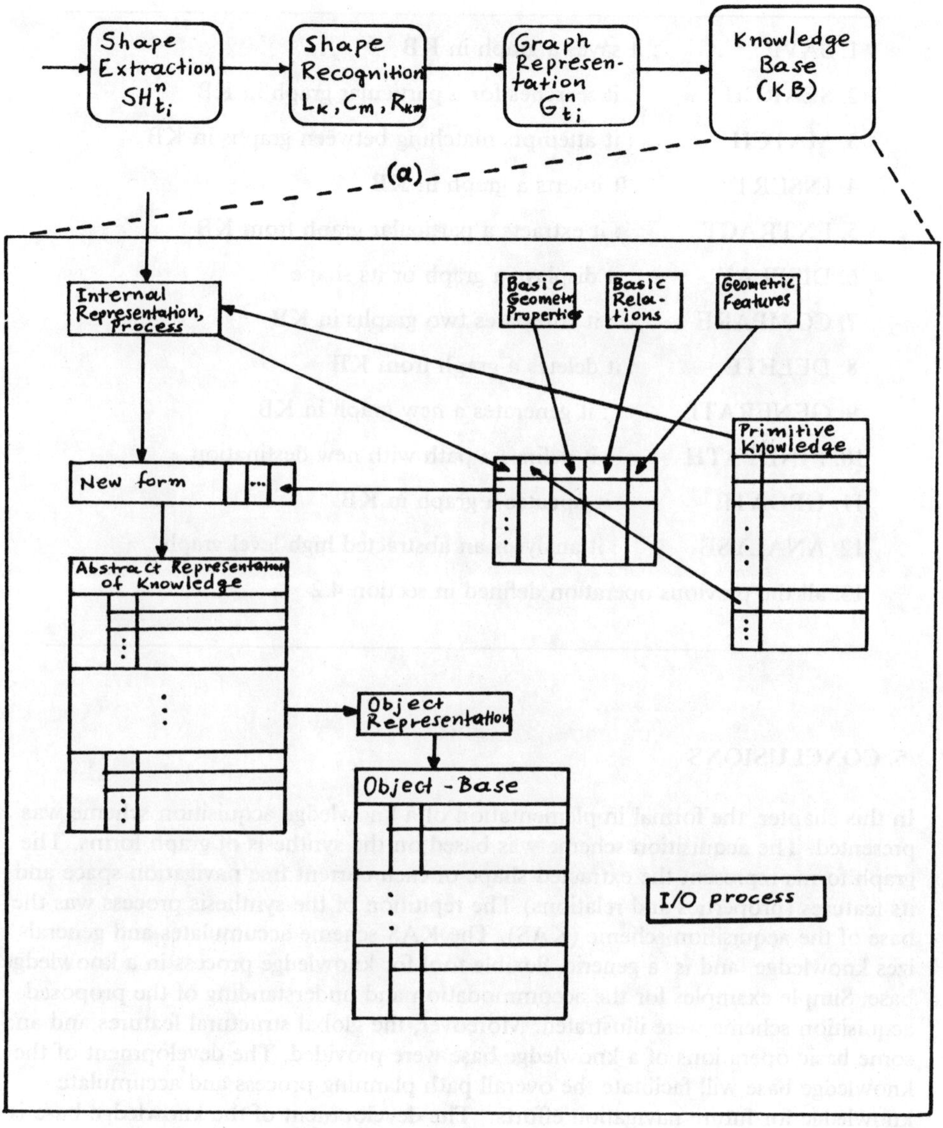

Figure 8. A global configuration of the KB.

TABLE - 1

1. SAVE	: it saves a graph in KB
2. SEARCH	: it searches for a particular graph in KB
3. MATCH	: it attempts matching between graphs in KB
4. INSERT	: It inserts a graph in KB
5. EXTRACT	: it extracts a particular graph from KB
6. DISPLAY	: it displays a graph or its shape
7. COMPARE	: it compares two graphs in KB
8. DELETE	: it deletes a graph from KB
9. GENERATE	: it generates a new graph in KB
10. FINDPATH	: it defines a path with new destination
11. UPDATE	: it updates a graph in KB
12. ANALYSE	: it analyses an abstracted high level graph
13. all the previous operation defined in section 4.2	

5. CONCLUSIONS

In this chapter, the formal implementation of a knowledge acquisition scheme was presented. The acquisition scheme was based on the synthesis of graph forms. The graph forms represent the extracted shape of each current free navigation space and its features (properties and relations) The repitition of the synthesis process was the base of the acquisition scheme (KAS). The KAS scheme accumulates and generalizes knowledge and is a generic, flexible tool for knowledge process in a knowledge base. Simple examples for the accommodation and understanding of the proposed acquisition scheme were illustrated. Moreover, the global structural features and and some basic operations of a knowledge base were provided. The development of the knowledge base will facilitate the overall path planning process and accumulate knowledge for future navigation efforts. The development of the knowledge base is in progress. In addition, the methodology of the proposed knowledge acquisition scheme has been successfully used in 2-D computer vision and image interpretation applications, [13].

REFERENCES

1. T.Lozano-Perez,*Spatial Planning: A configuration space approach*, IEEE Trans. on Computers, 32,2,1983,pp.108-120.

2. R.Brooks,*Solving the find-path problem by good representation of the free space*,IEEE Trans. on SMC, 13,3,1983,pp.190-197.

3. J.Schwartz and M.Shamir,*On the peano mover's problems*, pub. Comm.Pure App.Math, Adv.Math.,Robotics Research, 1983.

4. J.Barraquand and J.Latombe,*Robot Motion Planning:A distributed representation approach*,IEEE J. on Automation & Robotics, 1991, to be appeared.

5. W.A.Woods,*Important issues on knowledge representation*, Proc. of the IEEE, 74,10,1986,pp.1322-1334.

6. G.Kahn et. al.*Strategies for knowledge acquisition*, IEEE-Trans. on PAMI, 7,5,1985.

7. K.S.Fu,*Syntactic Pattern Recognition and Applications* Prentice Hall, 1982.

8. N.Bourbakis,*A heuristic collision free path planning for an autonomous navigation platform*,IJ.on Intelligence & Robotic Systems 1,4,1989,pp.375-387.

9. N.Bourbakis,*Autonomous Mobile Robots and Path Planning Strategies*,Chapter in Intelligent Robotics, 1991.

10. N.Bourbakis, *An knowledge acquisition scheme using attributed graphs*,Technical Report 1989, submitted.

11. L.Gonzenes,*Collision avoidance for robots in an environmenta l flexible assembly cell*,Proc. IEEE Conf. on Robotics,CA,1984.

12. B.W.Wah,*Computers for Symbolic Processing*,chapter 1, pp.1-73, Artificial Intelligence Processing, J.Wiley, 1990.

13. N.Bourbakis,*Image interpretation using knowledge acquisition* TR-1990, submitted.

Applications of Learning and Planning Methods
edited by N. G. Bourbakis

CHAPTER 14

Path Planning for Two Cooperating Robot Manipulators

Qing Xue
Department of Electrical Engineering
Indiana University and Purdue University
at Indianapolis
1201 East 38th Street
Indianapolis, IN 46205
P. C-Y. Sheu
Dept. of Electrical & Computer Engineering
Rutgers University
Piscataway, NJ 08855

Abstract

In this paper, we investigate the path planning problem for two planar robot manipulators which cooperate in carrying a rectangular object from an initial position and orientation to a destination position and orientation in a 2-D environment. To approach the problem, we model the robot arms, the carried object and the straight line connecting two robots bases together as a 6-link closed chain. The path planning problem for the 6-link closed chain is solved by the collision-free feasible configuration finding algorithm and the collision-free path finding algorithm. The former finds all collision-free feasible configurations (CFFC's) for the 6-link closed chain at each quantized interval for two of the six joint angles. The latter builds a connection graph of the CFFC's and the transitions between any two groups of CFFC's at adjacent joint intervals. Finally, a graph search method is employed to find a collision-free path for each robot joint. The proposed algorithms can deal with cluttered environments and can be proved to converge.

1. Introduction

The problem of collision-free path planning for robots has been studied for many years. Most of researches on the path planning have been focused on a single robot. Among them, the Configuration Space method [1] and the method for solving the Piano Movers' problem [2] have received most attention. These methods are further developed for different applications, particularly in view of reducing the computational complexity.

Recently, the problem of collision-free path planning for two cooperating manipulators has received more attention. Motions executed by two coordinating arms can be classified into two groups [3]: loosely coordinated motions and tightly coordinated motions. In loosely coordinated motions, two robot arms execute two independent working sequences for two unrelated tasks, but share a common working space. In tightly coordinated motions, two robot arms execute two related working sequences for a common task.

The problem of collision avoidance among two (multiple) loosely coordinated robots has been investigated in [4]. In this approach, the robots have master/slave relationship. The problem of planning a collision-free path for two closely cooperating robot arms is studied in [5]. The robot arms in the approach are restricted to be cylindrical robot arms. Another approach to the same problem with two robots working on the same task is proposed in [6]. Since the object and the manipulators in the approach are modeled as a combination of several spheres, this approach is an approximate approach which is difficult to deal with littered environment.

In this paper, we describe an approach for the path planning problem for two planar robot manipulators with three revolute joints which cooperate in carrying a rectangular object from an initial position and orientation to a destination position and orientation in a 2-D environment. This approach can deal with cluttered environments and convex polygonal obstacles. In addition, the approach can be proved to converge (that is, a collision-free path from the initial position and orientation of the carried object to the final position and orientation of the carried object can be found if it exists; otherwise the approach issues a message indicating that no collision-free path exists). In the approach, the robot arms, the carried object, and the straight line connecting the bases of the robots form a 6-link closed chain. The path planning problem for the 6-link closed chain is solved by the collision-free feasible configuration finding algorithm and the collision-free path finding algorithm. The former finds all collision-free feasible configurations (CFFC's) for the 6-link closed chain at each quantized interval for two of the six joint angles. The latter builds a connection graph of the CFFC's and the transition between any two groups of CFFC's at each pair of adjacent joint intervals first. Subsequently, a graph search method is used to find a collision-free path for each robot manipulator. For simplicity, the dynamics of the moving object is not considered. Furthermore, it is assumed that all information about the environment is available. Consequently, path planning is performed off-line.

2. Problem Formulation

In the approach for finding the paths of two closely cooperating planer manipulators which carry a rectangular object moving in a 2-D environment, we assume that each robot manipulator consists of a lower link, an upper link, an end-effector and three revolute joints. The joints are located between the base and the upper link, between the upper link and the lower link, and between the lower link and the end-effector, respectively. The four corners of the rectangular object are denoted as b_1, b_2, b_3 and b_4. It is assumed that the two robot manipulators grasp the rectangular object at points a and b which are located at the intersections of the boundary and the centerline of the rectangular object by their end-effectors (see Fig. 2-1). For convenience, point a is chosen as the reference position for the carried object. In addition, the angle between the centerline of the rectangular object and the x-axis of the world coordinate system is chosen as the orientation of the carried object. Finally, in a two-dimensional environment, it is assumed that there are p stationary obstacles which are represented by convex polygons.

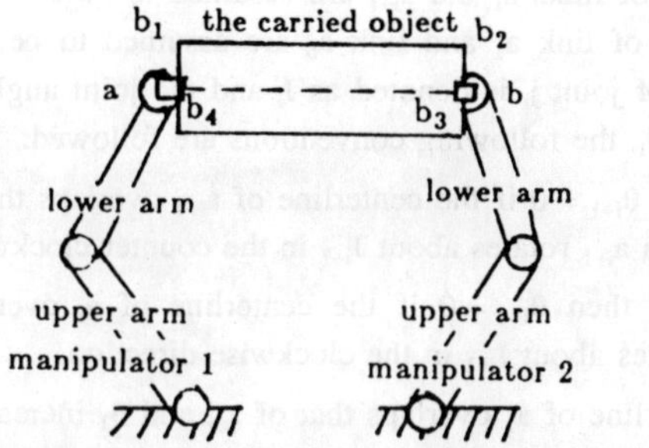

Fig. 2-1: Two cooperating robot manipulators

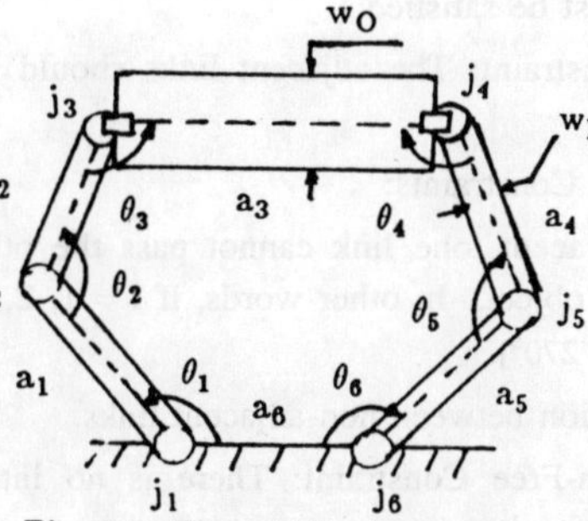

Fig. 2-2: The 6-link closed chain

In the following discussion, a straight line segment which connects the bases of the two robot manipulators is considered as a virtual link. Consequently, the virtual link, the links of the robot manipulators, and the carried rectangular object form a 6-link closed chain (see Figure 2-2). In the closed chain, the upper and the lower links of one manipulator, the carried object, the lower and the upper links of the other manipulator, and the virtual link are denoted as links a_1 to a_6, respectively.

Let the length of each link a_i, where $1 \le i \le 6$, be l_i. It is assumed that all links of the manipulators have the same width w_R and the carried rectangular object has width w_O. Furthermore, the centerlines of links a_i and a_{i+1} are assumed to be connected by the revolute joint j_{i+1} and the centerlines of link a_1 and link a_6 are assumed to be connected by the revolute joint j_1. The location of joint j_i is denoted as J_i and the joint angle of j_i is denoted as θ_i. To define the direction of θ_i, the following conventions are followed:

(1) If i = 1 or 2, then $\theta_{i+1} = 0$ if the centerline of a_{i+1} overlaps that of a_i; on the other hand, θ_{i+1} increases when a_{i+1} rotates about J_{i+1} in the counter clockwise direction.

(2) If i = 3, 4 or 5, then $\theta_{i+1} = 0$ if the centerline of a_i overlaps that of a_{i+1}, and θ_{i+1} increases if a_i rotates about J_{i+1} in the clockwise direction.

(3) $\theta_1 = 0$ if the centerline of a_1 overlaps that of a_6, and θ_1 increases if a_1 rotates about J_1 in the counter clockwise direction.

Our problem is to plan a path for each robot joint so that the carried object can be moved from a given starting position and orientation to a given final position and orientation without colliding with obstacles. To find the path for each joint in the 6-link closed chain, the following constraints must be satisfied:

(1) The Closed Chain Constraint: The adjacent links should always be connected to each other.

(2) The link Collision-Free Constraints:

  (a) If two links are adjacent, one link cannot pass the other link and any link cannot intersect the carried object. In other words, if i = 1, 2, 5, 6, then $\theta_i \in [0°, 360°]$, and if i = 3, 4, $\theta_i \in [90°, 270°]$.

  (b) There is no intersection between non-adjacent links.

(3) The Obstacle Collision-Free Constraint: There is no intersection between the 6-link closed chain and the obstacles.

For convenience, we define a gesture of the 6-link closed chain which satisfies the closed chain constraint to be a *configuration* and define a configuration which satisfies the closed chain constraint and the link collision-free constraints to be a *feasible configuration.* Furthermore, a feasible configuration which does not collide with any obstacles is defined to be a *collision-free feasible configuration* and is denoted as CFFC. In addition, it is assumed that the configurations of the 6-link closed chain at the given initial and final positions and orientations of the carried object (link a_3) are collision-free and feasible.

3. Finding Collision-Free Feasible Configurations

In order to find the collision-free feasible configurations (CFFC's) for a 6-link closed chain, the following proposition which describes the relationship between different configurations will be employed:

Proposition 3-1: In order to move a 6-link closed chain from one configuration to another configuration, the positions of at least three consecutive links in the closed chain need to be changed.

It is easy to show that the above is true since any three consecutive links a_i, a_{i+1} and a_{i+2} in a closed chain have one internal degree of freedom if one end of a_i and one end of a_{i+2} which are not connected to a_{i+1} are fixed.

For convenience, any link in which the position of one or both ends can be changed is called a *changeable link*. Therefore, all links, except a_6, in the 6-link closed chain are changeable links. By Proposition 3-1, any position change of a 6-link closed chain is caused by the position change of a set of three consecutive changeable links. Therefore, each set of three consecutive changeable links a_i, a_{i+1} a_{i+2}, where $1 \le i \le 3$, in a 6-link closed chain is defined to be a *basic changeable unit* BCU(a_i, a_{i+1}, a_{i+2}). Consequently, there are three basic changeable units BCU(a_1,a_2,a_3), BCU(a_2,a_3,a_4) and BCU(a_3,a_4,a_5) in a 6-links closed chain. Since there is one degree of freedom of motion in a BCU, there are three degrees of freedom of motion in a 6-link closed chain.

Our strategy to find all the CFFC's for a 6-link closed chain is to find all the CFFC's generated by a BCU for each quantized motion of the other two changeable links which are not in the BCU. For illustration, we arbitrarily choose BCU(a_2,a_3,a_4) and quantize the motion of links a_1 and a_5 of the closed chain in the following discussion. In the course of finding CFFC's, the constraints given in Section 2 are satisfied one by one in the following subsections.

3.1. Finding all the Configurations Generated by BCU(a_2,a_3,a_4)

Among the three changeable links in a BCU, we call a link with one fixed end a *semi-changeable link*. Accordingly, since one end of a_2 and one end of a_4 are fixed in BCU(a_2,a_3,a_4) as a_1 and a_5 are fixed, a_2 and a_4 are semi-changeable links.

The closed chain constraint requires that each pair of adjacent links be connected to each other. In Figure 2-2, it can be observed that a necessary condition to satisfy the closed chain constraint is $|J_2J_5| \le l_2 + l_3 + l_4$. For convenience, we assume that $|J_2J_5| \le l_2 + l_3 + l_4$ holds in the following discussion.

Since there is only one degree of freedom of motion for a_2, a_3 and a_4 in BCU(a_2,a_3,a_4), if the position of one link (called the active link) is changed, the positions of the other links (called the passive links) must be changed accordingly. Since the positions of the semi-changeable links a_2 and a_4 can be uniquely described by joint angles θ_2 and θ_5, respectively, it is convenient to choose a semichangeable link as the active link. Suppose a_4 is chosen as the active link, the motion range of θ_5 in which the closed chain constraint is satisfied needs to be found. In order to simplify the discussion, we first introduce the following definitions.

Definition 3.1-1: If one end of the centerline of a changeable link or a semichangeable link a_i is fixed at a point V, then all the possible positions for the other end of the centerline of a_i form a circle with radius l_i and center V. This circle is called the *link position circle* of a_i and is denoted as $LPC_{a_i}(V)$ or LPC_{a_i} (see Figure 3.1-1). □

In order to find the motion range of θ_5, we draw two circles I and II which have a common center J_2 and radii $|l_2+l_3|$ and $|l_2-l_3|$, respectively (see Figure. 3.1-2). If we only consider the closed chain constraint, based on reference [7], the motion range, $MR(\theta_5)$, of θ_5 should be a set of regions on LPC_{a_4} which lie within circle I, but outside circle II and this set of regions could contain one continuous region or two continuous regions which are symmetric with respect to line J_2J_5 that passes through J_2 and J_5.

The position of the active link a_4 can determine the positions of the passive links a_2 and a_3. From Figure 3.1-3, for a given position of a_4 in $MR(\theta_5)$, the possible positions of joint j_3 (which connects the passive links a_2 and a_3) should be on $LPC_{a_2}(J_2)$ and $LPC_{a_3}(J_4)$ at the same time. Hence, the possible positions for joint j_3 should be at the intersections of LPC_{a_2} and LPC_{a_3}. If $l_2 + l_3 + l_4 < |J_2J_5|$, LPC_{a_2} and LPC_{a_3} have two intersections. Consequently, two possible positions of j_3 can be found. In other words, a given position of a_4 in $MR(\theta_5)$ can produce two possible positions for the passive links a_2 and a_3 and therefore two configurations can be found. In each case, if we enumerate the possible values of θ_5 within its motion range, all the configurations of $BCU(a_2,a_3,a_4)$ can be found. If a_2 is chosen as the active link, the motion range $MR(\theta_2)$ and all the configurations of $BCU(a_2,a_3,a_4)$ can be found in a similar way.

3.2. Finding all the Feasible Configurations for $BCU(a_2,a_3,a_4)$

In this section, assuming a_4 is still chosen as the active link in $BCU(a_2,a_3,a_4)$, we shall to find a set of subregions of $MR(\theta_5)$ in which the link collision-free constraints are satisfied. These subregions of $MR(\theta_5)$ are called the *feasible regions* $FR(\theta_5)$ of θ_5. After $FR(\theta_5)$ is found, all the feasible configurations of $BCU(a_2,a_3,a_4)$ can be determined by restricting the motion of θ_5 to $FR(\theta_5)$.

In order to find the feasible region $FR(\theta_5)$ in which the link collision-free constraints are satisfied, the link collision-free constraint for adjacent links is considered first; subsequently we consider the link collision-free constraint for non-adjacent links.

3.2.1. Finding Subregions of $MR(\theta_5)$ which Satisfy the Link Collision-Free Constraint for Adjacent Links

In order to satisfy the link collision-free constraint for adjacent links in $BCU(a_2,a_3,a_4)$, θ_i, where i = 2, 5, should be in the range [0°, 360°] and θ_i, where i = 3, 4, should be in the range

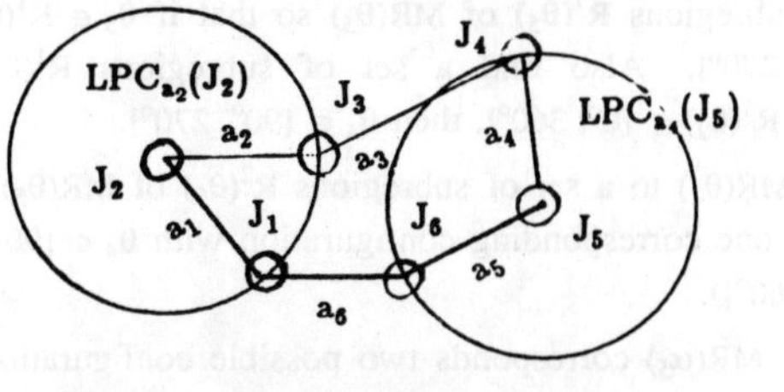

Fig. 3.1-1: The link position circle LPC$_{a_2}$(J$_2$) and LPC$_{a_4}$(J$_5$)

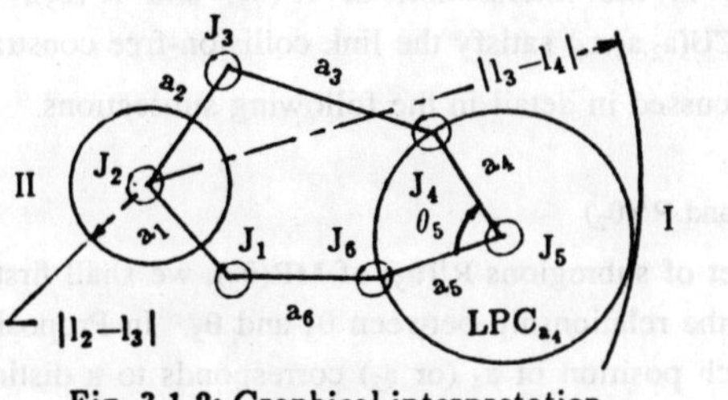

Fig. 3.1-2: Graphical interpretation of circles I and II

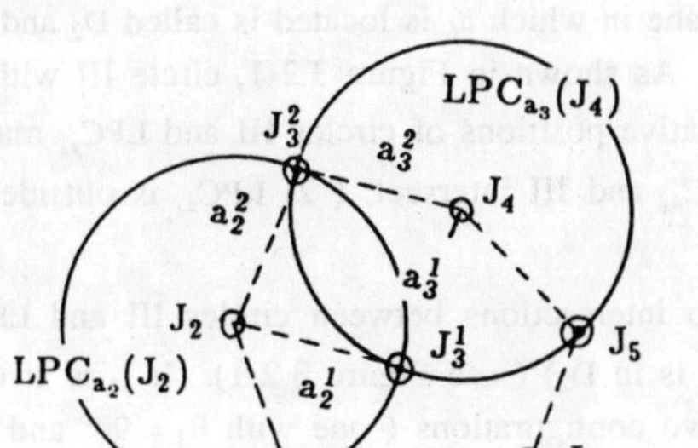

Fig. 3.1-3: One position of a$_4$ corresponds two positions of a$_2$ and a$_3$

[90°, 270°]. Therefore we need to find a set of subregions of MR(θ_5) in which the corresponding configurations have $\theta_5,\theta_2 \in$ [0°, 360°] and $\theta_3,\theta_4 \in$ [90°, 270°]. To find this set of subregions of MR(θ_5), the following three steps can be followed:

(1) Find a set of subregions $R^1(\theta_5)$ of MR(θ_5) so that if $\theta_5 \in R^1(\theta_5)$ and $R^1(\theta_5) \subseteq$ [0°, 360°] then $\theta_3 \in$ [90°, 270°]. Also find a set of subregions $R^1(\theta_2)$ of MR(θ_2) so that if $\theta_2 \in R^1(\theta_2)$ and $R^1(\theta_2) \subseteq$ [0°, 360°], then $\theta_4 \in$ [90°, 270°].

(2) Map $R^1(\theta_2)$ of MR(θ_2) to a set of subregions $R^2(\theta_5)$ of MR(θ_5) so that for a $\theta_5 \in R^2(\theta_5)$, there is at least one corresponding configuration with $\theta_4 \in$ [90°, 270°] and $\theta_2 \in$ [0°, 360°] ($R^1(\theta_2) \subseteq$ [0°, 360°]).

(3) Since each $\theta_5 \in$ MR(α_5) corresponds two possible configurations, when $\theta_5 \in R^2(\theta_5)$, one of two corresponding configurations may do not have $\theta_4 \in$ [90°, 270°] and $\theta_2 \in$ [0°, 360°]. In this step, we need to find $R^3(\theta_5)$ from $R^2(\theta_5)$ so that if $\theta_5 \in R^3(\theta_5)$, all the corresponding $\theta_4 \in$ [90°, 270°] and $\theta_2 \in$ [0°, 360°]. In other words, in $R^3(\theta_5)$, the configurations which do not have $\theta_4 \in$ [90°, 270°] and $\theta_2 \in$ [0°, 360°] are removed. Consequently, if θ_5 is in the intersection of $R^1(\theta_5)$ and $R^3(\theta_5)$, all the corresponding configurations of BCU(a_2,a_3,a_4) satisfy the link collision-free constraint for adjacent links.

These three steps are discussed in detail in the following subsections.

3.2.1.1. Finding $R^1(\theta_5)$ and $R^1(\theta_2)$

In order to find a set of subregions $R^1(\theta_5)$ of MR(θ_5), we shall first introduce Proposition 3.2-1, which establishes the relationship between θ_3 and θ_5. In Proposition 3.2-1 and the following discussion, as each position of a_4 (or a_2) corresponds to a distinct value of θ_5 (or θ_2), the following convention is adopted: If the changeable end of the centerline of a_4 (or a_2) is at a point P on LPC_{a_4} (or LPC_{a_2}), the joint angle of j_5 (or j_2) is denoted as θ_{5P} (or θ_{2P}).

Proposition 3.2-1: If we draw a line passing through J_2 and J_5, it separates the entire plane into two subplanes. The subplane in which a_6 is located is called D_2 and the other subplane is called D_1 (see Figure 3.2-1). As shown in Figure 3.2-1, circle III with center J_2 and radius $\sqrt{l_2^2+l_3^2}$ can be drawn. The relative positions of circles III and LPC_{a_4} may fall into one of the following three cases: (1) LPC_{a_4} and III intersect, (2) LPC_{a_4} is outside III, and (3) LPC_{a_4} is inside III.

(1) In Case 1, there are two intersections between circles III and LPC_{a_4} (say, u_1 and u_2 where u_1 is in D_1 and u_2 is in D_2) (see Figure 3.2-1). If j_4 is at u_1 or u_2, then each of the intersections forms two configurations (one with $\theta_3 = 90°$ and one with $\theta_3 = 270°$). If j_4 is outside III then each position of a_4 gives two configurations, and in both configurations $\theta_3 \in$ [90°, 270°].

(2) If LPC_{a_4} is outside III, then for each $\theta_5 \in$ MR(θ_5), the corresponding configurations have $\theta_3 \in$ [90°, 270°].

(3) If LPC_{a_4} is inside III, no $\theta_5 \in MR(\theta_5)$ has configurations with $\theta_3 \in [90°, 270°]$.

Suppose circle III and circle LPC_{a_4} have two intersections u_1 and u_2. Let the value of θ_5 when j_4 is at u_1 be θ_{5u_1}, and the value of θ_5 when j_4 is at u_2 be θ_{5u_2}. Since a_5 is the reference of θ_5, the regions of θ_5 corresponding to the position of j_4 that is outside III can be represented as: $\theta_{5u_1} \le \theta_5 \le \theta_{5u_2}$ if $\theta_{5u_1} \le \theta_{5u_2}$, or $0° \le \theta_5 \le \theta_{5u_2}$ and $\theta_{5u_1} \le \theta_5 \le 360°$ for $\theta_{5u_1} \ge \theta_{5u_2}$.

By definition, $R^1(\theta_5)$ should be equal to the intersection between $MR(\theta_5)$ and the regions of θ_5 in which $\theta_3 \in [90^0, 270^0]$, therefore we obtain:

(a) If III and LPC_{a_4} intersect at u_1 and u_2,

$$R^1(\theta_5) = \begin{cases} MR(\theta_5) \cap [\theta_{5u_1}, \theta_{5u_2}] & \text{if } \theta_{5u_1} \le \theta_{5u_2} \\ MR(\theta_5) \cap \left\{ [0°, \theta_{5u_2}] \cup [\theta_{5u_1}, 360°] \right\} & \text{if } \theta_{5u_1} > \theta_{5u_2}. \end{cases}$$

(b) If LPC_{a_4} is outside circle III, $R^1(\theta_5) = MR(\theta_5)$.

(c) If LPC_{a_4} is inside circle III, $R^1(\theta_5) = \Phi$.

From the discussion following Definition 3.1-1 in Section 3.1, $MR(\theta_5)$ may consist of one continuous region or two continuous regions which are symmetric with respect to the line which connects J_2 and J_5. From Figure 3.2-1, it can be observed that the regions of θ_5 in which $\theta_3 \in [90°, 270°]$ are symmetric with respect to line J_2J_5 also. Consequently, we can conclude that $R^1(\theta_5)$ consists of one continuous region or two continuous regions which are symmetric with respect to line J_2J_5 as well.

In order to find $R^1(\theta_2)$, we can draw a circle IV with center J_5 and radius $\sqrt{l_3^2+l_4^2}$. Similar to the case of finding $R^1(\theta_5)$, $R^1(\theta_2)$ can be found according to the relative positions of IV and LPC_{a_3}. Suppose circles IV and LPC_{a_3} have two intersections w_1 and w_2, $R^1(\theta_2)$ can be obtained by a group of equations which is similar to equations to find $R^1(\theta_5)$. Meanwhile, $R^1(\theta_2)$ has the same property as $R^1(\theta_5)$.

3.2.1.2. Finding $R^2(\theta_5)$

As mentioned above, if $\theta_2 \in R^1(\theta_2)$ then all the corresponding $\theta_4 \in [90°, 270°]$ and all the corresponding $\theta_2 \in [0°, 360°]$ ($R^1(\theta_2) \subset [0°, 360°]$). Since a_4 is chosen as the active link, we need to map $R^1(\theta_2)$ to region $R^2(\theta_5)$ of θ_5 so that if $\theta_5 \in R^2(\theta_5)$, at least one corresponding configuration has $\theta_2 \in R^1(\theta_2) \subset [0°, 360°]$ and $\theta_4 \in [90°, 270°]$.

In order to map $R^1(\theta_2)$ to region $R^2(\theta_5)$ of θ_5, we first introduce the following definition:

Definition 3.2-1: Given θ_5 (or θ_2), a_4 (or a_2) has a unique position in $BCU(a_2,a_3,a_4)$. However, for each unique position of a_4 (or a_2), a_2 (or a_4) and a_3 have two possible positions with which two configurations can be obtained, and the area occupied by the closed chain

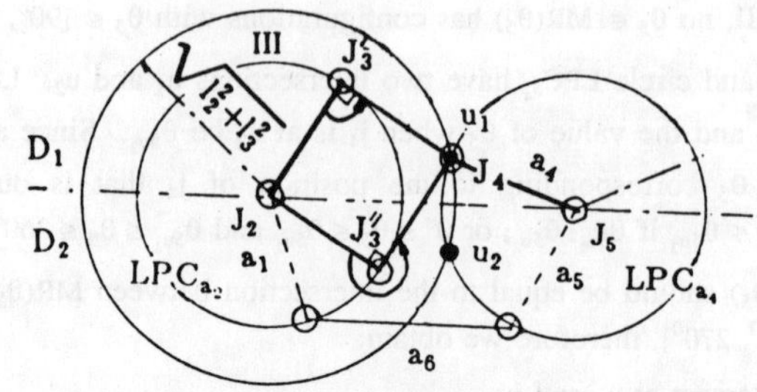

Fig. 3.2-1: Graphical interpretation of Proposition 3.2-1 (Two intersections between LPC_{a_4} and III)

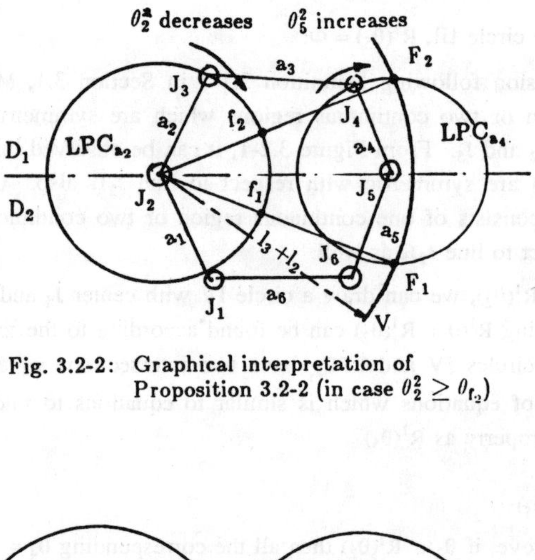

Fig. 3.2-2: Graphical interpretation of Proposition 3.2-2 (in case $\theta_2^2 \geq \theta_{f_2}$)

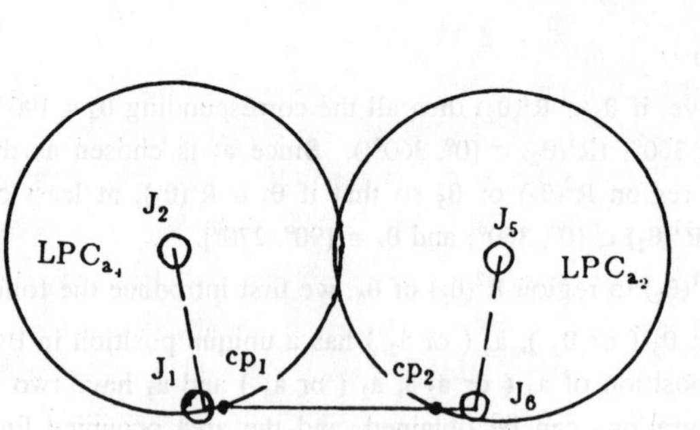

Fig. 3.2-3: Critical points cp_1 and cp_2

under one configuration is a subset of that under the other configuration. We use θ_i^1 (i = 2,3,4,5) to represent the joint angles in the configuration with which the 6-link closed chain occupies a smaller area and call this configuration as a *smaller configuration*. Similarly, we use θ_i^2 (i = 2,3,4,5) to represent the joint angles in the configuration with which the 6-link closed chain occupies a larger area and call this configuration as a *larger configuration*.

The following Proposition 2.3-2 gives the relationship between θ_2^i and θ_5^i (i = 1,2).

Proposition 3.2-2: Let V be a circle centered at J_2 with radius $l_2 + l_3$. Suppose LPC_{a_4} and V have two intersections F_2 and F_1 where F_2 is in D_1 and F_1 is in D_2 (D_1 and D_2 are the sub-planes separated by line J_2J_5) (see Figure 3.2-2), two line segments J_2F_1 and J_2F_2 which intersect LPC_{a_2} at f_1 and f_2 can be drawn. Then we can have the following relationship between θ_2^1 (θ_2^2), θ_5^1 (θ_5^2) and f_1 (f_2) (if V and LPC_{a_4} have one intersection or no intersection, we consider f_1 and f_2 coincide on line J_2J_5):

In case $\theta_2^i \geq \theta_{2f_i}$ (i = 1,2),

if θ_2^2 decreases, then θ_5^2 increases.

or if θ_2^1 decreases, then θ_5^1 decreases.

In case $\theta_2^i < \theta_{2f_i}$ (i = 1,2),

if θ_2^2 decreases, then θ_5^2 decreases;

or if θ_2^1 decreases, then θ_5^1 increases.

By Proposition 3.2-2, when a_2 monotonously moves in a continuous subregion of $MR(\theta_2)$ on LPC_{a_2} in which f_2 (for the larger configurations) or f_1 (for the smaller configurations) only can exist at the boundary, a_4 monotonously moves in a continuous region on LPC_{a_4} to form the larger configurations or the smaller configurations. Therefore a continuous subregion of $MR(\theta_2)$ on LPC_{a_2} in which f_2 or f_1 only can exist at the boundary can be mapped to two continuous regions on LPC_{a_4} (one is for the larger configurations, the other is for the smaller configurations). If f_2 or f_1 exists inside a region on LPC_{a_2}, in order to map this region to the regions on LPC_{a_4}, we first decompose the region into two subregions in which f_2 or f_1 exists at the boundary. Subsequently, each subregion is mapped to the regions on LPC_{a_4}. Finally, we make the union of all mapped regions on LPC_{a_4}. According to above analysis, in order to map $R^1(\theta_2)$ to the region $R^2(\theta_5)$ of θ_5, the following procedure is used.

(1) Find the points on LPC_{a_4} which correspond to the boundaries of the subregions of $R^1(\theta_2)$ on LPC_{a_2}. In other words, suppose there are m (m = 1 or 2) continuous subregions in $R^1(\theta_2)$ and each subregion can be represented as $[\theta_{2min^n}, \theta_{2max^n}]$ where n = 1,m, draw $LPC_{a_3}(min^n)$ and $LPC_{a_3}(max^n)$ which intersect LPC_{a_4} at q_1^n (or r_1^n) and q_2^n (or r_2^n) for the larger configurations (or the smaller configurations), respectively.

(2) Map the subregions of $R^1(\theta_2)$ on LPC_{a_2} to regions among F_1, F_2 and the points obtained in the previous step. In other words, the region $[\theta_{2min^n}, \theta_{2max^n}]$ on LPC_{a_2} should be mapped to $[\theta^2_{5i}, \theta^2_{2j}]^n$ for the larger configurations and $[\theta^1_{5k}, \theta^1_{2l}]^n$ for the smaller configurations on LPC_{a_4} (the superscription n is used to indicate that these regions are mapped from the n-th subregion of $R^1(\theta_2)$), where

(a) if $\theta_{2max^n}, \theta_{2min^n} \geq \theta_{2f_2}$, then $i = q_2^n$, $j = q_1^n$ $k = r_1^n$ and $l = r_2^n$;

(b) if $\theta_{2min^n} \leq \theta_{2f_2} \leq \theta_{2max^n}$, then $i = \begin{cases} q_1^n & \text{if } \theta_{5q_1^n} \geq \theta_{5q_2^n} \\ q_2^n & \text{if } \theta_{5q_2^n} > \theta_{5q_1^n} \end{cases}$, $j = F_2$ $k = r_1^n$ and $l = r_2^n$;

(c) if $\theta_{2f_1} < \theta_{2min^n}, \theta_{2max^n} \leq \theta_{2f_2}$, then $i = q_1^n$, $j = q_2^n$ $k = 2_1^n$ and $l = r_2^n$;

(d) if $\theta_{2min^n} \leq \theta_{2f_1} \leq \theta_{2max^n}$, then $i = q_1^n$, $j = q_2^n$ $k = \begin{cases} r_1^n & \text{if } \theta_{5r_1^n} \leq \theta_{5r_2^n} \\ 2_2^n & \text{if } \theta_{5r_2^n} < \theta_{5r_1^n} \end{cases}$, and $l = F_1$;

(e) If $\theta_{2max^n}, \theta_{2min^n} \leq \theta_{2f_2}$, then $i = q_1^n$, $j = q_2^n$ $k = 2_2^n$ and $l = r_1^n$.

(3) Intersect $MR(\theta_5)$ and the union of all the regions obtained in the previous step to obtain $R^2(\theta_5)$, i.e., $R^2(\theta_5) = \bigcup_{n=1}^{m}([\theta^2_{5i}, \theta^2_{2j}]^n \cup [\theta^1_{5k}, \theta^1_{2l}]^n) \cap MR(\theta_5)$ ($m = 1$ or 2).

3.2.1.3. Finding $R^3(\theta_5)$

As described earlier, each value of $\theta_5 \in MR(\theta_5)$ corresponds to two configurations (the smaller configuration and the larger configuration). After mapping $R^1(\theta_2)$ to $R^2(\theta_5)$, it is guaranteed that for each $\theta_5 \in R^2(\theta_5)$, we can have at least one configuration with $\theta_2 \in R^1(\theta_2)$. However, there is no guarantee that the other configuration has $\theta_2 \in R^1(\theta_2)$. Since we need a set of regions of θ_5 in which all the configurations satisfy link collision-free constraint for adjacent links, $R^2(\theta_5)$ should be modified in order to find a set of regions $R^3(\theta_5)$ so that if $\theta_5 \in R^3(\theta_5)$, then all the corresponding configurations have $\theta_2 \in R^1(\theta_2)$. To accomplish this, we first introduce the following definitions.

Definition 3.2-2: Each $\theta_5 \in MR(\theta_5)$ corresponds two configurations (the larger configuration or the smaller configuration). If on a subregion $M(\theta_5) \in MR(\theta_5)$, only the smaller configurations need to be considered, then $M(\theta_5)$ is called a *small-one-configuration region* and is denoted as $M(\theta_5)^1$. If on $M(\theta_5)$, only the larger configurations need to be considered, then $M(\theta_5)$ is called a *larger-one-configuration region* and is denoted as $M(\theta_5)^2$. Similarly, if both the larger configurations and the smaller configurations need to be considered on $M(\theta_5)$, it is called a *two-configuration region* and is denoted as $M(\theta_5)^0$. □

We define the operations between two-configuration regions, small-one-configuration regions and larger-one-configuration regions as follows:

Definition 3.2-3:

(a) The union (or intersection) of two two-configuration regions is a two-configuration region.

(b) The union (or intersection) of two (small- or large-) one-configuration regions is a (small- or large-) one-configuration region.

(c) The union of a two-configuration region $R(\theta_5)^0$, a small-one-configuration region $S(\theta_5)^1$ and a large-one-configuration region $T(\theta_5)^2$ is a region $M(\theta_5)$ which contains three subregions $R(\theta_5)^0$, $S(\theta_5)^1$ and $T(\theta_5)^2$ (For convenience, we denote the two-configuration subregion of $M(\theta_5)$ as $M^*(\theta_5)^0$ ($= R(\theta_5)^0$), the smaller or the larger -one configuration subregion of $M(\theta_5)$ as $M^+(\theta_5)^1$ ($= S(\theta_5)^1$) or $M^{++}(\theta_5)^2$ ($= T(\theta_5)^2$)) respectively.

(d) The intersection of a two-configuration region and a (small- or large-) one-configuration region is a (small- or large-) one-configuration region.

(e) The intersection of a large-one-configuration region and a small-one-configuration region is empty.

(f) The result of subtracting a two-configuration region from a (small- or large-) one-configuration region is a (small- or large-) one-configuration region. □

Since for each $\theta_5 \in R^2(\theta_5)$, it is not guarantee that both corresponding configuration have $\theta_2 \in R^1(\theta_5)$, we need to figure out which subregion of $R^2(\theta_3)$ has both types of configurations with $\theta_2 \in R^1(\theta_5)$ and which subregion only has the smaller configurations (or the larger configurations) with $\theta_2 \in R^1(\theta_5)$. By Definition 3.2-2 and Definition 3.2-3, this task is to find $R^3(\theta_5) = R^{3*}(\theta_5)^0 \cup R^{3+}(\theta_5)^1 \cup R^{3++}(\theta_5)^2$, where $R^{3*}(\theta_5)^0$ contains the subregions of $R^2(\theta_5)$ in which each θ_5 corresponds to two configurations with $\theta_2 \in R^1(\theta_2)$, $R^{3+}(\theta_5)^1$ (or $R^{3++}(\theta_5)^2$) contains the subregions of $R^2(\theta_5)$ in which each θ_5 corresponds to one, i.e., the smaller (or the larger) configuration that has $\theta_2 \in R^1(\theta_2)$.

Suppose $R^1(\theta_2)$ consists of two continuous subregions (the case for one region can be discussed similarly) which are symmetric with respect to line J_2J_5. Each subregion can be represented as $[\theta_{2min^n}, \theta_{2max^n}]$ where n = 1,2 (subregion is in D_2 if n = 1 and subregion is in D_1 if n = 2) In Section 3.2.1.2, the subregions of $R^1(\theta_2)$ are mapped to a set of subregions $R^2(\theta_5)$ which are between the points q_1^n (or r_1^n) and q_2^n (or r_2^n) (n = 1,2) that are the intersections between $LPC_{a_3}(min^n)$, $LPC_{a_3}(max^n)$ and LPC_4 for the larger configurations (or the smaller configurations). To find $R^{3*}(\theta_5)^0$, all the possible subregions of $R^2(\theta_5)$ in which each θ_5 corresponds to two configurations with $\theta_2 \in R^1(\theta_2)$ can be found as follows:

(1) By Proposition 3.2-2, in case $\theta_{2min^2} \le \theta_{2f_2} \le \theta_{2max^2}$, if j_3 which is on LPC_{a_2} moves from θ_{2min^2} to θ_{2max^2}, then θ_5^2 increase from $\theta_{5q_1^2}$ to θ_{5F_2} first, subsequently, it decreases from θ_{5F_2} to $\theta_{5q_2^2}$. Consequently each θ_5 in the subregion $I = [\theta_{5q_*^2}, \theta_{5F_2}]^2$, where $\theta_{5q_*^2} = \max\left\{\theta_{5q_1^2}, \theta_{5q_2^2}\right\}$, has two configurations with $\theta_2 \in [\theta_{2min^2}, \theta_{2max^2}] \subset R^2(\theta_2)$. Simi-

larly, if $\theta_{2min^1} \le \theta_{2f_1} \le \theta_{2max^1}$, then each θ_5 in the subregion $H = [\theta_{5F_1}, \theta_{5r_*^1}]^1$ where $\theta_{5r_*^1} = \min\left\{\theta_{5r_1^1}, \theta_{5r_1^1}\right\}$, has two configurations with $\theta_2 \in [\theta_{2min^1}, \theta_{2max^1}] \subset R^2(\theta_2)$.

(2) Recall that two continuous subregions of $R^1(\theta_2)$ can be mapped to four continuous regions on LPC_{a_4}. If θ_5 is in the intersection between any two of these four regions on LPC_4, it can correspond to two configurations with $\theta_2 \in R^1(\theta_2)$, where an intersection of these four regions on LPC_{a_4} could be either

(a) an intersection of two subregions in $R^2(\theta_5)$ which are mapped from one subregion in $R^1(\theta_2)$. Let K be the union of this type of intersections, ie., $K = \bigcup_{n=1}^{2}\left\{[\theta_{5i}^2, \theta_{2j}^2]^n \cap [\theta_{5k}^{n^1}, \theta_{2l}^1]^n\right\}$ (where i,j,k and l are determined by the procedure described in Section 3.2.1.2), or

(b) an intersection of two groups of subregions in $R^2(\theta_5)$ in which each group is mapped from a subregion of $R^1(\theta_2)$ (Let O be the union of this type of intersections, i.e., $O = \bigcap_{n=1}^{2}\left\{[\theta_{5i}^2, \theta_{2j}^2]^n \cup [\theta_{5k}^1, \theta_{2l}^1]^n\right\}$).

By the above analysis, we have

$$R^{3*}(\theta_5)^0 = I^* \cup H^* \cup K \cup O$$

where

$$I^* = \begin{cases} I & \text{if } \theta_{2f_2} \in [\theta_{2min^2}, \theta_{2max^2}] \\ \Phi & \text{if } \theta_{2f_2} \notin [\theta_{2min^2}, \theta_{2max^2}], \end{cases} \quad H^* = \begin{cases} H & \text{if } \theta_{2f_1} \in [\theta_{2min^1}, \theta_{2max^1}] \\ \Phi & \text{if } \theta_{2f_1} \notin [\theta_{2min^1}, \theta_{2max^1}], \end{cases}$$

By the definition of the smaller configurations, all the subregions of $R^2(\theta_5)$ which are not in $R^{3*}(\theta_5)^0$ and in which each θ_5 has one configuration with $\theta_2 \in [\theta_{2min^1}, \theta_{2max^1}]$ should belong to $R^2(\theta_5)^1$. Consequently, $R^{3+}(\theta_5)^1 = [\theta_{5i}^1, \theta_{2k}^1]^1 \cup [\theta_{5k}^1, \theta_{2l}^1]^1 - R^{3*}(\theta_5)^0$. For the same region, $R^{3++}(\theta_5)^2 = [\theta_{5i}^2, \theta_{2k}^2]^2 \cup [\theta_{5k}^2, \theta_{2l}^2]^2 - R^{3*}(\theta_5)^0$.

Now, by definition 3.2-2, $R^1(\theta_5)$ should be a two configuration region. Therefore, $R^1(\theta_5)$ can be denoted as $R^{1*}(\theta_5)^0$. Consequently, if $\theta_5 \in R^{1*}(\theta_5)^0 \cap R^3(\theta_5)$, then the link collision-free constraint about adjacent links can satisfied.

3.2.2. Finding Subregions of MR(θ_5) which Satisfy the Link Collision-Free Constraint for Non-Adjacent Links

To satisfy the link collision-free constraint for non-adjacent links, each pair of non-adjacent links cannot intersect each other. It can be observed from the 6-link closed chain that among non-adjacent links, only changeable links a_2, a_3 and a_4 can possibly intersect the fixed links a_1, a_5 or a_6. Since θ_3 and θ_4 should be in [90°, 270°], the changeable links a_2, a_3 or

a_4 can possibly intersect the fixed links a_1, a_5 or a_6 only if θ_5 or θ_2 becames small. In the following, we shall find the values of θ_{51} (or θ_{21}) by which one of a_3 and a_4 (or one of a_3 and a_2) touches a_1, a_5 or a_6 as θ_5 (or θ_2) becomes small. Subsequently, the value of θ_{21} should be mapped to the regions of θ_5. In other words, we shall to find a subregion $R^4(\theta_5)$ of $MR(\theta_5)$ in which each configuration has $\theta_5 \geq \theta_{51}$ and $\theta_2 \geq \theta_{21}$. Consequently, if $\theta_5 \in R^4(\theta_5)$, the link collision-free constraint for non-adjacent links can be satisfied.

Definition 3.2-4: An intersection between $LPC_{a_2}(J_2)$ and the centerline of one of the fixed links a_1, a_5 or a_6 is defined to be *a critical point* and is denoted as cp_1 (which one among a_1, a_5 and a_6 is intersected by $LPC_{a_2}(J_2)$ depends on the length of a_2). Similarly, an intersection between $LPC_{a_4}(J_5)$ and the centerline of one of the fixed links a_1, a_5 or a_6 is defined as a *critical point* as well and is denoted as cp_2 (see Figure 3.2-3).

It can be observed from Figure 3.2-4(a) that in case of $w_O \leq w_R$ (w_O and w_R are the widths of the carried object and robot link respectively), when θ_5 (or θ_2) becames small, link a_4 (or a_2) (which is robot link) first touches the fixed link a_1, a_5 or a_6 at $\theta_5 = \theta_{5cp_2}$ (or $\theta_2 = \theta_{2cp_1}$). On the other hand, in case of $w_O > w_R$ (see Figure 3.2-4(b)), when θ_5 (or θ_2) becames small, link a_3 (which is the carried object) first touches the fixed link a_1, a_5 or a_6 at $\theta_5 = \theta_{5R_2}$ (or $\theta_2 = \theta_{2R_1}$). Putting θ_{5cp_2} and θ_{5R} (or θ_{2cp_1} and θ_{2R}) together, if $\theta_5 \geq \theta_{51}$ and $\theta_2 \geq \theta_{21}$, where

$$\theta_{51} = \begin{cases} \theta_{5cp_2} & \text{if } w_O \leq w_R \\ \theta_{5R_2} & \text{if } w_O > w_R \end{cases}, \quad \theta_{21} = \begin{cases} \theta_{2cp_1} & \text{if } w_O \leq w_R \\ \theta_{2R_1} & \text{if } w_O > w_R \end{cases},$$

the link collision-free constraint for non-adjacent links can be satisfied.

To find $R^4(\theta_5)$ of θ_5 in which each configuration has $\theta_5 \geq \theta_{51}$ and $\theta_2 \geq \theta_{21}$, $\theta_2 \geq \theta_{21}$ should be mapped to the motion region of θ_5. We first find the value of θ_5 by which the corresponding configurations have $\theta_2 = \theta_{21}$. For this, a circle VI with center R (for $w_O > w_R$) or cp_1 (for $w_O \leq w_R$) and radius l_3 can be drawn. As shown in Figure 3.2-5, VI intersects LPC_{a_4} at points Q^1 and Q^2. Consequently, two configurations corresponding with $\theta_2 = \theta_{21}$ can be obtained. Among them, we assume the larger one with $\theta_5^2 = \theta_{5Q^1}$ and the smaller one with $\theta_5^1 = \theta_{5Q^2}$. If $\theta_{21} \leq \theta_{2f_1}$, by Proposition 3.2-2, each $\theta_2 \in [\theta_{2min}, \theta_{21}]$ (or $\leq \theta_{21}$), where θ_{5min} is the smallest angle in $MR(\theta_2)$, corresponds $\theta_5^2 \leq \theta_{5Q^2}$ (for the larger configurations) and $\theta_5^1 \leq \theta_{5Q^1}$ (for the smaller configurations) (see Figure 3.2-5). Consequently, region $W(\theta_5) = [\theta_{2min}, \theta_{21}]$ on LPC_{a_2} should be mapped to region $N_1(\theta_5)$ on LPC_{a_4}, where $N_1(\theta_5) = [\theta_{5Q^2}, \theta_{5Q^1}]$. By the definition of the smaller configuration, each smaller configuration in region $N_1(\theta_5)$ on LPC_{a_4} has $\theta_2 \in W(\theta_2)$ on LPC_{a_2}. Since $R^4(\theta_5) \subset MR(\theta_5)$ is a set of subregions in which corresponding $\theta_5 \geq \theta_{51}$ and $\theta_2 \geq \theta_{21}$, the smaller configurations on region $N_1(\theta_5)$ should not be included in $R^4(\theta_5)$. Consequently, if θ_{5max} is used to represent the largest angle in $MR(\theta_2)$, then

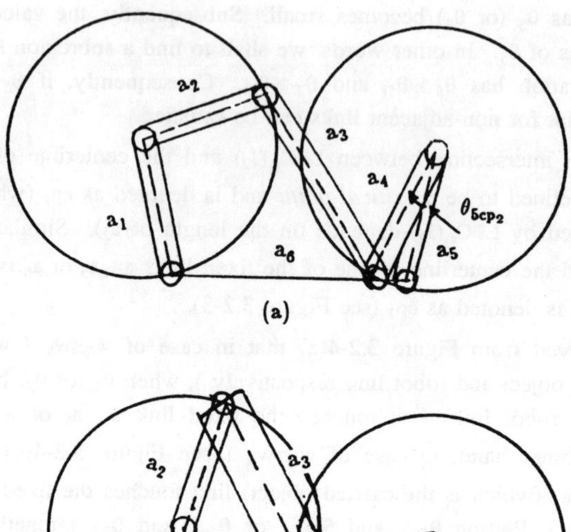

Fig. 3.2-4: An explanation of θ_{5cp_2} and θ_{5cp_2}

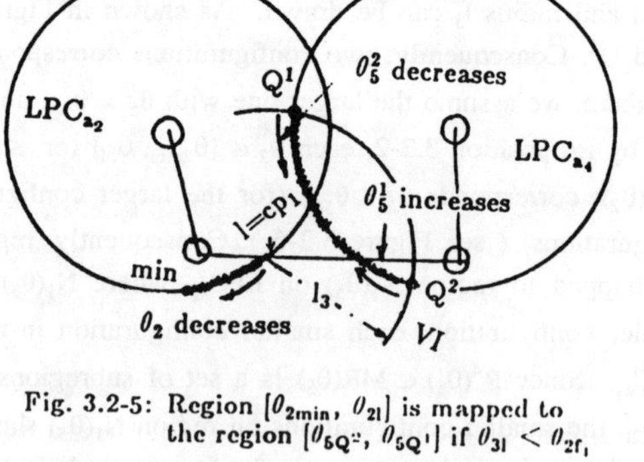

Fig. 3.2-5: Region $[\theta_{2min}, \theta_{2l}]$ is mapped to the region $[\theta_{5Q^2}, \theta_{5Q^1}]$ if $\theta_{3l} < \theta_{2f_1}$

$$R^4(\theta_5) = R^{4*}(\theta_5)^0 \cup R^{4++}(\theta_5)^2$$

where

$$R^{4*}(\theta_5)^0 \text{ (two configuration)} = [\theta_{5l}, \theta_{5max}] \cap MR(\theta_5) - N_1(\theta_5),$$
$$R^{4++}(\theta_5)^2 \text{ (larger–one–configuration region)} = N_1(\theta_5) \cap MR(\theta_5).$$

Similarly, if $\theta_{2l} > \theta_{2f_1}$, by Proposition 3.2-2, region $W(\theta_2)$ on LPC_{a_2} should be mapped to region $N_2(\theta_5) = [\theta_{5Q^1}, \theta_{5max}] \cup [\theta_{5min}, \theta_{5Q^2}]$, where θ_{5min} is the smallest angle of $MR(\theta_5)$. In this situation, $R^4(\theta_5)$ is still equal to $R^{4*}(\theta_5)^0 \cup R^{4++}(\theta_5)^2$, but

$$R^{4*}(\theta_5)^0 \text{ (two configuration)} = [\theta_{5l}, \theta_{5max}] \cap MR(\theta_5) - N_2(\theta_2),$$

$$R^{4++}(\theta_5)^2 \text{ (larger–one–configuration region)} = N_2(\theta_5) \cap MR(\theta_5).$$

Consequently, if $\theta_5 \in R^4(\theta_5) = R^{4*}(\theta_5)^0 \cup R^{4++}(\theta_5)^2$ the link collision-free constraint for non-adjacent links is satisfied.

Combined the results in Section 3.2.1 and in Section 3.2.2, we can have the feasible region $FR(\theta_5)$ in which the link collision-free constraints (for adjacent links and for non-adjacent links) and closed chain constraint are satisfied is the feasible region as follows:

$$FR(\theta_5) = R^3(\theta_5) \cap R^4(\theta_5) = FR^*(\theta_5)^0 \cap FR^+(\theta_5)^1 \cap R^{++}(\theta_5)^2$$

where

$$FR^*(\theta)^0 = R^{1*}(\theta_5)^0 \cap R^{3*}(\theta_5)^0 \cap R^{4*}(\theta_5)^0$$
$$FR^+(\theta)^1 = R^{1*}(\theta_5)^0 \cap R^{3+}(\theta_5)^1 \cap R^{4*}(\theta_5)^0$$
$$FR^{++}(\theta)^2 = R^{1*}(\theta_5)^0 \cap \left\{ W_1(\theta_5) \cup W_2(\theta_5) \cup W_3(\theta_5) \right\}$$
$$W_1(\theta_5) = [R^{3++}(\theta_5)^2 \cap R^{4*}(\theta_5)^0]$$
$$W_2(\theta_5) = [R^{3*}(\theta_5)^0 \cap R^{4++}(\theta_5)^2]$$
$$W_3(\theta_5) = [R^{3++}(\theta_5)^2 \cap R^{4++}(\theta_5)^2]$$

3.3. Finding All the Collision-Free Feasible Configurations

Without considering obstacles, in the last section we computed the feasible regions $FR(\theta_5) = FR^*(\theta_5)^0 \cup FR^+(\theta_5)^1 \cup FR^{++}(\theta_5)^2$ in which all the feasible configurations can be obtained. If there are obstacles in the environment, collision with obstacles should also be considered. In this section, we shall find the subregions of $FR(\theta_5)$ in which each feasible configuration is collision-free. In the discussion, the simplest case in which $w_O=0$, $w_R=0$ and there is only one obstacle in the environment is studied first. Subsequently, the result will be modified by adding a width to each link of the closed chain and by adding all the obstacles in the environment.

3.3.1. Finding All the CFFC's in a Single Obstacle Environment

In this subsection, we assume that there is a single obstacle environment and the width of the carried object and robot links are zero (i.e., w_O=0 and w_R=0). Obviously, if we can find subregions of FR(θ_5) in which all three changeable links a_2, a_3 and a_4 are collision-free with the obstacle, then all the CFFC's in this situation can be obtained. For this, following procedure is adopted.

(1) Find subregions of FR(θ_5) in which the corresponding position of a_4 does not have any collision with the obstacle. We call this type of subregion as a possible collision-free feasible region of a_4 and denote it as $PCFFR_{a_4}$.

(2) Find subregions of $PCFFR_{a_4}$ in which the corresponding positions of a_4 and a_2 do not have any collision with the obstacle. Similarly, we call this type of subregion as a possible collision-free feasible region of a_2 and a_4 and denote it as $PCFFR_{a_2,a_4}$.

(3) Finally, find subregions of $PCFFR_{a_2,a_4}$ in which the corresponding positions of all changeable links a_2, a_3 and a_4 are collision-free with the obstacle. This type of subregion is called as a collision-free feasible region and is denoted as CFFR.

Each step is discussed in detail in the following subsections.

3.3.1.1. Finding $PCFFR_{a_4}$

In this subsection and following discussions, we use $PCFFR^i_{a_4}$, where i = 0, 1 or 2, to represent a $PCFFR_{a_4}$ which is a two-configuration region, a small-one-configuration region, or a large-one-configuration region respectively. Suppose there is an obstacle O_1 in the environment which intersects LPC_{a_4} at points A and B (see Figure 3.3-1). To find $PCFFR_{a_4}$, we can draw rays from J_5 to A, B and all the corners of O_1 which is inside LPC_{a_4}. Each ray corresponds to a angle θ_{5d_j}, where d_j is the intersection between the ray and LPC_{a_4}. Since $PCFFR^i_{a_4}$'s are subregions in which a_4 is collision-free with the obstacle, we can obtain that

$$PCFFR^0_{a_4} = \left\{[\theta_{5A'},\theta_{5max}] \cup [\theta_{5min},\theta_{5B'}]\right\} \cap FR^*(\theta_5)^0,$$

$$PCFFR^1_{a_4} = \left\{[\theta_{5A'},\theta_{5max}] \cup [\theta_{5min},\theta_{5B'}]\right\} \cap FR^+(\theta_5)^1 \text{ and}$$

$$PCFFR^2_{a_4} = \left\{[\theta_{5A'},\theta_{5max}] \cup [\theta_{5min},\theta_{5B'}]\right\} \cap FR^{++}(\theta_5)^2,$$

where $\theta_{5A'} = \max_j \theta_{5d_j}$, $\theta_{5B'} = \min_j \theta_{5d_j}$, θ_{5max} is the largest angle of FR(θ_5) and θ_{5min} is the smallest angle of FR(θ_5).

3.3.1.2. Finding $PCFFR_{a_2,a_4}$

If θ_5 is in a $PCFFR_{a_4}$, a_4 does not collide with any obstacles. However we still don't know if the closed chain is in a CFFC as a_2 and a_3 may collide with the obstacles. Consequently, in this subsection we shall to find a subregion, $PCFFR_{a_2,a_4}$, of $PCFFR_{a_4}$, in which both a_2 and a_4 are collision-free. Similar with $PCFFR^i_{a_4}$, we use $PCFFR^i_{a_2,a_4}$ to represent a $PCFFR_{a_2,a_4}$ which is a two-configuration region, a small-one-configuration region, or a large-one-configuration region respectively. To find $PCFFR^i_{a_2,a_4}$, we assume $[\theta_{5C},\theta_{5D}]$ is a $PCFFR^2_{a_4}$ (the case that $[\theta_{5C},\theta_{5D}]$ is a $PCFFR^1_{a_4}$ can be discussed in a similar way) (see Figure 3.2-2). Supposing that the locations of j_4 and j_3 which form a larger configuration are denoted as J^2_4 and J^2_3 respectively. In case θ_{5C}, $\theta_{5D} > \theta_{5F_2}$+, (where θ_{5F_2} was defined in Proposition 3.2-2 in Section 3.2.1.2), since J^2_4 can be moved between C and D, by Proposition 3.2-2, in order to form a larger configuration J^2_3 must be on the locus of $LPC_{a_2}(J_2)$ between D' and C' (where C' is an intersection of $LPC_{a_2}(J_2)$ and $LPC_{a_3}(C)$, and D' is an intersection of $LPC_{a_2}(J_2)$ and $LPC_{a_3}(D)$). However, a_2 may have a collision with the obstacle if θ_2 is in the region $[\theta_{2D'},\theta_{2C'}]$. If a_2 has a collision with the obstacle in the entire region of $[\theta_{2D'},\theta_{2C'}]$, then there is no $PCFFR^2_{a_2,a_4}$ in $[\theta_{5C},\theta_{5D}]$. Otherwise, we can find a subregion $[\theta_{2s_1'},\theta_{2s_2'}]$ of $[\theta_{2D'},\theta_{2C'}]$ in which a_2 does not have any collision with the obstacle. Subsequently, by the similar method for finding $[\theta_{D'},\theta_{C'}]$, a corresponding subregion $PCFFR^2_{a_2,a_4} = [\theta_{5s_1},\theta_{5s_2}]$ in $[\theta_{5C},\theta_{5D}]$ of θ_5 can be obtained so that if θ_5 is in $[\theta_{5s_1},\theta_{5s_2}]$, there is no collision between links a_2, a_4 and the obstacle.

If $PCFFR_{a_4}$ is a two-configuration region $PCFFR^0$, this region can be decomposed into two one-configurations. The first one is denoted as $PCFFR^{01}_{a_4}$ in which only the smaller configurations are considered. The other one is denoted as $PCFFR^{02}$ in which only the larger configurations are considered. Based on $PCFFR^{01}_{a_4}$ and $PCFFR^{02}_{a_4}$, $PCFFR^{01}_{a\,a_4}$ and $PCFFR^{02}_{a_2,a_4}$ can be obtained by above method, where superscriptions o1 and 02 in $PCFFR_{a_2,a_4}$ have the similar meaning with that in $PCFFR_{a_4}$.

3.3.1.3. Finding CFFR

Our goal is to find collision-free feasible regions (CFFR's) which are subregions of $PCFFR_{a_2,a_4}$ and in which all the changeable links a_a, a_3 and a_4 are collision-free with the obstacle. Similar with $PCFFR^i_{a_4}$ and $PCFFR^i_{a_2,a_4}$, we use $CFFR^i$ to represent a CFFR which is a two-configuration region, a small-one-configuration region, or a large-one-configuration region respectively. Since $CFFR^i$ is a subregion of $PCFFR^i_{a_2,a_4}$, there should be some CFFC's that divide $PCFFR^{i}_{a_2,a_4}$ into several subregions and the subregion which is in one side of the

+ The case θ_{5C}, $\theta_{5D} \leq \theta_{5F_2}$ can be discussed in a similarly way. If $\theta_{5C} < \theta_{5F_2} < \theta_{5D}$, then we consider two regions $[\theta_{5C},\theta_{5F_2}]$ and $[\theta_{5F_2},\theta_{5D}]$.

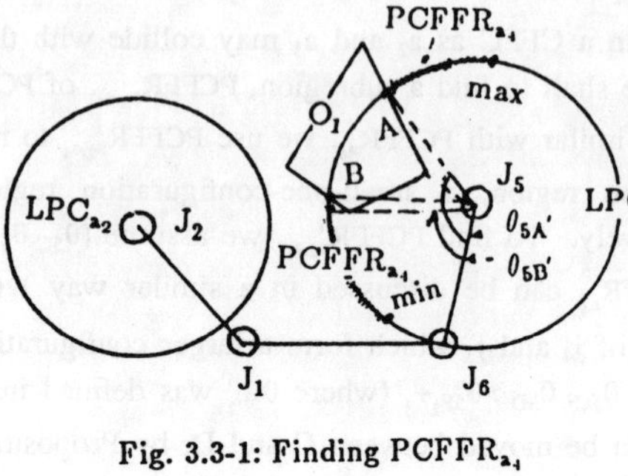

Fig. 3.3-1: Finding PCFFR$_{a_4}$

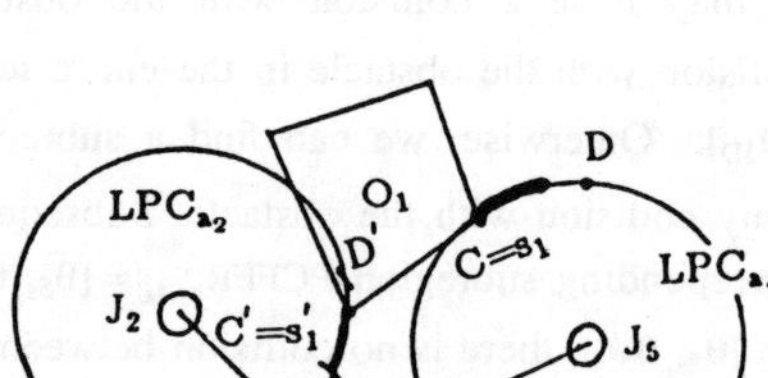

Fig. 3.3-2: Finding PCFFR$^2_{a_2,a_4}$

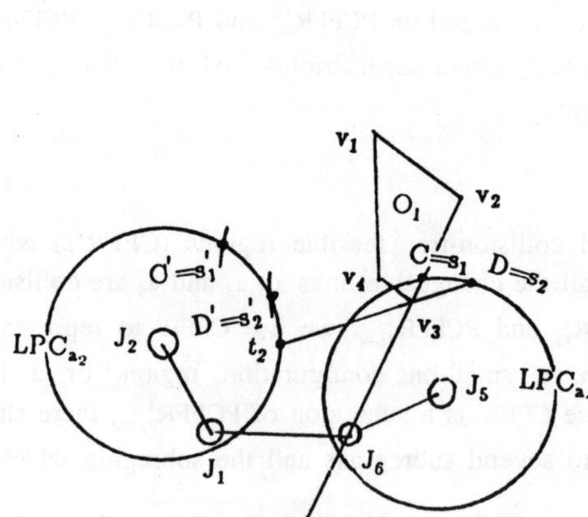

Fig. 3.3-3: A method to find critical CFFR

CFFC is a CFFR, but the subregion which is in the other side of the CFFC is not a CFFR. We call this kind of CFFC in a $PCFFR_{a_2,a_4}$ as a *critical* CFFC. Obviously, if we can find all the critical CFFC's, then we can find all the CFFR's. Consequently, all the CFFC's can be found. To find the critical CFFC's, the following lemma is introduced.

Lemma 3.3-1 : If a 6-link closed chain is in a critical CFFC of a $PCFFR_{a_2,a_4}$, then a_3 must pass through at least one corner or an edge of the obstacle.

Proof: If a 6-link closed chain is in a critical CFFC, either of a_2, a_3, or a_4 should touch the obstacle. Since link θ_5 is in a $PCFFR_{a_2,a_4}$, there is no collision between links a_2, a_4 and the obstacle. Therefore link a_3 should touch the obstacle to form a critical CFFC. Since the obstacle is a convex polygon and link a_3 is a line segment, link a_3 must pass through a corner or an edge of the obstacle in a critical CFFC. □

By Lemma 3.3-3, we can find the critical CFFC's by finding the CFFC's in which the location of a_3 touchs a corner or an edge of the obstacle. In the following, we shall discuss the method to find the critical CFFC's in which a_3 touches a corner of the obstacle. In Figure 3.3-3, it is assumed that v_3 is a corner of the obstacle and $[\theta_{s_1}, \theta_{s_2}]$ is a $PCFFR^2_{a_2,a_4}$ on LPC_{a_4}. In order to find $CFFR^2$ ($CFFR^0$ including $CFFR^{01}$ and $CFFR^{02}$ and $CFFR^1$ can be obtained in a similar way) we can draw a line s_1v_3 which may intersect $LPC_{a_2}(J_2)$ at t_1. We can also draw a line v_3s_2 which may intersect $LPC_{a_2}(J_2)$ at t_2. If both intersections do not exist or the intersection of the two arcs t_1t_2 and $s_1's_2'$ is empty, then there is no critical CFFC for the corner. Otherwise, we can find a case as shown in Figure 3.3-4 where the arc T_1T_2 is the intersection of t_1t_2 and $s_1's_2'$. (If there exists only one t_i, i = 1 or 2 , line t_is_i would separate the entire plane into two subplanes and separate LPC_{a_2} into two parts. If the intersection between $s_1's_2'$ and one part of LPC_{a_2} which is in the same subplane with line s_jv_3, where $i,j = 1,2, i \neq j$, is empty, then there is no critical CFFC for the corner. Otherwise, $T_1T_2 = s_1's_2'$.) In Figure 3.3-4, we draw two rays from T_1 and T_2 which pass through the corner v_3 and intersect LPC_{a_4} at K_1 and K_2. If there is a line segment with length l_3 which has one end on $[T_1, T_2]$ of LPC_{a_2}, one end on $[K_1, K_2]$ of LPC_{a_4}, and passes through v_3, we can find a feasible position of a_3 in the area shown in Figure 3.3-4. This feasible position of a_3 determines the critical CFFC in $PCFFR^2_{a_2,a_4}$. After checking each corner of the obstacle, the edges of the obstacle should also be checked. For this, link a_3 is overlapped with the edges of the obstacle. If there is a possible collision-free feasible configuration in which link a_3 and an edge of the obstacle overlap, it is a critical CFFC. After finding all the critical CFFC's in all the $PCFFR_{a_2,a_4}$'s, all the CFFR's for a single-obstacle environment can be obtained.

3.3.2. Modifications

In the previous subsections, the width of the robot arms and the (rectangular) object being carried are not considered. If $w_O \neq 0$ or $w_R \neq 0$, CFFR of θ_5 (which is computed by the above method) needs to be revised. Suppose $w_O = w_R \neq 0$, in case θ_5 is in a critical CFFC,

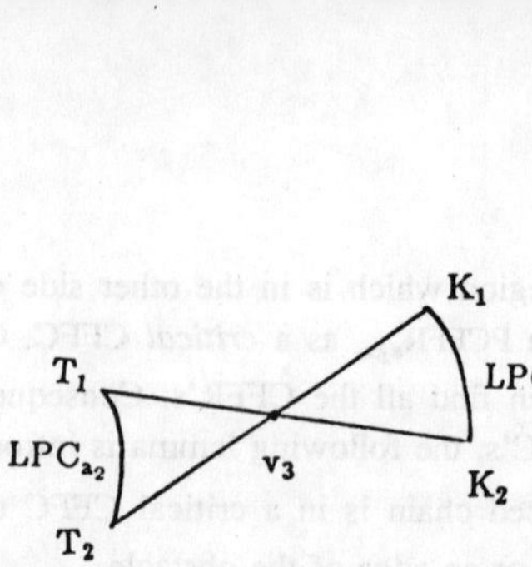

Fig. 3.3-4: A region for finding critical CFFR

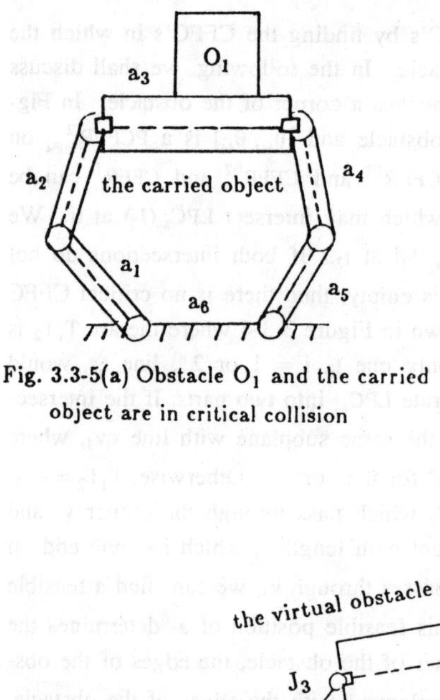

Fig. 3.3-5(a) Obstacle O_1 and the carried object are in critical collision

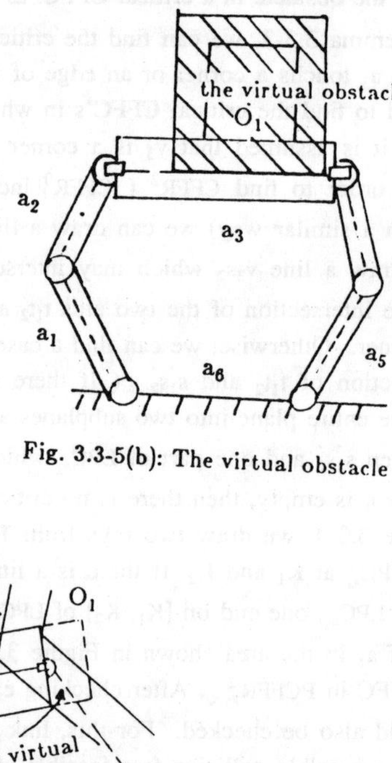

Fig. 3.3-5(b): The virtual obstacle

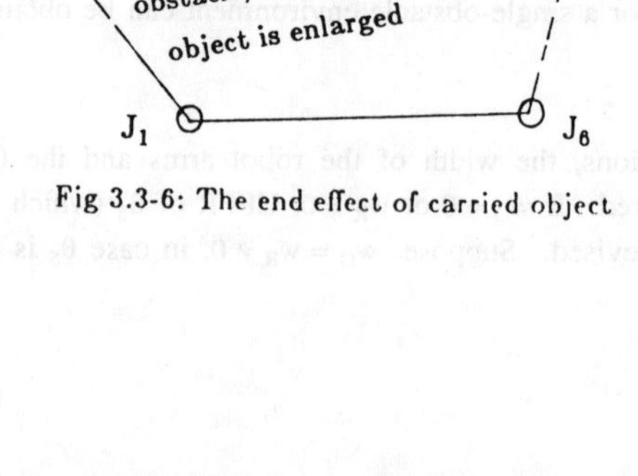

Fig 3.3-6: The end effect of carried object

the edge of the rectangular object which is outside the closed chain should pass through one corner or one edge of the obstacle instead of the centerline of a_3 (see Fig. 3.3-5(a)). If the edge of the rectangular object which is outside the closed chain passes through an edge or a corner of the obstacle, it can be regarded as if the centerline of the rectangular object passes through a corner or an edge of an expanded obstacle whose edges are parallel to the original obstacle with distance $\frac{1}{2}w_O$ and whose corners are the connections of its edges. This expanded obstacle is defined as a *virtual obstacle* of the original one (see Fig. 3.3-5(b)). Consequently, considering the width of the carried object and the robot arms, a virtual obstacle (instead of an original one) can be used in the above method.

It should be noted that if we use a virtual obstacle in the above method in finding a possible collision-free feasible region of a_4, two problems need to be resolved:

(i) Links a_2 and a_4 have width w_R, which normally is different from w_O.

(ii) If a virtual obstacle is used, in fact the carried (rectangular) object can be considered to have the shape shown in Figure 3.3-6. Therefore we need to cut off the effect of the extra parts at the two ends of the carried object (which is called the end effect) as shown in Figure 3.3-6.

Both of these issues can affect the boundary of a possible collision-free feasible region of a_4. These two problems can be solved in the following way. If there is a virtual obstacle which intersects LPC_{a_4} as shown in Figure 3.3-7, then two possible collision-free feasible regions of a_4 (i.e, $[\theta_{5A},\theta_{5max}]$ and $[\theta_{5min},\theta_{5B}]$) can be obtained:

(a) If $w_O<w_R$, θ_{5A} and θ_{5B} should be changed to θ_{5A}' and θ_{5B}' (see Fig. 3.2-7(a)), where

$$\theta_{5A}'=\theta_{5A}+\frac{w_R-w_O/2}{l_4}$$

$$\theta_{5B}'=\theta_{5B}-\frac{w_R-w_O/2}{l_4}$$

(b) If $w_O>w_R$, suppose $[\theta_{5A},\theta_{5max}]$ and $[\theta_{5min},\theta_{5B}]$ are $PCFFR_{a_4}^2$ in which only the larger configurations are considered (if these two regions are other types of $PCFFR_{a_4}$, the discussion is similar). we need to check the collision-freeness of the larger configuration with θ_{5A}^2 and the larger configuration with θ_{5B}^2. If the larger configuration with θ_{5A}^2 or the larger configuration with θ_{5B}^2 is not collision-free or is critical collision-free, θ_{5A} or θ_{5B} is the boundary of a $PCFFR_{a_4}^2$. If the larger configuration with θ_{5A}^2 or the larger configuration with θ_{5B}^2 is obstacle collision-free but not critical collision-free, the critical collision-free feasible configuration would not be resulted from the edge v_4v_3 (or the corner v_3) of the obstacle touching the edge b_1b_2 of the carried object which is outside the closed chain (see Fig. 3.2-7(b)). It is obtained by the edge b_2b_3 of the carried object touching the edge v_4v_3 (or the corner v_3) of the obstacle. Due to the end effect of the virtual obstacle, the boundary of a $PCFFR_{a_4}^2$ in this situation should be solved by a

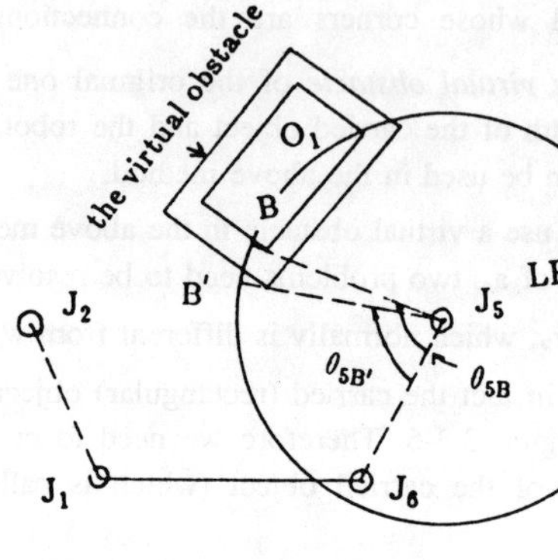

Fig. 3.3-7(a): Eliminating end effect when $w_O \leq w_R$

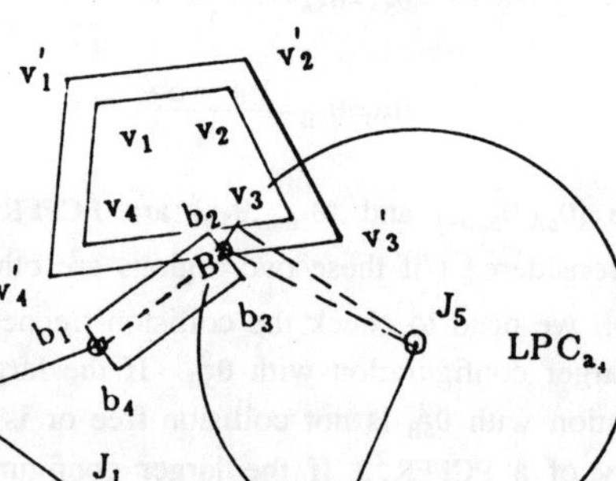

Fig. 3.3-7(b): Eliminating end effect when $w_O > w_R$

group of equations.

3.3.3. Finding CFFR's in a Multiple-Obstacle Environment

According to the above discussion, the critical CFFC's in $PCFFR_{a_4}$ or $PCFFR_{a_2,a_4}$ of θ_5 can be found if there is only one obstacle in the environment. Consequently, all the CFFC's in the possible collision-free feasible region with respect to the obstacle can be found. If there is more than one obstacle in the environment, the following theorem can be used to find all the CFFC's in the feasible region of θ_5.

Theorem 3.3-1: Suppose there are p obstacles $O_1,...,O_p$ in the environment e and this environment can be obtained by taking the union of p subenvironments e_j, j = 1,...,p, where each e_j has only one obstacle O_j. Assuming $U_j(i)$ (i = 01, 02 1, 2) is the union of the $CFFR^{i}$'s on LPC_{a_4} for e_j,j=1,...,p, the union U(i) of the CFFR's on LPC_{a_4} in the environment e is equal to $\bigcap_{j=1}^{p} U_j(i)$.

Proof: Since $U(i) = \bigcap_{j=1}^{p} U(i)_j$ (i = 01, 02, 1, 2), if J_4 is at a point which is inside U(i), the 6-link closed chain does not collide with any obstacle. If j_4 is at a point which is not inside U(i), we can always find an obstacle O_k such that the 6-link closed chain collides with this obstacle. Therefore, U(i) is equal to the union of $U(i)_j$. □

This theorem can be explained by Figure 3.3-8.

4. Finding A Collision-Free Path

After all the collision-free feasible configurations (CFFC's) for a 6-link closed chain have been found, the problem of finding a collision-free path for each joint angle of the robot arms needs to be solved. The 6-link closed chain described in Section 2 has three degrees of freedom. In Section 3, we tried to find all the CFFC's in $BCU(a_2,a_3,a_4)$ under a set of quantized values of θ_1 and θ_6 in term of the motion region of θ_5. Therefore, three degrees of freedom θ_1, θ_6 and θ_5 have been chosen as the independent variables. The collision-free path finding problem with three independent variables can be solved in the following four steps:

(1) Find the transitions between $CFFR^{i}$'s (i = 0,1,2) at fixed values of θ_1 and θ_6.

(2) Find the transitions between the $CFFR^{i}$'s (i = 0,1,2) at adjacent (quantized) values of θ_1 and θ_6.

(3) Build a *connection graph*, in which each vertex is a CFFR in $BCU(a_2,a_3,a_4)$ for specific θ_1 and θ_6, and in which each edge is a transition between a pair of CFFR's at adjacent (quantized) values of θ_1 and θ_6. If a connectivity subgraph which contains a CFFR at the starting position and a CFFR at the final position can be found, a collision-free path for the 6-link closed chain exists.

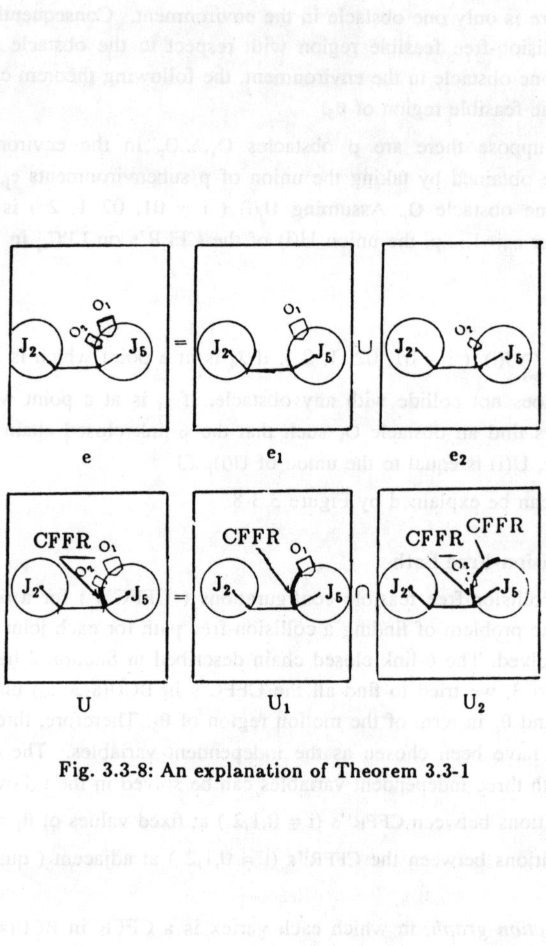

Fig. 3.3-8: An explanation of Theorem 3.3-1

(4) Search for a collision-free path for the 6-link closed chain in the connectivity subgraph.

4.1. Finding Transitions between CFFR^{i}'s (i= 0,1,2) in BCU(a_2,a_3,a_4)

For specific θ_1 and θ_6, there is one degree of freedom, say θ_5, in BCU(a_2,a_3,a_4). We say a transition exists between two CFFR^{i}'s (i = 0, 1, 2) in BCU(a_2,a_3,a_4) if we change the value of θ_5 continuously, a CFFC in one CFFR can be automatically transferred to a CFFC in another CFFR. In BCU(a_2,a_3,a_4), a CFFR could be a CFFR^0 which consists of CFFR^{01} and CFFR^{02}, CFFR^1 or CFFR^2. There are four possible types of transitions between a set of CFFR's: (1) transitions between two CFFR^{i}'s (i= 01, 02, 1,2), (2) transitions between CFFR^{01} and CFFR^{02}, (3) transitions between two different one-configuration regions CFFR^1 and CFFR^2 (or between CFFR^i and CFFR^{oj} where $i \neq j$, i,j = 1,2), and (4) transitions between CFFR^{01} and CFFR^1 (or between CFFR^{02} and CFFR^2). Since any two of the CFFR^{i}'s (i= 01,02, 1,2) are separated by obstacles, among these possible types of transition, we can observe that a transition exists between a pair of CFFR's in BCU(a_2,a_3,a_4) only in two cases: (a) CFFR^0 contains a points at which a_2 and a_3 align, and (b) CFFR^{01} and CFFR^1 (or CFFR^{02} and CFFR^2) are connected by a point on LPC$_{a_4}$.

4.2. Finding Transition between CFFR^{i}'s (i= 0,1,2) in Different BCU(a_2,a_3,a_4)'s

Besides the transitions between CFFR^{i}'s (i = 0, 1, 2) in the same BCU(a_2,a_3,a_4), transitions between CFFR^{i}'s (i = 0, 1, 2) in different BCU(a_2,a_3,a_4)'s need to be found. For each BCU, there are three kinds of CFFR's: CFFR^0 (which includes CFFR^{01} and CFFR^{02}), CFFR^1 and CFFR^2. For convenience, CFFR(0i) and CFFR^i (i = 1, 2) are denoted as CFFR$_i$; furthermore if a CFFR^{0i} and a CFFR^i are connected by a point on LPC$_{a_4}$, they are considered as one CFFR$_i$. Given two adjacent (quantized) values of θ_1's and two adjacent (quantized) values of θ_6's, we need to investigate three types of transitions between CFFR$_i$ and CFFR$_j$ (i,j = 1,2) which are in different BCU(a_2,a_3,a_4)'s:

(1) transitions between CFFR$_i$ and CFFR$_j$ (i,j = 1,2), which have the same (quantized) value of θ_1 and different but adjacent (quantized) values of θ_6,

(2) transitions between CFFR$_i$ and CFFR$_j$ (i,j = 1,2), which have different but adjacent (quantized) values of θ_1 and the same value of θ_6,

(3) transitions between CFFR$_i$ and CFFR$_j$ (i,j = 1,2), which have different but adjacent (quantized) values of θ_1 and different but adjacent (quantized) values of θ_6.

The first and the second types of transitions are symmetric, and if the quantized intervals of θ_1 and θ_6 are small enough, the third type of transitions can be approximated by a combination of a first type transition and a second type transition. Therefore we shall focus on the first type transitions; the others can be solved in a similar way. To find the first type transitions, we consider the transitions between two CFFR$_i$ (i = 1, 2) which are in different BCU's first, then the transitions between CFFR$_i$ and CFFR$_j$ ($i \neq j$, i,j = 1,2) which are in different BCU'S will be discussed.

Suppose we use $BCU(a_2,a_3,a_4)|_{\theta_1^k,\theta_6^g}$ to represent $BCU(a_2,a_3,a_4)$ at a specific value of θ_1 (θ_1^k in this case) and a specific value of θ_6 (θ_6^g in this case), and use $CFFR_i[\theta_5^{kl},\theta_5^{ku}]_{\theta_1^k,\theta_6^g}$ to represent a $CFFR^i$ (i = 1 or 2) in $BCU(a_2,a_3,a_4)|_{\theta_1^k,\theta_6^g}$ whose upper limit is θ_5^{ku} and lower limit is θ_5^{kl}. Assume there are two basic changeable units $BCU(a_2,a_3,a_4)|_{\theta_1^1,\theta_6^1}$ and $BCU(a_2,a_3,a_4)|_{\theta_1^2,\theta_6^1}$ which have different values of θ_1 but same value of θ_6. A transition between $CFFR_i[\theta_5^{1l},\theta_5^{1u}]_{\theta_1^1,\theta_6^1}$ in $BCU(a_2,a_3,a_4)|_{\theta_1^1,\theta_6^1}$ and $CFFR_i[\theta_5^{2l},\theta_5^{2u}]_{\theta_1^2,\theta_6^1}$ in $BCU(a_2,a_3,a_4)|_{\theta_1^2,\theta_6^1}$ can be obtained as follows. Since $BCU(a_2,a_3,a_4)|_{\theta_1^1,\theta_6^1}$ and $BCU(a_2,a_3,a_4)|_{\theta_1^2,\theta_6^1}$ have the same value of θ_6, they have the same link position circles LPC_{a_4} (see Figure 4-1). For each BCU, $CFFR_i$'s are separated by obstacles and therefore one $CFFR_i$ cannot be transferred to another $CFFR_i$ in a specific BCU. Consequently if there is a transition between $CFFR^i[\theta_5^{1l},\theta_5^{1u}]_{\theta_1^1,\theta_6^1}$ and $CFFR_i[\theta_5^{2l},\theta_5^{2u}]_{\theta_1^2,\theta_6^2}$, there is at least one $\theta_{5I}^i \in [\theta_5^{1l},\theta_5^{1u}] \cap [\theta_5^{2l},\theta_5^{2u}]$ such that the larger or the smaller configuration with θ_{5I}^i in both BCU's are collision-free.

Suppose $[\theta_5^l,\theta_5^u]_I = [\theta_5^{1l},\theta_5^{1u}] \cap [\theta_5^{2l},\theta_5^{2u}]$ is not empty, then there exists a θ_{5I}^i such that the larger or the smaller configuration with θ_{5I}^i in both BCU's are collision-free. We define a transition to exist between two larger (or smaller) configurations with θ_{5I} in $BCU(a_2,a_3,a_4)|_{\theta_1^1,\theta_6^1}$ and $BCU(a_2,a_3,a_4)|_{\theta_1^2,\theta_6^1}$ if we can find a continuous sequence of CFFC's which connects these two configurations, assuming θ_1^i is changed from θ_1^1 to θ_1^2 continuously.

To find this sequence of CFFC's, we assume $w_O = 0$ and $w_R = 0$ first. Again assuming J_2' and J_2'', $J_3^{i'}$ and $J_3^{i''}$ are the positions of joints j_2 and j_3 in the smaller (or the larger) configuration with θ_{5I} of $BCU(a_2,a_3,a_4)|_{\theta_1^1,\theta_6^1}$ and $BCU(a_2,a_3,a_4)|_{\theta_1^2,\theta_6^1}$, respectively, the link position circles $LPC_{a_3}(I)$, $LPC_{a_2}(J_2')$, $LPC_{a_2}(J_2'')$ and $LPC_{a_1}(J_1)$ can be drawn as shown Figure 4-2. Subsequently, $J_3^{i'}$ and $J_3^{i''}$, which are the intersection between $LPC_{a_3}(I)$ and $LPC_{a_2}(J_2')$ and the intersection between $LPC_{a_3}(I)$ and $LPC_{a_2}(J_2'')$, respectively can be obtained. For convenience we introduce two notations here. Suppose there are two points A and B, $\overline{AB}$ denotes a straight line between points A and B, $\widehat{AB}$ denotes a circular arc between points A and B (on some circle). From Figure 4-2, it is observed that in order to guarantee the obstacle collision-freeness for links a_1 and a_3 in the sequence of feasible configurations, the area $\Omega_i = \overline{IJ_3^{i'}}\widehat{J_3^{i'}J_3^{i''}}\overline{J_3^{i''}I}$ (i = 1, 2) and the area $\Psi = \overline{J_1J_2'}\widehat{J_2'J_2''}\overline{J_2''J_1}$ should be obstacle collision-free. To identify the collision-free positions for link a_2, we can draw the lines $\overline{J_2'J_3^{i'}}$ and $\overline{J_2''J_3^{i''}}$ (i = 1, 2). Since any lines between $\widehat{J_2'J_2''}$ on $LPC_{a_1}(J_1)$ and $\widehat{J_3^{i'}J_3^{i''}}$ on LPC_{a_3} should be in the area $\Phi_i = \widehat{J_2'J_2''}\overline{J_2''J_3^{i''}}\widehat{J_3^{i''}J_3^{i'}}\overline{J_3^{i'}J_2'}$ (see Figure 4-2), if Φ_i is obstacle collision-free then a_2 in transition should be obstacle collision-free. Consequently, in the case of $w_O = 0$ and $w_R = 0$, if the area Ω_i, the area Ψ and area Φ are all obstacle collision-free, then there is a collision-free transition between the two larger (or smaller) configurations with θ_I^i in $BCU(a_2,a_3,a_4)|_{\theta_1^1,\theta_6^1}$ and $BCU(a_2,a_3,a_4)|_{\theta_1^2,\theta_6^1}$. Otherwise there is no collision-free transition between these two collision-free feasible configurations.

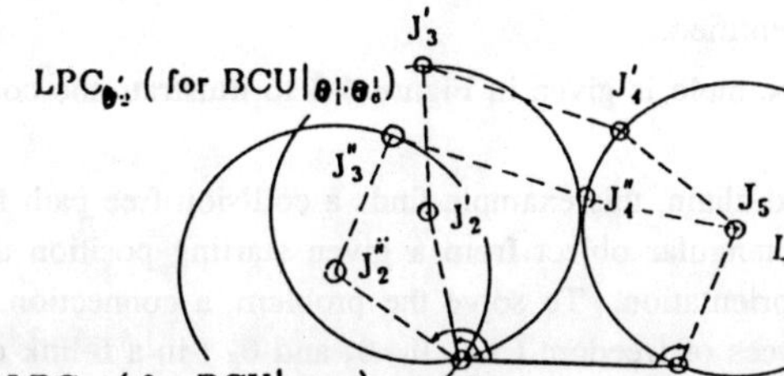

Fig. 4-1: BCU$|_{\theta_1'\theta_0'}$ and BCU$|_{\theta_1''\theta_0'}$ have the same LPC$_{a_4}$

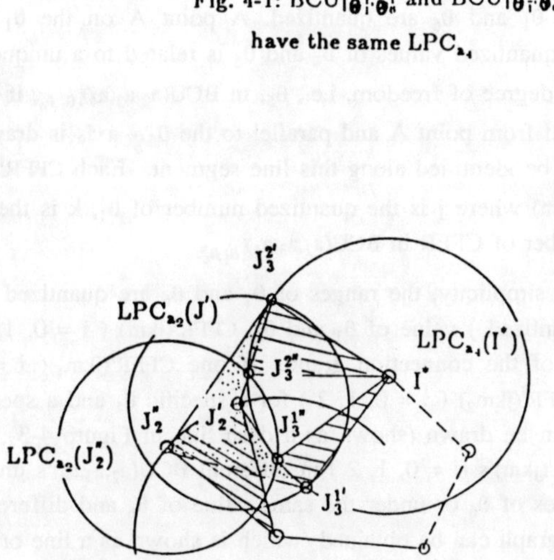

Fig.4-2: obstacle collision-freeness of a transition when $w_O = 0$ and $w_R = 0$

If $w_O \neq 0$ or $w_R \neq 0$ (which counts for most cases), the above method needs to be modified. In such cases, instead of $LPC_{a_4}(I)$ and $LPC_{a_1}(J_1)$, we can draw two circles VII and VIII, which have centers I and J_1 and radii $\sqrt{(\frac{w_O}{2})^2+l_3^2}$ and $\sqrt{(\frac{w_R}{2})^2+l_2^1}$ respectively.

In practice, the intersected region $[\theta_5^l,\theta_5^u]_I = [\theta_5^{1l},\theta_5^{1u}] \cap [\theta_5^{2l},\theta_5^{2u}]$ should be quantized and the transitions for each quantized value should be checked until one transition can be found or no transition can be identified.

As a summary, an example is given in Figure 4-3 to illustrate the construction of a connection graph.

Given a 6-link closed chain, this example finds a collision-free path for each joint of the robot which carries a rectangular object from a given starting position and orientation to a given final position and orientation. To solve the problem, a connection graph can be built. Since there are three degrees of freedom (i.e., θ_1, θ_5 and θ_6) in a 6-link closed chain, a connection graph can be described in a 3-D space for different values of θ_1, θ_5 and θ_6, where θ_1 and θ_6 are chosen as the axes in a horizontal plane and θ_5 is chosen as the vertical axis. In this example, variables θ_1 and θ_6 are quantized. A point A on the θ_1 - θ_6 plane which corresponds to specific quantized values of θ_1 and θ_6 is related to a unique $BCU(a_2,a_3,a_4)|_{\theta_1,\theta_6}$. Since there is only one degree of freedom, i.e., θ_5, in $BCU(a_2,a_3,a_4)|_{\theta_1,\theta_6}$, if a vertical line segment which is originated from point A and parallel to the θ_5 - axis is drawn, all the CFFR's in $BCU(a_2,a_3,a_4)|_{\theta_1,\theta_6}$ can be identified along this line segment. Each $CFFR^i$ (i = 0, 1, 2) can be denoted by $CFFR^i(jkm)$ where j is the quantized number of θ_1, k is the quantized number of θ_6, and m is the number of CFFR in $BCU(a_2,a_3,a_4)|_{\theta_1^j,\theta_6^k}$.

In Figure 4-3, for simplicity, the ranges of θ_1 and θ_6 are quantized into two intervals. For each specific (quantized) value of θ_1 and θ_6, $CFFR^i(jkm)$ (i = 0, 1, 2) can be found, which are the vertices of the connection graph. If one $CFFR^i(jkm_1)$(i = 0, 1, 2) can be transferred to anther $CFFR^i(jkm_2)$ (i = 0, 1, 2) for a specific θ_1 and a specific θ_6, an edge in the connection graph can be drawn (shown as a dash line in Figure 4-3). If a transition can be found between $CFFR^i(jkm)$'s (i = 0, 1, 2) in different $BCU(a_2,a_3,a_4)$'s under the same value of θ_1 and different values of θ_6 or under the same value of θ_6 and different values of θ_1, an edge of the connection graph can be obtained (which is shown as a line or a double dash line in Figure 4-3). After all the $CFFR^i$'s (i = 0, 1, 2) and the transitions among them are found, a connection graph can be built. If a connectivity subgraph which contains a CFFR at the starting position and a CFFR at the final position can be found, a collision-free path for the 6-link closed chain can be obtained from this subgraph.

In Figure 4-3, there is a connectivity subgraph shown in bold lines which contains $CFFR^1(113)$ at the starting position (with $\theta_1 = \theta_1^1$ and $\theta_6 = \theta_6^1$) and $CFFR^1(334)$ at the final position (with $\theta_1 = \theta_1^3$ $\theta_6 = \theta_6^3$). Consequently, we can find a path for the 6-link closed chain which passes through vertices $CFFR^1(113)$, $CFFR^1(122)$,..., $CFFR^1(333)$. After a path for the 6-link closed chain has been found, by the relationship between the joint angles in the 6-

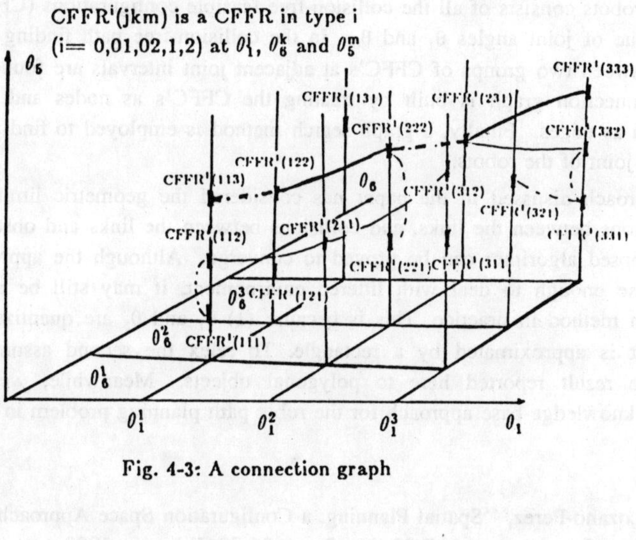

Fig. 4-3: A connection graph

link closed chain, a path for each joint can be obtained.

5. Discussion and Conclusion

We have described an off-line approach to plan a collision-free path for two robots which coordinate in carrying a rectangular object in a 2-D environment. The main idea of the approach is to map the free space in the Cartesian world space of the robots to the robot's joint space. In the joint free space, each joint configuration of the robots satisfies the closed chain constraint, the link collision-free constraints, and the obstacle collision-free constraint. A collision-free path for the robots is then computed in the robot's joint free space. Two major algorithms, namely the collision-free feasible configuration finding algorithm and the collision-free path finding algorithm, are employed in the approach. In the collision-free feasible configuration finding algorithm, the constraints are superpositioned one by one as the free space in the Cartesian space is mapped to the joint space of the robots. The joint free space of the robots consists of all the collision-free feasible configurations (CFFC's) for each quantized value of joint angles θ_1 and θ_6. In the collision-free path finding algorithm, the transitions between two groups of CFFC's at adjacent joint intervals are found first. Subsequently a connection graph is built by treating the CFFC's as nodes and the transitions between them as edges. Finally, a graph search method is employed to find a collision-free path for each joint of the robots.

The approach disussed in the paper has considered the geometric limitations of joint angles, collisions between the links, and collisions between the links and obstacles. In general, the proposed algorithm can be proved to converge. Although the approach described here is precise enough to deal with littered environment, it may still be regarded as an approximation method in practice. This is because (1) θ_1 and θ_5 are quantized, and (2) the carried object is approximated by a rectangle. To relex the second assumption, we are extending the result reported here to polygonal objects. Meanwhile, we are currently developing a knowledge-base approach for the robot path planning problem in a 3/D environment.

References

[1]. Tomas Lozano-Perez, ''Spatial Planning: a Configuration Space Approach,'' *IEEE Transactions on Computers*, vol. C-32, No. 2, pp. 26-38, February 1983.

[2]. J.T.Schwartz and M.Sharir, ''On the Piano Movers' Problem: I. The Special Case of a Rigid Polygonal Body Moving amidst Polygonal Barries,'' *Commun. Pure Appl. Math.*, vol. 36, pp. 345-398, 1983.

[3]. Y. F. Zheng and Fred R. Sias, Jr., ''Two Robot Arms in Assembly,'' *Proc. IEEE International Conference on Robotics and Automation*, pp. 1230-1235, 1986.

[4]. E. Freund and H. Hoyer, ''Pathfinding in Multi-Robot Systems: Solution and Applications,'' *Proc. IEEE International Conference on Robotics and Automation*, pp 103-111, San Francisco, California, 1986.

[5]. S. Fortune, G. Wilfong and C. Yap, ''Coordinated Motion of Two Robot Arms,'' *Proc. IEEE International Conference on Robotics and Automation*, pp. 1216-1223, 1986.

[6]. R. Zapata, A. Fournier and P. Dauchez, ''True Cooperation of Robot in Multi-Arms Tasks,'' *Proc. IEEE International Conference on Robotics and Automation*, pp. 1255-1260, 1987.

[7]. A.Midha, Zhen-Lu Zhao and I.Her, ''Mobility Conditions for Planar Linkage Using Triangle Inequality and Graphical Interpretation,'' *Transactions of the ASME* vol. 107, September 1985.

NOTES